The Sierra Club Handbook of
Seals and Sirenians

The Sierra Club Handbook of

SEALS AND SIRENIANS

Randall R. Reeves,
Brent S. Stewart, and
Stephen Leatherwood

Illustrations by
Pieter A. Folkens

SIERRA CLUB BOOKS SAN FRANCISCO

THE SIERRA CLUB, FOUNDED IN 1892 by John Muir, has devoted itself to the study and protection of the earth's scenic and ecological resources–mountains, wetlands, woodlands, wild shores and rivers, deserts and plains. The publishing program of the Sierra Club offers books to the public as a nonprofit educational service in the hope that they may enlarge the public's understanding of the Club's basic concerns. The point of view expressed in each book, however, does not necessarily represent that of the Club. The Sierra Club has some sixty chapters coast to coast, in Canada, Hawaii, and Alaska. For information about how you may participate in its programs to preserve wilderness and the quality of life, please address inquiries to Sierra Club, 730 Polk Street, San Francisco, CA 94109.

Library of Congress Cataloging-in-Publication Data

Reeves, Randall R.
The Sierra Club handbook of seals and sirenians / Randall Reeves, Stephen Leatherwood, Brent Stewart.
p. cm.
Includes bibliographical references and index.
ISBN 0-87156-656-7
1. Pinnipedia. 2. Sirenia. 3. Marine mammals. I. Leatherwood, Stephen. II. Stewart, Brent S. III. Title. IV. Title: Handbook of seals and sirenians.
QL737.P6R44 1992
599.74'5–dc20 92-946
CIP

Production by Robin Rockey and Janet Vail

Cover design by Abigail Johnston

Cover illustration by Pieter A. Folkens

Book design by Abigail Johnston
based on a design by Jon Goodchild

Illustrations by Pieter A. Folkens

Printed and bound in Hong Kong by Dai Nippon Printing Company, Ltd.

10 9 8 7 6 5 4 3 2 1

For Randi Roe Olsen
and
Pamela K. Yochem
and
Thomas Oscar and
Cheryl Lynne Leatherwood
and
Cynthia Lee Hill

Contents

Foreword

IT IS A RARE PLEASURE to have been offered the opportunity to haul out from the depths of my narrow niche (paleontology of marine mammals) to bask, however obliquely, in the bright light of this book. In short, I am proud to be associated in any way with it, and by extension with its (1983) companion volume on cetaceans.

Truly successful writing on esoteric subjects for a diverse readership of serious-minded nonspecialists, particularly on a subject as popular, emotion-laden, and fraught with scientific uncertainties as that of marine mammals, is the most demanding writing I know. It has to be right, brief, uncondescending, and interesting. Most of us researchers get by well enough in writing up our results for hardcore journals used only by our peers, but most of us are too snobbish or we think too busy to write for a wider audience, while we decry the low quality of what is produced to meet that demand. In the craft of writing, perhaps more than in any other, a few years of unwilling soon becomes an everafter of unable.

Such is not the case here. These authors have unquestionable credentials in and mastery of marine mammalogy, coupled with rare aptitude and willingness to share their special knowledge with a wide audience. Their combined record of research and publication is more than enough to assure them a place of more than average respect in the inner circle of marine mammalogists, but in addition they have persevered in polishing their innate skills in broader communication. At risk of slighting his coauthors, I feel compelled to say a special word about the one (Reeves) best known to me personally. I have known Randy since 1976, when he served with me, nominally as my assistant, while I chaired the Committee of Scientific Advisors to the Marine Mammal Commission under the U.S. Marine Mammal Protection Act. Any successes that we achieved during that period can be attributed largely to his ability to grasp complex, controversial, often ill-defined subjects quickly and correctly and to encapsulate them in clear, persuasive prose. Because I know the person behind the prose, I know that you can believe what you read in this book.

That brings me to the inevitable question of why this book (and its companion) justifies cutting still more trees, when there are so many books, some excellent, about marine mammals on the market. The modern era might be said to have begun in 1958 with V. B. Scheffer's *Seals, Sea Lions, and Walruses. A Review of the Pinnipedia* (Stanford, CA: Stanford University Press). This is still a valuable source, though outdated in this dynamic field, and covering pinnipeds only, as do the fine books by J. E. King (*Seals of the World,* 2d ed., 1983) and most recently W. N. Bonner (*The Natural History of Seals,* 1989), D. Renouf (*The Behaviour of Pinnipeds,* 1991), and M. Riedman (*The Pinnipeds. Seals, Sea Lions, and Walruses,* 1990). Other books are in part more inclusive taxonomically but regional in focus, for example, D. Haley's *Marine Mammals of Eastern North Pacific and Arctic Waters* (2d ed., 1986) and R. T. Orr and R. C. Helm's *Marine Mammals of California* (1989). The only publication of comparable substance and scope (though omitting the polar bear and the marine otter) is *The Handbook of Marine Mammals,* edited by S. H. Ridgway and R. J. Harrison (1981–89), and it is ten times more costly, more bulky (four volumes versus two) and has some of the unevenness in style, content, and illustration seemingly inherent in many-author compendia. In contrast, the present work is that of three authors who not only know the subject scholarly but have rubbed against most of the subjects literally. Further, in addition to fresh, well-chosen photographs, these two books benefit immeasurably, both in pleasure and information, from the artistry of best-in-the-business Larry Foster and Pieter Arend Folkens. If one had the money, time, and space for a dozen or more volumes, plus posters, then well and good, but if one is limited in any of these three commodities, say in packing light for a busy, pricey, natural history cruise anywhere in the world, then these two Sierra Club handbooks are first choice.

CLAYTON E. RAY
Smithsonian Institution
Washington, DC

Preface

IN 1978 TWO OF US (Stephen Leatherwood and Randall Reeves) began compiling a nontechnical book on marine mammals of the world. Initially, we expected to cover all species in a single volume, but it quickly became clear that this was impractical. *The Sierra Club Handbook of Whales and Dolphins,* published in 1983, became the first of two installments. The present book, covering all the marine mammals except the cetaceans, completes the set. It should be regarded as a companion volume to the whale and dolphin book.

Reaction to the *Handbook of Whales and Dolphins* has been favorable, and it is partly that warm reception that has emboldened us to complete this one. Since 1982, when the manuscript on cetaceans was delivered to Sierra Club Books, Reeves and Leatherwood have continued their wide-ranging research and writing on marine mammals, and Brent Stewart, in addition to conducting his other research, has started and finished his Ph.D. dissertation on northern elephant seals. Most importantly, however, the state of knowledge about marine mammals has improved enormously. There is now much more to say about these animals than there was a decade ago.

However eagerly we may have greeted the new information and insights, we also recognized how much more difficult they would make the job of writing a readable, judicious, and accurate summary. The burgeoning interest in marine mammals has meant that researchers have fanned out around the globe, probing the lives of seals, sirenians, and other aquatic air breathers. Scientists have addressed their chosen research questions using classic techniques of direct observation, but now also equipped with high technology as a powerful new ally. They have used such tools as radio and satellite transmitters and receivers, high-resolution sonar, time-depth recorders, computer-based automatic pattern recognition systems, remote television cameras and miniature video recorders that can be attached to animals, magnetic resonance imaging systems, and ultrasound devices, to mention but a few. Improvements in the ways animals are captured, handled, and kept healthy in captivity and in the design and operation of special experimental research facilities have meant that much of the work with wild populations could be augmented in one way or another with studies of captives under controlled conditions. Findings by scientists in many different disciplines (for example, medical hematology, physiology, behavioral ecology) are providing a better understanding of pinnipeds. Under this multidisciplinary approach, old truisms have given way to new truths. And, of course, as always happens in attempts to unlock Nature's secrets, many of the mysteries, rather than disappearing, have only deepened.

What we offer in this book, then, is not some final word on the subject, but rather a kind of progress report. Once again, as with the whale and dolphin book, we have depended on colleagues to shore up what we know from the literature and our own experiences in the field and laboratory. We are humbled by the generous spirit in which they have responded to our calls for help. While accepting full responsibility for this book, including errors of omission and commission, we want to record our thanks to the scientists who helped us get things right. For reviewing selected portions of the manuscript we thank Paul K. Anderson, John Bengtson, Annalisa Berta, Mônica Borobia, Michael M. Bryden, Martin Cawthorn, Daniel P. Costa, Enrique Crespo, Randall W. Davis, D. William Doidge, Daryl P. Domning, Michael Donoghue, Francis H. Fay, Charles W. Fowler, John Francis, Kathryn J. Frost, William G. Gilmartin, Mads-Peter Heide-Jørgensen, Alex Hiby, Wybrand Hoek, Karl W. Kenyon, Kit Kovacs, John K. Ling, Tom Loughlin, Lloyd F. Lowry, Arthur W. Mansfield, Rob Mattlin, Helene Marsh, Richard Merrick, Daniel K. Odell, Peter J. H. Reijnders, John Reynolds, David E. Sergeant, Robert E. A. Stewart, Ian Stirling, Frank S. Todd, and Pamela K. Yochem. We owe special thanks to Clayton E. Ray for reading the entire manuscript, to W. Nigel Bonner for reading the pinniped section, and to F. H. Fay and Lawrence G. Barnes for help with the Introduction. Pieter Folkens wishes to thank Todd Telander for his valued assistance, including his original work on the polar bear and sea otter illustrations. Marjorie Crosby contributed by offering us a computer-equipped sanctuary in which to work undisturbed. Kathy Kangas helped in various ways with preparation of this book. The Reeves family–Randi, Justin, and Brendan–remained supportive throughout the book's lengthy gestation, and we thank them. The many colleagues who contributed photographs are identified in the captions; we are deeply in their debt. Finally, for sharing with us their enthusiasm, experience, knowledge, and camaraderie in the field over the years, we thank George A. Antonelis, Frank T. Awbrey, John Bengtson, Kirk Connally, Murray D. Dailey, Robert L. DeLong, Roger D. and Suzanne E. Hill, Carl L. Hubbs, Tom R. Loughlin, Edward D. Mitchell, Sam H. Ridgway, Alexey Yablokov, Pamela K. Yochem, and Alexander Zorin. Danny Moses championed our cause at Sierra Club Books and made sure that this book reached publication.

The demise of the Union of Soviet Socialist Republics (Soviet Union) occurred while this book was already in production (1991–92). Our apologies to readers, and to citizens of the new republics, for any confusion or slight caused by the now-archaic references to the U.S.S.R.

RANDALL R. REEVES
BRENT S. STEWART
STEPHEN LEATHERWOOD
Solana Beach, California
June 1991

Conversion Factors

To change	into	multiply by
Centimeters	Inches	0.3937
Meters	Feet	3.281
Meters	Yards	1.094
Kilometers	Statute miles	0.6214
Kilometers	Nautical miles	0.5396
Nautical miles	Kilometers	1.853
Square kilometers	Square nautical miles	0.29
Square nautical miles	Square kilometers	3.43
Fathoms	Feet	6.0
Fathoms	Meters	1.829
Kilograms	Pounds	2.205
°Centigrade	°Fahrenheit	1.8, then add 32

Part I

GENERAL INTRODUCTION

General Introduction

This is a book about all the marine mammals other than the cetaceans (whales, dolphins, and porpoises). It includes the pinnipeds (seals, sea lions, and walruses), sirenians (manatees, dugong, and sea cow), otters (sea otter and marine otter only), and polar bear. We acknowledge that the thread connecting the animals in this handbook is a loose one. To call them all *marine* mammals is in fact somewhat misleading. The Baikal seal and the Ladoga and Saimaa ringed seals are confined to freshwater systems, as is the Amazon manatee. Certain populations of harbor seals and manatees split their time among salt-, brackish-, and freshwater environments, while the Caspian seal is confined to a large, inland saline sea. The animals covered also differ in the degrees to which they are adapted to an aquatic life. Seals are amphibious, whereas sirenians are fully aquatic. Sea otters only come ashore occasionally in some areas; they give birth, nurture their young, mate, molt, and forage in the water. Marine otters (as well as river otters in many coastal portions of their range) den along rocky coastlines and forage in the marine environment. Hippopotamuses, beavers, and muskrats (see Appendix 1 for scientific names) may spend more of their time in the water than polar bears do, yet we do not lump these semiaquatic herbivores with the marine mammals. The polar bear gains its status as a marine mammal because it spends most of its life at sea, where it preys on pinnipeds and cetaceans. Even though they den and give birth mainly on land (many in the Beaufort Sea den on pack ice), polar bears are intimately linked to the sea ice by their dependence on ringed seals.

Our choices of which species to include are arbitrary to some extent, stemming in part from the traditional tendency of game managers and zoologists to see marine mammals as peripheral to the mammalian mainstream. It really was not until 1971, when the U.S. Congress was preparing the Marine Mammal Protection Act (MMPA; passed in 1972), that a formal definition of marine mammals became necessary. Our decision to include the polar bear, sea otter, and marine otter in this book along with the pinnipeds and sirenians follows the MMPA's precedent.

In 1976 the U.S. Marine Mammal Commission, whose existence was mandated by the MMPA, prepared a list of marine mammal species, including recommended common names as well as currently accepted scientific names. We have adopted this list, with only a few changes. For all species, we have mentioned the basis for their common and scientific names.

We seriously considered preparing a map or series of maps for this book. In the end, however, we decided to provide only a table showing breeding ranges by major region for each of the 42 species covered (Appendix 3). There are simply too many details concerning distribution and movement, and thus too many place names, mentioned in the text for maps with any meaningful resolution to be accommodated in the small, portable format of a Sierra Club handbook.

Introduction to the Animal Groups

Animals in the four different groups treated in this book are introduced briefly below. As there are more than 30 pinniped species, our introductory comments about them are most detailed and lengthy. The introductions to the other groups are abbreviated since they include many fewer living species – four sirenians, two otters, and one bear. For them, lengthy introductions would duplicate the species accounts.

The Pinnipeds

CLASSIFICATION AND NOMENCLATURE: The term pinniped is derived from the Latin *pinna*, meaning "fin," "wing," or "feather," and *pedis*, meaning "foot," thus, "fin-footed" animal. Pinnipedia has been used to designate both an order and a carnivoran suborder, but it has less firm status in taxonomy today. Like the term pachyderm applied to elephants, rhinoceroses, and hippopotamuses, it is antiquated and colloquial but retains a certain utility. There is no ambiguity in the word. Everyone who uses it understands that it encompasses all the amphibious aquatic Carnivora whose front and hind limbs are flippers. All exhibit a high degree of aquatic specialization, manifested in a similar basic body plan, advanced swimming and diving abilities, and various other physical, behavioral, and ecological characteristics, discussed below.

Some authors have treated the pinnipeds as a separate order of mammals, the Pinnipedia, equal in rank to the Rodentia (rodents), Artiodactyla (cloven-hoofed mammals), and Cetacea (whales, dolphins, and porpoises), for example. In support of this position, they cite the striking similarity of the various pinniped species, which equals or exceeds that of rodents, artiodactyls, or cetaceans, and the marked differences between pinnipeds, as a group, and the fissiped carnivores (those with separated, as opposed to webbed, toes). However, most authors today do not agree that the pinnipeds belong to a separate order. Rather, they place them in the order Carnivora, along with the terrestrial families of bears (Ursidae),

Profile of a ringed seal. This northern phocid is one of the most abundant seals. (Baffin Island, Canada, September 1965: Fred Bruemmer.)

otters and weasels (Mustelidae), cats (Felidae), and dogs (Canidae), to mention but a few. Within such a construct, the living pinnipeds are classified into two reasonably well defined carnivoran families, Phocidae and Otariidae; a third, Odobenidae, is less well defined.

The phocids are often called true seals, as well as earless or hair seals. (The term *hair seal* was also used by early naturalists and sealers for sea lions to distinguish them from fur seals.) Among the phocids, there are three species of monk seal in tropical and warm temperate regions of the Northern Hemisphere, two elephant seals with antitropical distributions, four species distributed mainly in the Antarctic, and ten species confined to the temperate and polar regions of the Northern Hemisphere. Phocids lack ear flaps (pinnae); the only external evidence of the ear is a small opening of the ear canal located just behind the eyes. The fore flippers are short and haired, with a claw on each of five digits. The hind flippers are of intermediate length, are oriented posteriorly, and cannot be rotated forward. Like the fore flippers, they have claws and hair, with thin webbing connecting the digits. These characteristics of the flippers are reflected in the phocids' methods of locomotion (see below). Phocid and odobenid testes are not scrotal, but lie separately outside the body cavity in the hypoder-

Harbor seals and other phocids almost always lack ear flaps. (Otter Harbor, San Miguel Island, California, July 1989: Brent S. Stewart.)

mal layer of the skin. Members of both groups have one or two pairs of teats.

The otariids, named after *Otaria,* the type genus, include two distinct subgroups of eared seals: fur seals and sea lions. Both have the following characteristics: external ear flaps; long, hairless or only partially haired fore flippers, containing splayed digits (the first longer and stronger than the rest), a hard cartilaginous leading edge, and very small nails located well away from the flipper edge; and relatively large hind flippers with larger nails on digits two to four and rudimentary claws on digits one and five. The hind flippers can be rotated beneath the body. Otariids have two pairs of teats, and the testes are contained in an external scrotum. The two subgroups differ in that the guard hairs of fur seals are surrounded by higher densities of short underfur. Most fur seals also have a

Pups of two otariid species, a California sea lion (left) and a northern fur seal (right). (San Miguel Island, California, August 1985: Brent S. Stewart.)

The thick-maned neck of bull sea lions is a conspicuous secondary sex characteristic. (Top: New Zealand sea lion, Campbell Island, New Zealand, January 1990: Roger Moffat. Bottom: South American sea lion, Punta Norte, Valdés Peninsula, Argentina: Claudio Campagna.)

relatively pointed snout compared with the broader, blunter nose of most sea lions.

There are five monotypic genera of living sea lions: *Zalophus* and *Eumetopias* of the Northern Hemisphere and *Neophoca, Phocarctos,* and *Otaria* of the Southern Hemisphere. They have been named sea lions because of the thick mane and somewhat lionlike roar of adult males.

The modern fur seals are assigned to two genera: the monotypic North Pacific genus *Callorhinus* and the polytypic genus *Arctocephalus,* represented in the North Pacific by one species and in the Southern Hemisphere by seven others. (The Galápagos fur seal lives in an archipelago that is partly in the Northern Hemisphere.) The nomenclature of the *Arctocephalus* fur seals has been confused for two reasons. First, many early works gave what is now understood to be a single species several, often ambiguous, names. Second and more recently, range expansion and recovery by some species have led to sympatric occurrences and even hybridization, raising doubts about the integrity of what were once considered well-defined species. Although the nomenclature of the modern fur seals is now fairly stable compared with what it was several decades ago, the phylogenetic relationships among the various species remain uncertain. Some early explorers, naturalists, and whalers called these seals sea bears, referring to the bearlike appearance, particularly of adult males, as they walked on land.

The third group, the odobenids or walruses, are in some respects morphologically intermediate between the phocids and otariids, but their basic morphology and the fossil record show them to be derivatives of sea lion–like animals. (This traditional view has been strongly disputed in recent years by some workers.) Although there is only one living species, the familiar walrus of high northern latitudes, the group was more widespread, extending into low latitudes, and much more diverse, including much of the otariid adaptive zone, in past geologic time. In fact, the extant walrus is an extreme variant and thus not a particularly representative odobenid. The walrus has no external ear flaps but can turn its rear flippers forward for locomotion on land. Claws on the fore flippers are rudimentary. Those on the hind flippers are largest on the middle three toes and slightly smaller on the outer two. The upper canines have developed into highly specialized, curved tusks. As indicated above, the testes are not scrotal, although they may appear partially scrotal in some cases, especially in very lean individuals.

The three major groups of pinnipeds are further distinguished by their differing modes of locomotion. On land or ice, the phocids move by hunching the body, like oversized caterpillars, or by wriggling from side to side. The fore flippers are used by some species, but in others neither the front nor the hind flippers touch the surface of the substrate, leaving the sternum and pelvis to take the pressure alternately. The otariids and walrus walk on their fore flippers, which are extended at right angles to the body, and either rotate their hind flippers forward at the ankle to walk on them or drag the hindquarters along with the rear flippers following passively. The phocids swim with a slashing, back-and-forth or figure-eight motion of the rear of the body, especially the hind flippers, and use the fore flippers largely for control. In contrast, the otariids extend the head and neck for steering as they swim with a strong sculling, or breaststroke, motion. The walrus uses a combination of these two methods, although its principal swimming power comes from alternating strokes of the hind flippers.

Phocids, such as this harbor seal, cannot rotate their hind flippers forward to "walk" on land or ice as otariids do. (Sable Island, Nova Scotia, Canada, January 1967: Fred Bruemmer.)

The phocids, otariids, and odobenids are also distinguished from one another by various internal anatomical characteristics, such as the shape of the tip of the tongue, the size of the canine teeth, the number of lower incisor teeth, the presence or absence of grooves on the incisors, and a number of skull features, particularly the configuration of the bones comprising the base of the cranium and the area around the ears.

There is little controversy about the distinctions among living pinniped species. Virtually all investigators recognize 33 of them.

A pregnant Galápagos sea lion walks along a sandy beach. Otariids use hind and fore flippers for locomotion on land. (June 1989: Frank S. Todd.)

An antarctic fur seal rockets from the water as it leaves its haul-out beach. (South Georgia, South Atlantic, January 1985: Marc Webber.)

A thirty-fourth, the Caribbean monk seal, survived into the mid-twentieth century but is now probably extinct.

EVOLUTION: The phylogenetic position of the pinnipeds within the Mammalia is currently a topic of vigorous debate. Researchers agree that the pinnipeds derived from terrestrial carnivores, specifically the Canoidea, a heterogeneous group that includes, among others, the mustelids, canids, and ursids. There is, however, disagreement about which of these groups gave rise to the pinnipeds.

California sea lions bodysurfing in breakers. (San Nicolas Island, California, January 1985: Stephen Leatherwood.)

A century-long debate continues about whether the true seals, fur seals, sea lions, and walruses are all descendants of a recent common ancestor *(monophyletic hypothesis)* or two different ancestral groups *(diphyletic hypothesis)*. The diphyletic hypothesis, supported by combinations of fossil, ecological, morphological, physiological, and biogeographical evidence, contends that sea lions and fur seals (otariids) and walruses (odobenids) derived from an ursid-like ancestor and the true seals (phocids) from a mustelid-like ancestor. Most molecular and biochemical data support the monophyletic hypothesis, although ambiguities about whether the ursids or the mustelids gave rise to all pinnipeds confuse interpretations. Some researchers have reported the greatest similarities between pinnipeds and ursids; others, between pinnipeds and mustelids.

To be valid, the theory for diphyly would require that the two lineages (phocid and otariid) appear to have converged to a striking extent, as the eared and earless seals show many similar adaptations to aquatic life. Proponents of diphyly argue that two groups of terrestrial mammals returning to the water would experience strong selection pressures for similar solutions to the problems of feeding, diving, and conserving heat and energy in that environment. To a proponent of monophyly, the similarities of living forms represent not convergence but similar retention of shared ancestral characteristics. The monophyly argument is itself complicated by biogeographical considerations, as it requires that the two (or three) groups of modern pinnipeds originated from the same group of terrestrial carnivores at various times on different continents and coasts.

The monophyly versus diphyly debate will continue. We are in no position to take sides in the ongoing controversy. However, in our opinion, an objective reexamination of assumptions and evidence is a healthy process, to be welcomed by the scientific community.

Otariids and odobenids were present in the North Pacific and phocids in the North Atlantic by the early Miocene, at least 22 million years ago. That the center of origin for the otariids and odobenids was the North Pacific is undisputed. Walruses appear first in the North Pacific, then in the North Atlantic, in the latest Miocene, about 6 million years ago. Although the phocids are generally thought to have arisen in the North Atlantic in the early Miocene, perhaps as much as 25 million years ago, our understanding of phocid origins is hampered by wide gaps in the fossil record and a lack of intermediate forms. (Phocids are not known from the North Pacific until quite late, in the Pleistocene, less than 1.5 million years ago.)

Sea lions are considered more derived, or recent, than fur seals. There is no evidence of otariids, living or extinct, in the North Atlantic (California sea lions escaped from captivity notwithstanding). Considering their widespread occurrence in the Southern Hemisphere, this could mean that dispersal from the North Pacific began after the Central American Seaway last closed in the Pliocene, some 10 to 8 million years ago. Otariids, and possibly phocids, may have dispersed from the Northern to Southern Hemisphere by way of the east coast of South America, with animals moving south and east from the North Pacific through the Central American Seaway; alternatively, they could have dispersed by moving south directly along the west coast of South America. The latter theory has the greatest support among biogeographers at present. Otariid fossils dating from 8 to 6 million years ago have been found in Peru, but there are no otariid fossils from the South Atlantic. Two possible routes have been available periodically for phocid and odobenid movements between the North Pacific and North Atlantic basins. The Central American Seaway is the southern route. The northern route is a two-pronged polar one, the so-called Northwest Passage across North America and the Northeast Passage across Eurasia.

Distribution and Migration: Pinnipeds can be highly accommodating subjects for study during periods when they are hauled out on land or ice. However, it has proven extremely difficult to learn about their distribution and movements while away from their island, mainland, or ice haul-out grounds. Most species are hard to detect and observe at sea, so novel techniques of study are being developed and used. Time-depth recorders combined with radio

One of the authors (Stewart) and colleague Pamela Yochem tagging weaned elephant seal pups. (San Nicolas Island, California, January 1985: Stephen Leatherwood.)

Crabeater seals often are encountered hauled out on ice floes in small groups of 10 to 15, although groups of several hundred are seen occasionally along the Antarctic Peninsula. (East side of Antarctic Peninsula, February 1987: Brent S. Stewart.)

transmitters have provided much useful data on pinnipeds at sea. The developing use of recording instruments linked to earth-orbiting satellites holds much promise. Once perfected, such gadgetry will permit documentation of behavior, physiology, and movements while pinnipeds are living at sea.

The worldwide, aggregate distribution of pinnipeds is centered along continental fringes and around oceanic islands in temperate and polar latitudes. Their requirement for consistently abundant, catchable food sources links these animals to areas of high productivity, for example, areas where nutrient enrichment occurs due to upwelling, indrift, or ice melting. The shelves surrounding continents and their associated archipelagoes, offshore banks, or seamounts, and regions where oceanic current systems collide and mix are often congregating sites for pinnipeds. So also are river mouths when anadromous fish congregate there.

Another requirement is a suitable substrate on which to haul out (that is, emerge from the water) for resting, molting, mating (in some species), and, above all, giving birth and nursing their offspring. The distribution of sea ice essentially defines the distributions of some species. For example, the ringed seal in the Arctic and the Weddell seal in the Antarctic use land-fast ice for giving birth and nursing. Baikal seals breed on the continuous ice that covers Lake Baikal in winter. Harp, hooded, ribbon, spotted, bearded, and crabeater seals give birth on the floating pack ice, and their annual reproductive cycles are intimately related to its formation, movement, and disintegration. Among the phocids,

A mixed aggregation of California sea lions and northern elephant seals on the beach in Tyler Bight, San Miguel Island, California. The two species can be readily distinguished by the elephant seal's much larger size. (May 1979: Brent S. Stewart.)

only the monk seals and the northern elephant seal live where no appreciable ice forms; they rely entirely on beaches, islets, or rocky ledges for hauling out.

All seals, because of their adaptations to the heat-sapping aquatic environment (see "Thermoregulation," below), face the problem, at least seasonally or periodically, of dissipating excess heat. This imperative may help explain the generally antitropical character of the group's world distribution. In general, phocids (with the significant exceptions of monk seals, harbor seals, and elephant seals) inhabit higher latitudes than otariids; the latter are mainly centered in subpolar and temperate latitudes. Of course, thermoregulation factors may be far less responsible for this pattern of distribution than are other energetic considerations, in particular the availability of energy-rich food to lactating females. Otariids, unlike phocids, appear incapable of acquiring large reserves of fat to sustain lactation. As a consequence, mothers need to give birth and nurse their young in areas within reasonable swimming distances of reliable food sources. The tropics and subtropics have relatively few areas where oceanographic conditions, especially upwelling, create high coastal productivity that could support large breeding concentrations of pinnipeds.

To some extent, the present-day distribution of pinnipeds has been shaped by human activity. For example, the absence of monk seals in the Caribbean Sea and Gulf of Mexico and throughout most of the Mediterranean Sea is an artifact of overexploitation, human disturbance, and coastal development. The overall pattern

The aggressive behavior exhibited by otariid bulls on rookeries extends from the beach into the adjacent shallows. (California sea lion off Los Islotes, Baja California, Mexico, October 1981: Howard Hall.)

of pinniped distribution, however, which leaves broad gaps in the tropical and subtropical zones, is mainly the result of natural limiting factors.

There are two basic kinds of breeding- and molting-season aggregations. Ice seals congregate in areas called grounds, or patches, which may vary in their precise location from year to year. Other seals traditionally form colonies, or rookeries, on the same beaches or rocky shores. These broad differences in congregating modes reflect major differences in breeding strategies between polygynous and non-polygynous species. Seals occupy their breeding and molting sites on a fixed schedule, and the events of each season begin, peak, and terminate with remarkable precision. From detailed observations of some species, notably the northern fur seal, seals often have been characterized as returning to their natal rookeries annually. However, from additional detailed study, some species, such as the northern elephant seal, are known to show less individual tenacity to specific sites. The habit of a seal population of returning annually to its traditional haul-out site can be so firmly entrenched that the animals keep coming back even in the face of intensive hunting pressure or severe harassment. Seals of most populations appear to disperse widely when they leave the ice or beach, although some populations of harbor seals and sea lions are thought to be fairly sedentary.

Young pinnipeds of most species, and particularly young males, are capable of dramatic deviations from their population's normal migratory route and schedule. They regularly wander hundreds or thousands of kilometers outside the normal limits of the range of the species, turning up in highly unlikely places.

Although some are circumpolar in the Northern or Southern Hemisphere, there is no truly cosmopolitan species of pinniped. This situation contrasts with that of the cetaceans, in which many whale and dolphin species have distributions spanning all or several ocean basins. Among pinnipeds, there is a high degree of endemism. Even in high latitudes of the Northern Hemisphere, where barriers to faunal dispersal have been ephemeral, a surprising number of species are effectively confined to either the North Pacific or the North Atlantic. Ribbon seals and spotted seals have not managed to invade the North Atlantic, nor have harp seals and hooded seals negotiated the Northwest or Northeast Passage often enough to establish populations in the North Pacific. The harbor seal, probably the most widely dispersed pinniped in the world, has disjunct populations in the North Atlantic and North Pacific. It exhibits considerable clinal variation, as reflected by graded differences in color, body size, behavior, and reproductive physiology. When the same species occurs in more than one ocean basin or in enclosed seas or lakes as well as the ocean, separate races or subspecies have generally emerged.

LIFE HISTORY: The pinnipeds have not made a total break with the land (or, in some cases, with ice, which serves as a suitable substitute for land). The life cycle of each species reflects the extent to which it is still tied to land or ice, although all nutritional requirements are met by the resources of the sea (or lakes). The activities that bring the seals ashore (or onto ice) are mainly pupping, nursing, mating, molting, resting, and escaping predators or other dangers. The briefer periods ashore usually punctuate longer periods at sea, although the durations of these periods vary among species.

Pinniped females of all species give birth to a single pup; twinning is rare. Pupping intervals vary between and within species. In many populations that have been growing rapidly, the interval between births is 1 year, while in other species with slowly growing or relatively stable populations, females may skip 1 or more years before giving birth again. Sexually mature females are usually receptive to mating for only a few days during a short breeding season. Females are mated again after they deliver their pups, but the interval between parturition and mating varies greatly among species, from 5 to 8 days in hooded seals and most fur seals to 20 to 40 days in elephant seals and California sea lions. The walrus and some phocids mate in the water. Fur seals and most temperate-region sea lions mate on land, but some sea lions living in warmer climates mate in the surf or in water near the shore. In most pinniped species whose reproductive physiologies have been studied, once the egg has been fertilized, development proceeds for only

Although a few copulations occur in the water, most take place unceremoniously on the beach. For the California sea lions in the foreground, the mating season is about 6 months away. (San Nicolas Island, California, January 1985: Stephen Leatherwood.)

7 to 10 days and then stops for several weeks or months (periods of 6 weeks to 5 months have been reported, varying by species). The pause in embryonic development that occurs between the time of fertilization and the time at which the blastocyst attaches to the uterine mucosa is generally called "delayed implantation" or "embryonic diapause." Delayed implantation, although suggested, has not yet been demonstrated for leopard and northern elephant seals and California and New Zealand sea lions. Recently, a free-floating blastocyst was found in an Australian sea lion, suggesting that implantation is delayed in this species. During the embryonic diapause, the blastocyst floats freely, dormant, in the uterus. Insufficient secretion of gonadotropic hormones by the pituitary causes the delay in implantation, although the specific hormones involved may vary among species. The actual timing of resumed development of the blastocyst and its attachment to the uterine mucosa varies among species. The timing of molt, associated changes in circulating hormones (implantation evidently follows the molt in most phocids), and the general physical condition of the female influence the duration of delayed implantation and suspended development. Delayed implantation, which is also known to occur in some other mammals (for example, weasels, otters, mink, roe deer), allows the seal to combine the activities of giving birth and mating in one season, while giving the female a needed respite from the energy demands of pregnancy during two energetically costly periods—the time when she is recovering weight lost during intensive lactation and

the time of molt. Ultimately, though, delayed implantation in these groups is the integrated result of hormonal, anatomical, ecological, and evolutionary influences.

Including the period of embryonic diapause, gestation in all pinnipeds, except the walrus and perhaps the Australian sea lion, lasts 10 to 12 months. The duration of significant fetal growth, however, depends on the delay in mating once a female has given birth. Fetal growth rates in species with short gestation periods are presumably faster than in those with longer gestation periods.

In most species, pupping occurs in spring or summer and is usually confined to a short season. However, some seals, including northern elephant seals, harp seals, and Caspian seals, give birth and mate mainly in winter. The walrus mates in winter but gives birth in spring, over one year later. In some species, such as the harbor seal, the pupping season starts earlier and lasts slightly longer for populations living in relatively mild climates. There is much variation among species in the period of pup dependency. In the seals that give birth on pack ice, it can be remarkably short (an average of only 4 days in the hooded seal!). In others, particularly otariids and the walrus, it can span a year or more. Seal milk is rich in fat, moderately high in protein, low in water content, and very low in lactose. The richness of the milk assures rapid pup growth and minimizes the time required of the mother to nurture her pup. The relative composition of the milk changes through the nursing period in response to increasing energetic stress on the female. Phocid pups gain weight much more rapidly than otariid pups. Otariid milk is less rich, and otariid mothers go to sea to forage, even during the pup's first few weeks of life, whereas most phocid mothers fast and remain ashore (or on the ice) during

A California sea lion giving birth. Most pinniped births are head first. (San Nicolas Island, California, June 1984: Brent S. Stewart.)

A mother South African fur seal turns to greet her newborn pup, whose umbilical cord is still attached. The mother will learn to recognize her pup by its call and scent. (Cape Cross, Namibia, November 1988: Fred Bruemmer.)

lactation. Females of a few phocid species, particularly harbor seals, evidently feed at sea while their pups are still dependent. When doing so, they are generally accompanied by their pups. However, at least on Sable Island in southeastern Canada, mothers leave their pups on shore while they forage nearby for brief periods. Mother ringed seals leave their pups on the ice in specially constructed, snow-covered lairs while foraging under the ice.

Females of some species can produce their first young at 3 or, less commonly, 2 years of age, but in most the average age at first reproduction is at least 5 years. Males are usually sexually mature at 4 to 6 years of age, but particularly in the polygynous species they often do not become socially mature and mate successfully

A large Galápagos sea lion pup nuzzles one of its mother's four teats. (Isla Plaza, Galápagos Islands, 23 July 1980: William T. Everett.)

Newborn gray seals have a white fetal coat that molts within a few weeks after birth. This lanugo is replaced by a spotted coat like that of adults. (Sable Island, Nova Scotia, Canada, January 1978: Fred Bruemmer.)

for several more years. Few seals live longer than 30 years in the wild, although some individuals have been known to attain ages of 40 or more years.

Because of the importance of the skin, hair, and fur in waterproofing and thermoregulation, pinnipeds molt once or twice during the first year of life and once each year thereafter. This process is a critical and energetically demanding part of each animal's annual cycle, and it structures the timing of haul-out behavior after the breeding season. The animals evidently eat little or nothing during much of the molting process. Blood, as a supplier of nutrients to the epidermis and dermis, is essential for the growth of new skin, fur, and hair and thus the completion of the molt. Its circulation to the skin is substantially reduced to conserve heat and oxygen when the seal is in the water, whether resting, swimming, or diving. Therefore, retreat from the water is important for rapid and safe replacement of the pelage. While on land or ice, most animals bask in the sun to warm the body surface and promote blood flow to the skin. Some species (for example, the elephant seals and walrus) may huddle together to enhance blood flow to the skin while minimizing heat loss and the metabolic cost of maintaining body temperature.

Ages of pinnipeds are usually estimated by counting the layers of tooth dentin, deposited in the pulp cavity until it fills up, or in the cementum (which is deposited continuously on the outside of the tooth's root). These cyclic layers are visible in prepared sections of the teeth examined under standard or electron microscopes. Canine teeth are usually examined, but for some species (for example, the crabeater seal) cementum layers on the roots of postcanine teeth are clearer for determining age and for studying

For most seals, molt is a prolonged process in which the hair is shed gradually. In monk seals and elephant seals, such as this young southern elephant seal, the molt is catastrophic, with the hair and epidermis shed in large clumps during a brief molting season. (Macquarie Island, southeast of Australia, February 1981: Frank S. Todd.)

the foraging, suckling, and reproductive histories of females, as reflected in the fine-scale layering. (Layers are formed at a fine, microscopic scale due to pulses of hormone production. Different hormones dominate according to the seal's principal activity at a given time, such as feeding, fasting, molting, or breeding.)

Pairs of dark and light annual layers also are formed in the claws, so, except when worn on the tip, they can be used for estimating age as well. Other layers that may reflect age are also present in ear bullae and the mandibles.

FOOD AND FEEDING: The ability to use a great diversity of prey, a characteristic of most pinnipeds, may be a factor in their wide dispersal. The bearded seal, which feeds mainly on bottom-dwelling organisms, is a relatively specialized feeder. However, even this seal exhibits considerable flexibility in its food habits. The walrus, often viewed as a specialist that feeds on benthic bivalve mollusks, may in fact eat more different kinds of organisms than any other pinniped. Its prey range in size from tiny amphipods and cumaceans (both groups of small crustaceans) to adult bearded seals.

One or another pinniped preys on almost every major form of nonplant marine life occurring on or along the continental shelves and in the epipelagic and mesopelagic zones far offshore. Crabeater seals eat mostly krill (although recent studies in some areas indicate that fish are seasonally quite important). Walruses and some of the larger otariids prey at least occasionally on seals

An immature male Galápagos sea lion feeds on a male bluechin parrot fish. (Bartholomé Beach, Galápagos, 1979: Jack S. Grove.)

and birds; some leopard seals may be seal- or penguin-eating specialists, at least seasonally. Many seals eat a variety of fishes, mollusks (particularly squid), and crustaceans (particularly shrimp). Deep divers like the elephant seals evidently forage on the deep-scattering layer, a complex of light-sensitive organisms that move upward in the water column at night and downward during the day.

Few details of how pinnipeds actually capture their prey are known. Little direct observation of wild seals feeding underwater has been possible, so, for most species, insight comes from inferences based on the behavior of captives, the identification of stomach contents, and anatomical structures. Divers have watched Hawaiian monk seals search the caves and crevices of coral reefs for individual prey such as lobsters and octopus. Also, monk seals and harbor seals have often been observed bringing a large fish or crustacean to the surface, where they shake it vigorously before swallowing it. Walruses in captivity vacuum the floors of their tanks with a powerful suction, and their feeding grounds in the wild appear to have been plowed. Their standard mode of exploiting bivalve mollusks must involve a systematic process of rooting in sediments for the exposed soft parts of their prey. Seals that consume large quantities of small schooling fishes must apply much of their energy to corralling and driving the schools into positions of vulnerability, such as against the surface or the sea floor, then taking many individual fish in a single feeding pass through the school.

Most of the time and under most circumstances, seals meet their entire requirement for water by metabolizing prey or, during fasts, their own blubber. However, otariids are known to drink at least occasionally. Bulls in particular may need to supplement their metabolic supply of water while engaged in breeding-season battles. Not only do they fast, but at times they may experience a greater need to get rid of excess heat by panting, and to dilute and eliminate toxic nitrogenous wastes by urinating.

Many seals swallow stones and pebbles, but there is no conclusive evidence as to why they do. Some stones may be ingested accidentally during feeding or play, but most probably are swallowed deliberately. California sea lions have been observed to spend considerable time and effort searching for and eating small rocks from the sea floor. Stones could provide necessary bulk in the stomachs of fasting seals, help grind up the seal's food, or perform some unexplained role in the seal's buoyancy or balance.

Pinniped dentition (see Appendix 2) is reduced from the typical mammalian pattern by the loss of the back molars; there are not more than six upper and four lower incisors in the cranium and mandibles, respectively. (The walrus, with six upper and six lower incisors, most of which are vestigial, may be the only exception.) Dental formulas for pinnipeds are expressed as I (incisors), C (canines), and PC (postcanines). The teeth posterior to the canines are essentially undifferentiated in pinnipeds. Consequently, rather than premolars and molars as are recognized for many mammals, the cheek teeth of pinnipeds are designated collectively as postcanines. The dental formulas are not always diagnostic because there is greater individual variation in the number of teeth in pinnipeds than there is in most terrestrial carnivores.

The teeth of most otariids have two or fewer cusps, although some Southern Hemisphere fur seals have tricuspid teeth. The teeth of phocids have three, four, or more cusps. The most striking example is the crabeater seal, whose teeth have a series of comblike cusps. Presumably, these cusps facilitate the straining of the crabeater's primarily small, shrimplike prey, in a manner reminiscent of filter-feeding in baleen whales.

DIVING: Pinnipeds depend on a variety of anatomical and physiological adaptations to feed in the water. They dive long and sometimes deep, repeatedly. Other mammals, even humans, have at least the beginnings of these abilities, which are simply much more developed in diving mammals. In the case of seals, the nostrils, already closed in the resting state, are sealed further by water pressure during diving. The external auditory canal probably is closed during dives by a combination of water pressure on the soft tissue and contraction of muscles around the canal. Strong laryngeal muscles prevent water from entering the throat when the mouth is open underwater.

A California sea lion diving near a blue shark off the Coronado Islands, Baja California, Mexico. Otariids and phocids use different methods of locomotion underwater, but both are adept swimmers and divers. (July 1989: Mark Conlin.)

Seals have a relatively larger blood volume than other mammals. Most phocids exhale just before diving, so there is little gas remaining in their lungs during diving. Research has shown that in at least some pinnipeds the lungs collapse entirely below about 25–70 m, forcing all air into cartilage-supported structures that are poorly vascularized, thus preventing nitrogen from being forced into the bloodstream and preventing blood vessels from rupturing into air spaces during dives. In phocids, the alveoli collapse entirely. An excess of dissolved nitrogen can be painful or even fatal to a mammal. To prevent the bends, a seal must avoid absorbing too much nitrogen or ascend at a slow enough rate to allow the nitrogen to come out of solution. Phocids dive continually over long stretches of time (up to several months) with only brief periods at the surface, and they maximize the amount of time at depth. Thus, it is often unrealistic for them to ascend slowly, and, as a consequence, adaptations to prevent nitrogen from being absorbed predominate.

This crabeater seal carries a sensor to record heart function and diving patterns. A researcher, kneeling, retrieves the stored data with a cable link to a field computer. (Northwestern Weddell Sea, Antarctic, February 1987: Brent S. Stewart.)

The concentration of circulating red cells is high in the blood (hemoglobin) and muscles (myoglobin) of pinnipeds. Thus, they can carry and store great quantities of oxygen in these tissues for use during routine, long dives. During dives, otariids store a greater proportion of oxygen in their lungs than do phocids. The myoglobin stored in muscles gives the meat of pinnipeds its characteristic dark red color.

During dives, the heart rate may decline from as much as 120 beats per minute to only 4 to 6 beats per minute, a condition known as bradycardia. (This is the experimental extreme, which probably is seldom, if ever, realized under normal free-diving conditions.) Blood flow is substantially reduced or eliminated to all but a few essential tissues and organs, such as the brain and heart. Even so, the tissues of the brain and heart appear to be quite tolerant of low oxygen availability. Most oxygen evidently is conserved for use by muscle tissue during dives. Redistribution of blood, heavy oxygen loading of myoglobin in muscle, and reduction in the use of oxygen (that is, a reduced metabolic rate) by most if not all tissues and organs permit pinnipeds to make long dives repeatedly for several hours, days, and even months, with little or no appreciable buildup of lactate in tissues. Thus, little time is needed, and indeed spent, at the surface between dives to recharge the muscle oxygen stores and to rid tissues of metabolic by-products. This permits pinnipeds to spend most of their time efficiently feeding at depth when at sea. Metabolism during most, possibly all, dives is evidently aerobic. However, the challenge to physiologists studying the diving energetics of pinnipeds is to clarify the relative importance of the biochemical and physiological mechanisms permitting them to remain submerged and active for periods longer than those predicted based on their resting metabolic rates measured to date.

The diving performances of the various pinniped subgroups differ. Phocids remain at sea for long periods (weeks or months), during which they are active most of the time, some virtually continuously. It is not clear whether they sleep at all while at sea. Their main activity is diving, and with due respect to the deep-diving sperm and bottlenose whales, phocid seals may be the best in the world at it. Northern elephant seal bulls have been documented to dive to approximately 1600 m and for about 80 minutes. One southern elephant seal dived to 1180 m, and one dive by a southern elephant seal female lasted for about 2 hours. Otariids generally

A small computer being glued to the back of a bull northern elephant seal. Using such instruments, scientists have found that elephant seals dive deeper than 1500 m and that males feed as far north as the Aleutian Islands. (San Miguel Island, California, February 1990: Jon Francine.)

dive less deeply; they seldom dive deeper than 100 m and for more than 10 minutes.

Walruses are able to dive for longer than 20 minutes and to depths of more than 100 m. However, they generally make their living in an environment that is shallower than 80 m, and their routine dives are 10 minutes or shorter.

THERMOREGULATION: The average internal body temperature of all seals is in the range of 36.5° to 37.5°C (97.7° to 99.5°F). However, the different species are exposed to a wide range of conditions, from glaring tropical sun to bone-chilling polar winds, and from the tepid shallows of tropical and temperate areas to the deep, frigid waters at the edges of sea ice. Moreover, individuals often experience extremes of temperature and chill factor within a short period. To maintain their core body temperature, they must be able to conserve heat when in the water or when their damp body is exposed to strong, cold winds, and to rid themselves of excess heat on windless, cloudless, hot days.

Water is denser and more viscous than air and has a higher thermal conductivity and a much higher thermal capacity. Without special adaptations, these would impose an insupportable drain on a seal's energy resources. Pinnipeds have adapted to this aspect of their environment by a variety of means. The surface area where heat loss might occur has been reduced by the large size and fusiform shape of the body and the reduced blood flow to the appendages. Bones and muscles associated with the appendages are largely buried within the body. Most species are well haired. The pelage acts in three ways to separate the seal's skin from the environment. Sebaceous glands at the base of the hair follicles secrete an oily film, the hairs trap air that helps keep the skin dry, and the hair coat itself constitutes an insulative barrier. A thick blubber layer underlying the skin functions as an insulator as well as a fat storage depot.

When ashore, sea lions, fur seals, and walruses conserve heat by lying with the flippers beneath them and by crowding together against the wind and chill. In some species, dominance can be expressed by displacing subordinates from preferred positions (those out of the wind, dry, or already warmed by another individual).

Pinnipeds can reduce or stop blood flow to the skin (peripheral vasoconstriction), particularly that covering the flippers, as a way of conserving heat. They also can shunt blood through the capillaries when heat loss is the goal. Rafting sea lions and fur seals often swim with their flippers waving in the air to take advantage of convective cooling. Many of the postures assumed by seals on warm sandy beaches, as well as their habit of flicking sand onto their backs, can be explained as attempts to dissipate heat or wet the skin. The need to stay cool can help shape the hierarchy of breeding territories of otariid bulls. Sites in the splash zone, in

Elephant seals flip damp sand onto their backs, presumably to keep cool. They also flip sand when agitated, apparently as displacement. (Año Nuevo Island, California, February 1974: Fred Bruemmer.)

shade, or on wet sand are preferred over those with full exposure to the sun and well away from water. A prominent feature of monk seals is their lethargy while hauled out on sandy beaches. Any unnecessary exertion, it seems, could contribute to overheating.

LENGTHS: In 1958, marine mammal biologist Victor B. Scheffer provided definitions for standard length (in a straight line, snout to tail tip of an unskinned seal, lying belly up) and curvilinear length (shortest surface distance from snout to tail tip along the back, belly, or side). The latter was to be used only when standard length was unavailable or unobtainable. All too often, unfortunately, researchers have failed to report how measurements were taken, so data from all sources cannot necessarily be pooled or compared. The reader should also be cautioned that in this book, maximum measurements are composites. In other words, maximum length and maximum weight do not necessarily refer to the same specimen.

SENSES: All seals have coarse, continuously growing vibrissae, or whiskers, on the face. The most obvious of these, the mystacial vibrissae along the sides of the snout, are supplied at their base with a complex system of nerves and muscles. These whiskers clearly serve a tactile function. It seems likely that, at least in some species, they provide critical information as the seal closes in on its prey. The walrus, which has as many as 300 vibrissae on each side of the muzzle, roots for invertebrates in bottom sediments, where a keen sense of touch is crucial. The function of the few vibrissae

The catlike face of a young Weddell seal. The vibrissae on the snout and above the eyes are important tactile organs. (Paulet Island, northeastern Antarctic Peninsula, 22 February 1990: Hans Reinhard.)

present above the eyes and behind the nostrils in some species is uncertain.

Some pinnipeds are obviously thigmotactic (that is, they seek out contact with their fellows). Walruses, fur seals, sea lions, and elephant seals often huddle together in wriggling, seething heaps while hauled out. At such times, body contact appears to be a delight, notwithstanding the grumbling, gouging, and biting. Such cozy behavior may be prompted primarily by the need to share and conserve body heat. How much, if at all, the sensations of touching one another help establish or maintain relationships is unclear.

There is no evidence that touch plays an important role in seal courtship. Mothers of most species nuzzle their pups, although it is not clear to what extent they do so to establish the pup's identity (by smell or touch), to provide reassurance to the pup, or simply to reinforce the maternal bond.

Seals do have a sense of smell, although it is apparently of far less use to them in the water, where the nostrils are tightly closed, than in the air. While hauled out on land or ice, detection of odors is important to many seals. Here, the sense of smell is acute. The response of many seals to approaching predators on land or ice is in part based on olfactory cues, which may be detected at a considerable distance. Mothers use smell, along with vocal, gustatory, and visual cues, to locate and identify their own pups. Males probably assess a female's readiness for mating by smelling her. It has been suggested for some species that the odors of secretions from various glands in the skin convey information from one seal to another.

Pinnipeds have an acute sense of hearing in air and water. It is probably their most important sense in both environments. Those species studied to date have been found capable of hearing sounds in frequencies up to 28 to 32 kHz in air and above 70 kHz in water. (By comparison, humans are not able to hear sounds above 20 kHz.) Seals also have excellent directional hearing (that is, the ability to determine the source of a given sound) in both mediums. In water, this directional hearing probably helps them close on prey until they are near enough to see or feel the victim, and they may well use listening to avoid would-be predators, especially killer whales. In air, hearing helps females returning from feeding forays, or otherwise separated from their calling pup, to locate it in the horde of seals on a rookery. Hearing in air was long thought to be best in arctic phocids and the walrus. However, laboratory experiments have shown that the northern fur seal and the California sea lion have greater hearing ranges and sensitivities than those phocids studied to date. Even the best of the pinnipeds probably do not hear as well in air as most fissiped carnivores.

The structures of the middle and inner ear are extremely complex, and hearing mechanisms are difficult to describe simply. It has generally been thought that pinnipeds hear underwater by resonant action (involving the ossicles and resonance through the skull) and conductive reaction (sound transmitted to the skull through those tissues oriented normal to the sound), and in air by simple ossicle vibrations, as in other mammals. The bones of the inner ear are isolated from the surrounding bone, a placement that presumably minimizes the confusion that otherwise would result from the general vibration of the skull by waterborne sound. However, Soviet scientists have recently proposed, based on experimental evidence, that sound conduction in water occurs through the air-filled meatus and ear canal as it does in air.

Most of the airborne sounds produced by pinnipeds originate in the vocal cords, although some are amplified or resonated in the proboscis (as in elephant and hooded seals) or air sac (as in ribbon and bearded seals and walruses). Polar seals are among the most loquacious of animals, in terms of both the frequency and variety of their underwater sounds. This suggests that vocalizations may be useful underwater in locating breathing holes as well as in maintaining territories, socializing, and keeping in contact with pups and neighbors. Among the polar seals, antarctic species also have extensive repertoires of in-air sounds, while arctic seals are more quiet. This suggests that in the Arctic in-air sounds carry too high a risk of detection by polar bears, humans, and other potential predators. The eerie songs (for example, the trills and sweeps of the bearded and Weddell seals) of polar seals often dominate the local acoustic environment.

The walrus has a sparse vocal repertoire on ice or land but produces some strange calls, including a resonant metallic gong, underwater. Otariids make few sounds underwater but are vociferous on land. Much of their roaring and bellowing is meant to intimidate rivals for territory or mates. The phocids that do not associate with ice, such as harbor, monk, and most gray seals, are relatively quiet both underwater and in air.

Some scientists have speculated that the clicks of certain species are used in echolocation. Other are exploring the echoranging potential of nonclick sounds made by some, notably polar, seals. To date, however, there is no conclusive evidence that seals echolocate.

A wide-eyed Australian sea lion approaches a diving photographer off Kangaroo Island, South Australia. Pinnipeds have good vision underwater. (February 1988: Furney Hemingway.)

Seals have large orbits and large eyes. Furthermore, the size of certain structures is taken as an indication that they are very sensitive to light, even at depths approaching 1000 m, where the principal source of light is, of course, bioluminescence. Seals are astigmatic on land, which means that they can see only fairly well even in bright light, but have good to excellent vision in water. The minimum angle of resolution of a northern sea lion, for example, was similar to that of a human with 20/20 vision. At least some species have the structures necessary for color vision, but the only experimental evidence for such a capability, as far as we know, is for a captive spotted seal and several captive California sea lions.

Otherwise healthy seals have been found to be blind in one or both eyes. This suggests that a combination of the senses, and not just sight, is used in navigating, finding food, and avoiding predators.

There is no agreement about whether most pinnipeds use, or even have, a sense of taste. There are some gustatory organs (olfactory lobe, taste buds, and connecting nerves) in harbor seals, but their level of function has yet to be demonstrated. Pinnipeds tend to swallow their food whole or in large chunks, without chewing, so taste would be of little importance to them. Prey selection is likely based on such factors as texture, size, catchability, and, above all, food value. It is not clear that taste would be a useful way of assessing any of these.

EXPLOITATION: No species of pinniped has been spared from human exploitation, although the antarctic phocids, owing to the remoteness of their ranges, have been only slightly exploited. Some species, such as elephant seals and fur seals, were hunted to very low numbers and have shown remarkable resilience by recovering dramatically under protection. Others, such as monk seals, were driven to similarly low levels and either became extinct (the Caribbean monk seal) or continued to survive in dangerously low numbers (Mediterranean and Hawaiian monk seals). Finally, populations of a few species have sustained centuries of exploitation with no evidence of any catastrophic reduction in their overall range or abundance (bearded seals, ringed seals, harp seals, hooded seals).

As will become evident from the species accounts that follow, pinnipeds have been killed in myriad ways: They have been clubbed to death on beaches and ice floes; harpooned from canoes, kayaks, and the edges of breathing holes in ice; netted and trapped; shot from ice edges, beaches, and vessels of all kinds. Probably the most common reason for killing them has been to obtain food, oil, fur, and leather, either for subsistence use or commercial sale. Fishermen generally dislike seals, regarding them as pests and serious competitors for fish, damagers of fishing gear, and hosts for parasites. As a result, campaigns to exterminate seals, or at least reduce their populations, have been common. Government agencies traditionally have responded sympathetically to the fishermen's complaints by offering bounty payments for killed seals, supplying ammunition to hunters, and even organizing official culls. Considering the past crusades against them, the pinnipeds are still a surprisingly diverse and abundant group of animals.

There is a well-established tradition of managed seal hunting, and it is clear that some populations can sustain substantial cropping for many decades, or even centuries. Often-cited examples are the South American fur seal in Uruguay, the northern fur seal

A young Polar Eskimo hunter from Moriussaq, Greenland, with a freshly killed ringed seal. Subsistence hunters throughout the Arctic kill more than 100,000 ringed seals annually. (Wolstenholme Fiord, northwestern Greenland, 3 June 1988: Stephen Leatherwood.)

The problem of entanglement of seals in fishing equipment and debris is widespread, as evidenced by the wound on the neck of the Caspian seal in the right foreground. (Southern Caspian Sea, January 1984: Sasha Zorin.)

in the Bering Sea, the southern elephant seal at the island of South Georgia, and the harp seal in the northwest Atlantic. These seals have been exploited for very long periods. However, with pinnipeds, as with many mammals, it is increasingly apparent that any strategy to manage harvests must take into account other factors affecting the exploited populations. Seals are losing access to important haul-out and breeding sites as coastal development proceeds rapidly. Large numbers die in working nets and fishing trash (discarded nets, traps, and plastic or nylon debris). Seals are shot because of real or perceived competition with and damage to fisheries. At the same time, their own food supplies are being reduced

This weaned northern fur seal pup became entangled in a fragment of trawl-net webbing on the rocks of St. Paul Island, Pribilofs. Had the net not been removed by researchers, the young seal's chances of survival would have been poor. (August 1986: Brent S. Stewart.)

by the expanding fisheries. Finally, some seals are dying outright and their reproductive capacity is being impaired by outbreaks of disease. The role of contaminants, introduced to marine and freshwater systems by humans, may be significant in reducing the animals' resistance to disease and other forms of stress.

In virtually every situation where a negative impact on seal populations has been suspected, scientific controversy has ensued. Ecological complexity seems to ensure the difficulty of proving, incontrovertibly, that there is a cause-and-effect relationship between any given human activity and changes in a seal population. Often, such changes are influenced in unknown ways by natural fluctuations in water temperature, current patterns, and local productivity, so the contributions of natural phenomena and our own actions have thus far been difficult to distinguish.

In short, pinnipeds are conspicuous competitors with humans for food and space. They are victims of our expanding and often insensitive use of the sea. Their chances for survival in many places depend on more than just the termination of deliberate killing in a commercial hunt. In the long term, their survival anywhere depends, as does ours, on a healthy world.

The Sirenians

The order Sirenia, a group of herbivorous aquatic mammals popularly known as sea cows, contains only four living species in two genera: three manatees and one dugong. The Amazon manatee

A diver being approached by a manatee in Homosassa Springs, Florida. In areas where manatees are predictably curious, a modest tourist business has developed. (5 February 1989: Howard Hall.)

is entirely confined to a freshwater environment, while the other two manatees, the West Indian manatee and the West African manatee, inhabit both marine and fresh water. The dugong is wholly marine-adapted, although it occasionally may appear in brackish or freshwater habitats for brief periods.

Another marine dugongid, Steller's sea cow, was exterminated by North Pacific sealers and sea otter hunters in the eighteenth century. Thus, the Sirenia have the unhappy distinction of including one of only two marine mammals to have disappeared in modern times, the other one being the recently extinct Caribbean monk seal. If present trends continue, however, further sirenian extinctions can be expected. Local extirpations have occurred, and more seem inevitable in the near future.

Like the cetaceans, the sirenians complete their life cycle in water, and the absence of any external evidence of hind limbs reflects their successful adaptation to such a life. The essential characteristics of sirenians were established at least 40 million years ago. Their nearest living relatives in the animal kingdom include elephants and possibly hyraxes. A highly specialized eating apparatus, described briefly in the species accounts that follow, and large, dextrous forelimbs are among their most distinctive external features. All members of the order are nonruminant herbivores, with an enlarged hindgut endowed with a rich microflora to facilitate the digestion of cellulose and other fibrous carbohydrates. Sirenians have axillary mammary glands, one at the base of each flipper. Their bones are dense and heavy. The teeth of manatees are continuously replaced horizontally throughout the animal's life. As each tooth moves forward with the migration of the entire row, its roots are resorbed. By the time it reaches the front of the row and is lost, it has been replaced by a newly erupted tooth at the rear of the row. Apart from the few teeth that erupt before or soon after birth (corresponding to deciduous premolars), all the manatees' teeth are true molars.

Their placid disposition, slow movement, nearshore distribution, and relatively poor hearing have made sirenians exceedingly vulnerable to exploitation, casual destruction, and habitat usurpation by humans. Explorers and adventurers relied on manatees and dugongs as ready sources of meat in some parts of the tropics. Native communities in virtually all reaches of their range continue to hunt the much-depleted populations despite national or local protection laws. The introduction and proliferation of increasingly mechanized vessels and of netting made from synthetic fibers have proven disastrous for sirenians. Entanglement in nets and contact with other kinds of fishing gear are serious problems for most populations. In addition, manatees in Florida often die from collisions with fast-moving vessels or become debilitated by lacerations from propellers. Sirenians are unaggressive, and their food requirements, unlike those of most other marine mammals, do not conflict with those of humans. They seem capable of habituating to, or at least tolerating, many kinds of human activities. However, uncontrolled development and intensive recreational use of waterways are incompatible with their survival.

Two captive dugongs are shown with their backs exposed as their pool refills at Jaya Ancol Oceanarium, Djakarta, Indonesia. From this perspective, one is struck by the cetaceanlike appearance of these shallow-water herbivores. (March 1983: Tas'an.)

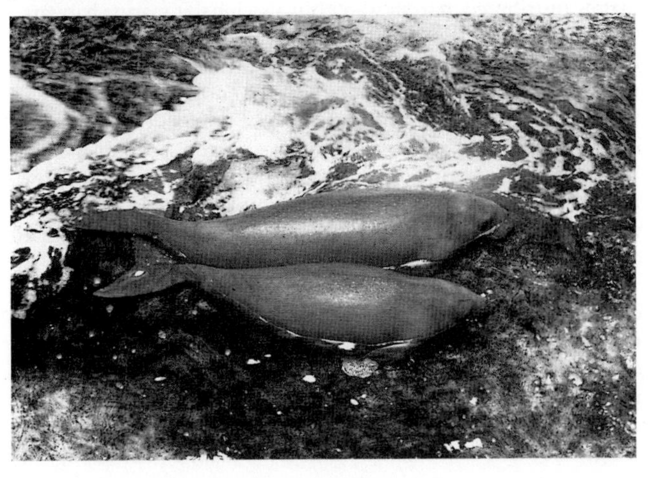

The prospects today of observing sirenians in the wild are generally bleak, except in a few exceptional places. Manatees are fairly accessible in parts of Florida where they congregate in winter, and sizable herds of dugongs inhabit Shark and Moreton bays, Australia. Most sightings outside these areas are the result of good luck or considerable diligence.

There has been some talk and a little action concerning the "semidomestication" of manatees, with a view to exploiting their impressive weed-eating capabilities and eventually marketing their meat. A few animals have been maintained successfully in captivity, but attempts at semidomestication have not lived up to expectations. Considering that all populations are now either endangered or threatened with extinction, such grand designs may have lost much of their practical appeal.

The Otters

The otters comprise a subfamily of the weasel and skunk family (Mustelidae) known as Lutrinae. The earliest fossil recognizable as an otter dates from about the mid-Miocene, 15 million years ago. Some molecular evidence suggests that otters diverged from the weasel lineage about 20 million years ago. All lutrines are torpedo-shaped, have short legs with some webbing between the toes, and have dense coats.

This fairly diverse group of small carnivores has a global distribution from cold temperate latitudes to the tropics. All of its members are associated to a considerable extent with the interface between water and land. Some species live exclusively in and along freshwater rivers and lakes. At least three, the European river otter, the North American river otter, and the African clawless otter are mainly distributed in fresh water but also inhabit marine habitats. Only two species are confined to marine habitats. The sea otter of the North Pacific, the largest of the Lutrinae, is the only living member of a genus that has been adapted entirely to a marine environment since at least the beginning of the Pleistocene epoch, some 2 million years ago. In some areas, sea otters spend their entire lives at sea. The marine otter of western South America, which is morphologically very similar to its freshwater congeners, is thought to have adapted to an exclusively marine niche more recently than any other living mammal, possibly excepting the polar bear.

The sea otter and the marine otter have exquisite pelts, and episodes of unregulated hunting for these pelts have nearly caused the demise of both species. Although the sea otter has made an encouraging recovery in much of its range around the less accessible northern rim of the Pacific, remnant populations along the coasts of California, Oregon, Washington, and British Columbia are in danger. So, too, are the small numbers of marine otters along the coasts of Peru and Chile. Both these furbearers must contend not only with the curse of wearing a commercially desirable coat but also with the animosity of fishermen intent on exploiting the same shellfish resources on which the otters depend.

One other mustelid deserves mention. The sea mink lived along the shores of the Bay of Fundy and Gulf of Maine in the northeastern United States and southeastern Canada. Little is known about its behavior and ecology. The sea mink survived until as recently as the 1890s but is now extinct.

The Polar Bear

The polar bear is one of four living members of the carnivore family Ursidae distributed widely in high latitudes of the Northern Hemisphere. Although brown bears occasionally visit the arctic coasts and even get stranded on ice floes, their range and that of

In summer, some polar bears remain with the disintegrating pack ice; others go ashore on the continental mainland or larger northern islands. Along the southwestern shore of Hudson Bay, some polar bears move many kilometers inland. (Churchill, Manitoba, October 1975: Fred Bruemmer.)

polar bears are generally mutually exclusive. American black bears are endemic to North America, ranging southward from the northern boreal forests to Mexico. Asiatic black bears inhabit deciduous forests and brushy areas north to the timber line in Asia.

All bears have a large head, a powerfully built torso, and a short tail. The ears are typically rounded and erect. There are five digits on all four limbs, and the strong, recurved claws are used for tearing and digging. The massive skull has a formidable dentition, including large, powerful canines.

Ursids are believed to have arisen in Europe during the early Miocene, perhaps 20 million years ago. The two extant genera, *Ursus* and *Tremarctos,* diverged in the Pliocene. The modern polar bear, the most recently derived ursid, had evolved from a brown bear–like animal by the late Pleistocene. The oldest polar bear fossil is only about 100,000 years old.

In the Arctic, the polar bear's presence influences the behavior of other mammals. Female ringed seals give birth and care for their young in snow-covered lairs, at least in part as a strategy for avoiding predation by polar bears. All arctic ice seals must remain alert while hauled out, lest they be surprised by a bear. Arctic foxes roam the sea ice in search of morsels left behind by bears. Inuit hunters seem to have copied certain of the polar bear's techniques for stalking and capturing seals. Finally, any visitor to the North is well advised to take precautions against the possibility of bear attacks. In many coastal areas, climatic severity and polar bears are the most palpable threats to human survival.

Further Reading

PINNIPEDS: Allen (1880), Barnes et al. (1985), Beddington et al. (1985), Bonner (1982, 1989a, 1989b), Bonner and Laws (1985), Boyd (1991), Burns et al. (1985), Burt (1971), Busch (1985), Croxall and Gentry (1987), Dierauf (1990), Fay and Fedoseev (1984), Food and Agriculture Organization of the United Nations (1982), Gentry and Kooyman (1986), Geraci and St. Aubin (1990), Haley (1986), Harwood et al. (1989), Huntley et al. (1987), King (1983), Kooyman (1989), Lavigne and Kovacs (1988), Laws (1984), Lentfer (1988), Malouf (1986), Orr and Helm (1989), Renouf (1991), Repenning et al. (1971, 1979), Ridgway and Harrison (1981a, 1981b), Riedman (1990), Ronald et al. (1983), Ronald and Mansfield (1975), Scheffer (1958), Siegfried et al. (1985), Trillmich and Ono (in press), Yablokov and Olsson (1989).
SIRENIANS: Domning (1978, 1982b, 1984–91), Marsh (1981), Reynolds and Odell (1991), Ridgway and Harrison (1985).
OTTERS: Chanin (1985), Kenyon (1969), Van Blaricom and Estes (1988).
POLAR BEAR: Bruemmer (1989), Larsen (1978), Stirling (1986, 1988).

Part II

PINNIPEDS

SEALS, SEA LIONS, & WALRUS

Walrus

Family: Odobenidae

Walrus

Odobenus rosmarus
(Linnaeus, 1758)*

NOMENCLATURE: *Odous* or *odonto* (Greek) means "tooth"; *bainō* (Greek) means "I walk," referring to the fact that tusks are sometimes used to help the animal move about, thus, "tooth-walker." *Rossmaal* or *rossmaar* (Norwegian) is derived from the Scandinavian word for walrus.

Two subspecies are recognized within the one living species. The Pacific walrus, *O. rosmarus divergens,* and the Atlantic walrus, *O. rosmarus rosmarus,* are geographically isolated and have slight differences in cranial morphology and tusk characteristics. Walruses found along the north coast of Asia, particularly in the Laptev Sea, are thought to have little contact with either the Pacific or Atlantic forms, and some Soviet investigators recognize them as a third subspecies, *O. rosmarus laptevi.*

---

*The presence or absence of parentheses around the name of the author who first described the species reflects a biological convention conveying a precise and necessary meaning. The name in parentheses indicates that, although the species name has remained the same since the date of naming, the species has since been assigned to another genus.

Historically, the walrus has been called sea horse or sea cow by English-speaking peoples. The Russians call it *morzh;* the French, *morse;* the Danes, *hvalros;* the Norwegians, *hvalross;* the Germans, *walross;* and the Dutch, *walrus* or *walros.* Native peoples of the far North call it *aivuk* (Yu'pik) or *aivik* (Inuit).

DESCRIPTION: The walrus is familiar to most people; it is the only living pinniped with external tusks. Its head is small relative to the bulky torso and is dominated by a broadly flattened muzzle with stiff, colorless whiskers (vibrissae). The paired vertical nostrils are situated immediately above this bristly pad, and the often "bloodshot" eyes are set high on the sides of the head. Although walruses appear in some respects more closely akin to the fur seals and sea lions (otariids) than to the true seals (phocids), they lack external ear flaps (pinnae). In adult males the massive neck and shoulders are covered with wartlike nodules or tubercles.

Walrus skin is extremely thick, up to 6 cm on an adult male's neck. The male's wrinkled, heavily scarred hide is covered by a sparse coat of cinnamon brown hair; a female is well haired throughout life. Pups (usually called calves) have short, gray to gray-brown hair; their flippers become black within 2 weeks after birth. Depending on whether it is in or out of the water, an individual's body color can vary from almost white (through vasoconstriction) to pink (through vasodilation) to reddish brown or gray. Males shed most of their hair in June and July, and it is replaced in July and August. The molt in females is apparently less well defined. In general, the annual molt in walruses is a more protracted process than it is in most other sparsely haired pinnipeds.

The fore flippers do not have exposed nails. They enable the walrus to hunker along on land or ice and even to clamber onto rocky ledges. The hind flippers can be rotated forward to allow an ungainly form of walking on solid surfaces.

Adult male Pacific walruses are about 20 percent longer and weigh 50 percent more than adult females. Pacific walrus males are, on average, 3.15 m long and weigh 1215 kg; females, 2.6 m and 812 kg. The largest males can weigh more than 1500 kg; large females, more than 1000 kg. Pups are usually less than 1.4 m long and weigh less than 85 kg at birth. Atlantic walruses are somewhat smaller. Walruses are exceeded in size among the pinnipeds only by the elephant seals.

The two upper canines develop early in life in both males and females into pointed, curved tusks that extend well outside the mouth. Individual tusks have been known to reach 1 m in length and to weigh more than 5 kg. Many are broken, and most are worn on the outer curvature near the tip. One function of the tusks of males is for fighting and display. Threat displays are usually adequate for establishing dominance, but similar-sized bulls are known to engage in brutal fights. The tusks also serve as picks, enabling their bearer to anchor itself and rest in the water next to an ice floe or to lift itself out of the water onto the ice. John Burns, an Alaskan mammalogist, once watched a mother walrus use her tusks to demolish a large piece of ice to free her pup, who had fallen into a crevice. The notion that the tusks are used for digging up clams and other prey in bottom sediments is almost certainly wrong, considering the patterns of wear on the tusks and facial bristles as well as observations of captive walruses feeding. There are 16 functional teeth in addition to the tusks: an incisor and 3 premolars (postcanines) in each upper row and a canine and 3 premolars (postcanines) in each lower row.

DISTRIBUTION: Walruses in modern times have occurred only in cold temperate, subarctic, and arctic latitudes of the Northern Hemisphere. Pleistocene fossils of the living species occur south to California, South Carolina, and France. The family Odobenidae was diverse and abundant off California and Mexico during Miocene and into Pliocene time, some 20 to 5 million years ago. At least one species ranged as far south as southern Florida during the Pliocene.

Pacific walruses are found during winter in the open pack ice of the Bering Sea, mainly between eastern Bristol Bay and an area southwest of St. Lawrence Island, and in the Gulf of Anadyr. Their northward migration through Bering Strait and into the Chukchi Sea begins in April. The population's distribution has been changing in recent years, as thousands now summer around the Diomede Islands, King Island, and Arakamchechen Island, areas where they were historically scarce or absent. In the Chukchi Sea they are scattered during the summer between Point Barrow in the east and the mouth of the Kolyma River on the East Siberian Sea in the west. During summers of light ice coverage they abandon the pack ice and haul out on shore at Wrangel and Herald islands and at traditional sites along the northern Chukchi Peninsula. In heavy ice years they remain with the ice and do not come ashore. Many males (at least 12,000) do not join the northward migration, but pass the summer on or near Round Island, in northern Bristol Bay; another 6000 to 8000 males summer in the Gulf of Anadyr and in Bering Strait. During the return migration in the fall, large groups haul out to rest at Big Diomede Island and the Punuk Islands and at some coastal sites on the Siberian mainland.

In the Atlantic, large herds formerly inhabited the Kara, Barents, and White seas and the shores of Novaya Zemlya, Franz Josef Land, Svalbard, and Bear Island. Their range is still vast, but the populations are much smaller today. Some walruses still winter in the southeastern Barents Sea and migrate in spring through Karskye Vorota Strait into the Kara Sea, returning by the same route in the fall. Sable Island (off Nova Scotia) and the Magdalen Islands (in the Gulf of St. Lawrence) were used by large numbers

Although usually associated with pack ice, walruses also can be found during summer and autumn along ice-free rocky and gravel beaches, such as this one in Kearny Cove, Devon Island, Canada. (October 1977: Wybrand Hoek.)

of walruses during historic times, but the species was essentially extirpated south of Labrador early in this century. The centers of abundance of the Atlantic walrus are in Hudson Strait, northern Hudson Bay, and northern Foxe Basin, and along portions of the Greenland, Devon Island, Ellesmere Island, and Baffin Island coasts. Relatively large numbers are present in Smith Sound, Jones Sound, and their adjacent channels and embayments. During winter (November to May or June), a few hundred walruses are present off central west Greenland (66° to 72° N) in the Davis Strait pack ice. Some winter in high arctic polynyas (areas of open water in the sea ice) such as the North Water, Hell Gate–Cardigan Strait, and Penny Strait–Queen's Channel, but there is also a northward migration in Davis Strait and Baffin Bay and a westward migration through Lancaster Sound during spring breakup. The migratory behavior of Atlantic walruses is poorly understood, and some local populations are probably nonmigratory.

The Laptev walrus is distributed mainly in the Laptev Sea, the eastern Kara Sea, and the western East Siberian Sea.

NATURAL HISTORY: Walruses are extremely gregarious. They often are found in aggregations of many hundreds. During the nonbreeding season, males and females sometimes are segregated, but on certain traditional summer and fall haul-out grounds both sexes and various age classes are well represented. The compact huddling of walruses when hauled out does not necessarily mean that their social relations are altogether amicable. Individuals compete aggressively for the more favorable spots, using tusks and body size to win or maintain social status.

Sexual maturity in most Pacific females is currently attained at around 8 years of age, with the first parturition at 10. During the 1950s and 1960s as this population was recovering from overexploitation, females matured about 2 years earlier. Males mature later. They are capable of mating successfully by age 10. However, they seem socially incapable of competing for mates under natural conditions until they are about 15. Walruses are polygynous, but since mating occurs on or near sea ice mainly from December to March, there has been little opportunity to study the details of their reproductive behavior and social relationships. Several investigators have suggested that the mating system of the Pacific walrus is reminiscent of a lek. Males display elaborately, using vocal and visual signals, on grounds occupied traditionally during the mating season. The competing males remain in the water, spaced about 7 to 10 m apart. They appear to defend territories by visual displays and sometimes by fighting. One difference between the walrus's mating system and the classic leks of many birds and ungulates is that the constant movement of the ice and shifting access to food require a high degree of mobility during the period of courtship. Also, the females do not congregate solely to present themselves for mating but to take advantage of suitable winter ice conditions in particular areas.

Copulation is thought to occur underwater. Attachment of the fertilized egg to the uterine wall is delayed for 4 to 5 months, and pups are born approximately 15 months after conception. Birth takes place between mid-April and mid-June, usually in May. Pups are suckled for as much as 2 to 2.5 years; no other pinniped has such a consistently prolonged period of lactation and maternal care. Females in their prime give birth in alternate years, becoming pregnant again in the winter following the latest birth (8 to 9 months postpartum), with 15 to 16 months of care and suckling still to go before another pup is born. Nursing females sometimes assume an upright posture in the water, with the head thrown back so that the tusks are parallel to the surface. The pup hangs upside-down, with its rump and hind flippers exposed as it feeds. Walruses also nurse on land or ice, but aquatic nursing appears to be much more common in this species than in other pinnipeds. A pup often rides on its mother's back, especially when they flee from danger. A mother will also use her fore flippers to grasp the pup and dive with it when threatened. The solicitude shown by mother walruses for their young has impressed many hunters.

As we know them today, walruses are usually associated with pack ice, on which they haul out to rest. There is some evidence that before they were heavily persecuted by humans, islands and beaches near shallow shellfish banks in many cold temperate and sub-

A walrus pup awakening from a nap. Walrus mothers are solicitous toward their pups, usually nursing them for at least two years. (Near Nome, Alaska, June 1976: Kathryn J. Frost.)

arctic areas were popular walrus habitats. They still haul out on land and rocky ledges seasonally in some parts of their range. A walrus can break through ice at least 20 cm thick with its head and maintain breathing holes with its tusks. In the North Pacific, where walruses are strongly migratory, floating cakes of ice allow the animals to ride passively much of the way to and from their summer feeding grounds. Some individuals probably migrate more than 3000 km in a year. Their usual swimming speed is 7 km per hour; they can reach at least 35 km per hour for short periods.

Walruses are primarily benthic (bottom) feeders and usually dive for periods of up to 10 minutes. Thus, they require access to shallow (80 m or less) banks and coastal margins. Feeling their way along the bottom with their sensitive whiskers, they apparently suck clams from their shells. Remarkably, as many as 3000 to 6000 tiny clam siphons can be ingested in a single feeding bout. Shell fragments are rarely swallowed, but substantial quantities of sand and gravel are often found in their stomachs. Slow-moving fish, snails, worms, soft-shelled (molting) crabs, and shrimp supplement bivalve mollusks in their diet. Birds are eaten occasionally. Winter (March) observations of feces-stained ice near and under large groups of walruses in the pack ice of northern Baffin Bay suggest that walruses can forage in deep water (200 to more than 500 m deep) when necessary. However, the maximum confirmed diving depth is 113 m, and close to 25 minutes is the longest submergence time recorded for a walrus.

Some walruses become regular predators of other marine mammals. According to Native hunters, the individuals that attack seals and small whales often acquire a peculiar appearance: powerful chest and shoulder muscles; long, thin tusks; and amber stains on their chest and tusks from their prey's blubber.

Walruses make several kinds of sound in air. Pups often bellow loudly when disturbed, and adults snort, cough, or roar at one another while attempting to establish dominance. The male's "song," a stereotyped, repetitive pattern of pulsing sounds produced underwater during late winter and early spring, apparently plays a role in courtship. Elastic membranes from the male's pharyngeal pouch (air sac) help the animal float while resting at the surface and produce a peculiar bell-like sound underwater.

Polar bears and killer whales prey on walruses, but such predation can exact a high cost from the predator. Adult walruses fight back, sometimes even managing to inflict serious injuries on their attackers. Ian Stirling, a Canadian biologist, once observed what he interpreted as a group threat display given by walruses to a polar bear:

> . . . 20 or more walruses suddenly surfaced simultaneously in the pool nearest to the bear and swam rapidly for several meters towards her. The water was churned up in their wake. As the group suddenly stopped, two large individuals of unknown sex at the front appeared to do a rapid forward-rolling surface dive during which they smacked their hind flippers on the water. The resultant sounds were loud like gunshots and the bear immediately turned and ran away 100 m or so to the north at a gallop (Stirling, 1984).

Walruses can live to ages of at least 40 years. As would be expected given their low rate of reproduction (at least compared with other pinnipeds), walruses have a relatively high survival rate. Mercury and organochlorine pesticide levels in walrus tissue from northern Baffin Bay are low compared to levels found in most other pinnipeds.

Many walruses have been held in captivity since 1608, when two pups were taken to England from Bear Island by a walrus hunting expedition. Nearly all captive walruses have been young ones whose mothers were killed by hunters, but in recent years a few have been born in captivity.

HISTORY OF EXPLOITATION: Walruses have been exploited commercially for their valuable ivory tusks, tough hides, and blubber oil since at least as long ago as the ninth century. In addition, northern peoples have traditionally depended on them as sources of food, ivory, and leather for several millennia. Two hides from female Pacific walruses are adequate to cover a 30-foot umiak, and the hides of pups make light, strong, practically unkinkable harpoon lines far superior to any made from synthetic or natural fibers. The elastic membranes from the pharyngeal pouches of males reportedly were used to cover drums in some parts of the eastern Canadian Arctic. As winter dog food, walrus meat and hide have no equal.

Norwegian, Russian, and British hunters relentlessly pursued walruses in the northeastern Atlantic for more than three centuries. By the 1850s the animals had been virtually exterminated at Bear Island, and by the 1880s they had been severely depleted at Svalbard. New hunting grounds were opened at Novaya Zemlya and in the Kara Sea in the 1860s, the Norwegians taking a record 2261 walruses there in 1887. Powered vessels made it possible to reach Franz Josef Land in the early 1920s, and Norwegians took thousands of walruses there after 1923. Hunting during World War I is thought to have dealt a final blow to the population around Svalbard, which only recently has shown signs of recovery. Fur trappers began hunting walruses on the coast of the Laptev Sea in the late nineteenth century, and walruses were hunted intensively along the Taymyr Peninsula in the 1930s. In 1953 and 1954, 1200 Laptev walruses were killed, most on Peschanyi Island.

Beginning in the tenth century, Norse colonists in south and west Greenland carried on what was to them a vital trade with Europe, exporting walrus ivory in exchange for essential items such as iron and wood. They made annual long-distance voyages to obtain walrus and narwhal tusks and polar bear hides in the Nordrsetur, or northern hunting grounds around Disko Bay. The scale of this early commercial walrus hunting is suggested by the crusade tithe and Peter's Pence of 635 kg of walrus ivory paid by the colonies in 1327. Although the Norse colonies in Greenland seem to have vanished by 1500, the commercial pursuit of walruses in Davis Strait did not end permanently. European whalers hunted walruses to fill out their cargoes, and trading companies (particularly the Hudson's Bay Company in Canada and the Royal Greenland Trading Company in Greenland) carried on a brisk trade in walrus products. Walrus hides were in demand at various times for the manufacture of bicycle and automobile tires, transmission belts, buffing wheels, and luggage. The hides of walruses killed in the Gulf of St. Lawrence during the eighteenth century were exported to the United States for carriage traces and to England for making glue.

In addition to the substantial commercial hunting in Davis Strait and Baffin Bay by Scottish whalers and Norwegian sealers, Greenlanders made increasing catches of walruses during the first half of the twentieth century. Danish biologist Erik Born has estimated that close to 12,000 walruses were harvested in west Greenland (not including Thule district) from 1900 to 1978. A Norwegian expedition made the last known commercial catch of walruses in Davis Strait, taking more than 1200 animals in 1951.

Intensive commercial hunting of walruses had begun in the Bering Sea region by the middle of the eighteenth century. The Russian American company secured 28,000 pounds of ivory, equivalent to 3000 or 4000 bulls, in a single year at the Pribilof Islands (where walruses are now virtually absent). Most of the commercial exploitation of the eighteenth and early nineteenth centuries was directed at bulls. Ivory was obtained by the Russian merchants in tribute and barter from Native communities as well as by direct harvesting. U.S. commercial whalers began hunting bowhead whales in the Bering Sea and Bering Strait region in 1848, and from that year to 1914, when commercial bowhead whaling in the western Arctic ceased, they took a total of at least 140,000 Pacific walruses. Much of the kill consisted of adult females, and this had a severe impact on the population.

Soviet commercial hunting of Pacific walruses in the present century took place from 1931 to 1960, reaching a peak of more than 8000 taken per year by 1937–38. Commercial walrus hunting resumed in the 1980s, and ship-based hunting continues, under quota, in the Bering and Chukchi seas. The United States banned commercial hunting as early as 1941, restricting the catch to meet subsistence needs of Native peoples. This policy continues, although the Native catch is itself now largely a commercial enterprise. Commercial walrus hunting has been banned in Canada since 1931.

CONSERVATION STATUS: Walrus Island and Northeast Point on St. Paul Island are the only former haul-out sites in the Pacific that have not yet been reoccupied by walruses. Northern sea lions occupied Walrus Island, a small rock outcrop in the Pribilofs, after walruses had been extirpated there during the late 1800s. Walruses are socially dominant among arctic pinnipeds. The Pacific walrus population is believed to have doubled between about 1960 and 1980, reaching a maximum population of more than 200,000. By 1978 its reproductive output was declining as the population approached the environmental carrying capacity. The combined annual kill of walruses in Alaska and the U.S.S.R. at least doubled during the early 1980s (to more than 10,000 animals per year, including sinking losses), and it is believed that the population is now declining.

The populations in the North Atlantic have not made a similar recovery. There may be only some 1500 to 2000 walruses in the northeastern Atlantic today. Protection from commercial hunting since the early 1950s has allowed them to begin recolonizing the waters around Svalbard, where as many as 550 have been seen in one herd in recent years (at White Island). The Northeast Greenland National Park, encompassing adjacent territorial waters, provides some protection for the small population of walruses north of Scoresby Sound. East Greenlanders kill only 10 to 20 walruses a year. Although locally depleted or eliminated in parts of the eastern Canadian Arctic, several thousand walruses remain in both Hudson Strait–Hudson Bay–Foxe Basin and in Davis Strait–Baffin Bay–Smith Sound (and their adjoining bays and inlets). The walrus population in Greenland's Thule district may be stable, but that off central west Greenland remains depleted.

Some 3000 Laptev walruses are present in the summer along the Taymyr Peninsula, and the total population of the subspecies is estimated at 4000 to 5000. Although protected from commercial hunting, the Laptev walrus is threatened by pollution and industrial activity.

Native peoples in all the polar nations continue to hunt walruses with high-powered rifles and motorboats. Some 200 are killed and secured in Greenland's Thule district each year, on average, providing about 80 tons of meat, blubber, and hide used mainly to feed sled dogs. People eat a small percentage of the meat, occasionally contracting trichinosis, a deadly parasitic disease, as a result. In Canada, the reported annual catch during the late 1970s and early 1980s averaged close to 550. The advent of motor toboggans for winter transportation has reduced dependence on dog teams in many areas; the diminished need for walrus flesh to feed dogs has, to some extent, relieved hunting pressure.

In some areas, the commercial value of the tusk ivory is an important incentive for the hunt. The distinction between commercial and subsistence hunting has been obscured by the growing importance of ivory as a cash crop in the increasingly cash-oriented Native economies. Many of the shot walruses are not recovered, and many of those that are secured are used only for their ivory. In some hunts in which firearms are used, at least one walrus and sometimes as many as three or four are killed and lost for every one secured. In Alaska particularly, the penis bone (baculum) of killed walruses is saved along with the tusks, to be sold as a novelty in tourist shops. It is the largest such bone of any living mammal, up to 62 cm long.

Any form of permanent human occupation, be it a Native village, an industrial site, or a military installation (for example, Distant Early Warning stations), appears to be incompatible with the terrestrial hauling out of walruses. Thus, some traditional haul-out grounds are no longer available to the animals.

FURTHER READING: Born (1984), Fay (1981, 1982, 1985), Fay and Fedoseev (1984), Fay et al. (1984, 1989, 1990), Lowry and Fay (1984), Miller (1975, 1976, 1985), Reeves (1978), Salter (1979), Vibe (1950).

Fur Seals and Sea Lions

Family: Otariidae

Northern Fur Seal

Callorhinus ursinus
(Linnaeus, 1758)

NOMENCLATURE: *Callorhinus* comes from the Greek *kallos* meaning "a beautiful object" and *rhinos* meaning "skin" or "hide," referring to the quality of the seal's fur. The name *ursinus,* which means "bearlike" in Latin, was applied because all fur seals were first called sea bears by Europeans.

DESCRIPTION: Pronounced sexual dimorphism in size begins at birth. Although the average length at birth is approximately 60 cm for both sexes, male neonates average 5.4 kg and females 4.5 kg. Adult females grow to an average of 1.4 m and weigh 30 to 50 kg; males grow to an average of 2.1 m and weigh 175 to 275 kg.

Individuals of both sexes have a relatively small head and a short, pointed snout. The rear flippers are extremely long and highly vascularized; they are waved in the air on warm, windless

A mixed aggregation of northern fur seals and subadult male northern sea lions. (Commander Islands, western Bering Sea, August 1980: Sasha Zorin.)

days to dissipate heat. Pups are black at birth. By weaning, at about 4 months of age, they are dark brown to black on the head and back, with a white or silvery chest and belly as well as some white on the sides of the face. Adult males are dark brown to black, with gray guard hairs on the back of the neck. Adult females have grayish black backs and silver to gray bellies.

DISTRIBUTION: The primary rookeries are on the Pribilof Islands (St. Paul and St. George) in the eastern Bering Sea and on the Commander Islands in the western Bering Sea. Small numbers of fur seals breed on Robben Island in the Sea of Okhotsk, on the Kuril Islands north of Japan, on Bogoslof Island in the eastern Aleutians, and on San Miguel Island off southern California. A few seals haul out seasonally on Southeast Farallon Island and, rarely, on San Nicolas Island, California.

Most females, pups, and juveniles leave the Bering Sea by late November and migrate south as far as southern California in the east and Japan in the west. They remain offshore along the continental shelf until March, when they begin returning north to the rookeries. Substantial numbers are taken in the high-seas squid driftnet fishery between 40° and 50°N. Adult males leave the rookeries from late August through early October and are believed to remain near the Aleutians during the winter. Little is known of their whereabouts during the nonbreeding season.

NATURAL HISTORY: Adult males arrive at the rookeries from late May to early June. They establish territories by threatening each

At summer rookeries, such as this one on the Pribilof Islands, northern fur seal bulls defend breeding territories containing 40 or more females. (St. Paul Island, Pribilofs, July 1972: Fred Bruemmer.)

other with stereotyped vocal and visual displays. Fights occur occasionally but are generally brief. One male will lunge at another and attempt to bite the relatively vulnerable fore flippers rather than the broad and well-protected neck and chest. Each territorial male aggressively guards and herds groups of mature females, up to 40 or more, until they come into estrus. Adult males can lose nearly 20 percent of their body mass due to their 1 to 2 month fast during the breeding season.

Females arrive at the rookeries throughout June. Some, particularly young individuals, arrive during early July and early August. They give birth 2 days after arriving at the rookery and remain ashore for another 8 to 10 days before coming into estrus and being mated. They then begin a series of foraging trips to sea. The first lasts from 4 to 5 days; subsequent trips generally lengthen to 8 to 10 days. While at sea, lactating females dive and feed most often at night, at depths averaging 68 m (maximum documented: 207 m in the Bering Sea and 230 m in southern California waters); the dives last, on the average, 2.6 minutes. Lactating females at the Pribilof Islands usually forage within 160 km of the rookeries, occasionally as far as 430 km away. Their visits ashore to nurse their pups last 1 to 2 days. Pups are weaned in October and November, about 125 days after birth, and go to sea soon afterward. Some may remain at sea through the next breeding season. From 5 to 12 percent of the pups die during the first month of life, many from hookworm infection, which causes severe anemia.

Females are sexually mature when 4 years old; nearly 30 percent of 4-year-olds and 65 percent of 5-year-olds are pregnant. Approximately 57 percent of the mature females give birth each year. Males are sexually mature when 4 to 5 years old, but few breed before they are 8 or 9. Bulls have a brief reproductive life; few breed for more than two seasons. Nearly half the pups die during their first year. Natural mortality of 2- and 3-year-olds averages 10 to 20 percent per year; for adult males, 32 to 38 percent; and for mature females, 10 to 11 percent. Maximum longevity for northern fur seals is about 26 years.

There is some movement of seals among rookeries. Immigrants from the Pribilof, Commander, and Robben islands contributed to the growth of the San Miguel Island colony. Similarly, the recent repopulation of the Kuril Islands evidently resulted from immigration from eastern and western Bering Sea colonies.

A sleek mother fur seal nurses her pup on a crowded rookery. (St. Paul Island, Pribilofs, July 1972: Fred Bruemmer.)

Throughout their range in the eastern Pacific, northern fur seals feed on more than 50 fish and 10 cephalopod species. The diet consists primarily of small schooling fish (mostly walleye pollock, herring, anchovy, capelin, and hake) and squid (primarily *Loligo opalescens* and several species of *Onychoteuthis*). Some commercially important species are eaten by fur seals throughout their range, particularly squid, hake, anchovy, sand lance, capelin, and walleye pollock. Lactating females consume nearly 1.6 times as much food as do nonlactating females. In summer, lactating females from the Pribilof Islands population eat an estimated 146,500 tons of fish and squid, while nonlactating females eat approximately 43,100 tons. Nearly 66 percent of the prey biomass is fish, the remainder squid.

Sharks, killer whales, and northern sea lions prey on fur seal pups; 3 to 7 percent of neonates on St. George Island were killed by male sea lions in 1974 and 1975. Perhaps as many as 5 percent of all northern fur seals become entangled in plastic debris and fishing net fragments at sea each year; an unknown proportion of those die.

HISTORY OF EXPLOITATION: Northern fur seals were killed by Aleuts living on the Aleutian Islands and by Indians along the coasts of Washington and British Columbia in prehistoric times. The Chumash and Nicoleño Indians who lived on the southern California Channel Islands from 10,000 years ago until the early 1800s also hunted fur seals; however, it is not clear if the seals bred there.

Northern fur seals were discovered in 1741–42 on the Commander Islands by G. W. Steller. The rookeries on St. George Island were found by the Russian sealer and explorer Gerasim Pribilof in 1786; St. Paul Island and its fur seal colony were discovered a year later. Aleuts soon were taken to the islands to help harvest the fur seals for pelts. Permanent Aleut settlements were established by 1820. Harvesting of fur seals was intensive and unregulated until 1799, when the Russian American Company assumed control over the harvests on the Pribilof Islands. This did not stop the harvests from declining. In 1847, the killing of females was discontinued, and the harvest of males on rookeries and haul-out grounds was limited to several thousand per year. The herds subsequently increased. After some 250,000 seals were killed in 1868, the U.S. government, which had just assumed responsibility for the fur seal industry, began restricting and managing harvests to ensure a profitable trade in pelts on a sustained-yield basis. Within only a few years, the revenues from sale of these furs matched the purchase price of the entire Alaskan territory, and the sale of harvested skins continued to be an important source of U.S. government revenue for many years.

Unregulated commercial sealing at sea, however, accounted for more than 800,000 seals, mostly mature females, between 1868 and 1911, again causing steep declines in the herds. Indeed, by 1909 fewer than 300,000 seals remained in the Pribilof herd. This prompted an international treaty, the North Pacific Fur Seal Convention, signed in 1911 by the United States, Japan, Russia, and Great Britain for Canada, prohibiting the killing of seals at sea. Furthermore, the U.S. government prohibited commercial sealing on the Pribilofs between 1912 and 1917. The United States and Russia were responsible for enforcement of the convention. An arrangement was made for sharing pelts taken on land. The United States, Russia, and Japan gave Canada a portion of the pelts that they harvested. Annual harvests of young males from 1958 through 1974 at the Commander and Robben islands were between 2000 and 10,000 and 3000 and 9000, respectively. The Pribilof herd increased to about 2.5 million by the late 1950s, even though 50,000 to 60,000 subadult males, mostly 3-year-olds, were killed each year under the convention from 1940 through 1956. Then the U.S. government began an experimental harvest of females; this was expected to stimulate productivity of the herd by increasing pregnancy rates and survival and by decreasing the age of first reproduction because of reduced competition among the surviving seals. Nearly 300,000 females were harvested at St. Paul and St. George islands between 1956 and 1968. Subsequently, births at St. Paul, the main harvesting site, declined, as did the harvest of subadult males, a response that was expected to be temporary. Births continued to decline, however, by an average of 7.8 percent per year from 1975 through 1981, with little change since then. Births also declined at similar rates at St. George, and in 1984, 35 percent fewer births occurred there than in 1977. Births at St. Paul did not change substantially from 1982 through 1989, but continued to decline at St. George at least through 1988.

CONSERVATION STATUS: In 1976, the world population of northern fur seals was estimated at 1.75 million, with nearly 900,000 in the Pribilof herd and 350,000 on Soviet islands, mostly (256,000) on the Commander Islands. There were approximately 278,000 births on St. Paul Island (about 80 percent of the Pribilof herd) in 1975 and approximately 33,000 and 69,000 on Robben Island and the Commander Islands, respectively, in 1974. The Pribilof herd was estimated at 870,000 in 1983; in 1989, nearly 171,000 pups were born at St. Paul and, in 1988, nearly 25,000 were born at St. George. In 1983, 11 pups were born on Bogoslof Island, and 90 were born there in 1989.

A small, recently established breeding colony of northern fur seals was discovered at Adams Cove on San Miguel Island in 1968, and another at nearby Castle Rock in 1972. Births have increased steadily at exponential rates, with minor exceptions, since then. Births declined by 60 percent in 1983, evidently because many females were in poor condition in late 1982 through early 1983 and either did not conceive or aborted fetuses when prey availability declined substantially during El Niño. Births began increasing soon after, and nearly 1400 pups were born at the two San Miguel Island rookeries in 1988.

The North Pacific Fur Seal Convention, which protected herds from harvest at sea, ended in 1984 when the U.S. government failed to ratify it. No commercial harvests have been allowed on St. Paul Island since then, but Aleuts can kill up to 2000 subadult males annually for subsistence. The U.S. government designated St. George Island as a Research Sanctuary in 1973, and only small harvests by Aleuts for subsistence are permitted there. Now, fewer than 2500 seals are killed annually on the Pribilofs by Aleuts for subsistence.

The reasons for the dramatic decline of the Pribilof Islands fur seal herd are still unknown, although reduced survival of juveniles seems to be the proximate cause. Commercial fishing in the North Pacific and Bering Sea has increased dramatically since the early 1960s, reducing prey available to seals. Seals also have died after becoming entangled in working fishing gear and in plastic marine debris and net fragments.

In June 1988, the U.S. National Marine Fisheries Service ruled that the Pribilof Islands herd was below its optimum sustainable population and designated it a depleted stock under the MMPA. Consequently, it is illegal to kill northern fur seals except for authorized scientific research or, in the case of Native peoples, for subsistence. However, a multiyear exemption recently was granted to the commercial fishing industry, allowing it to kill up to 50 northern fur seals incidentally each year during fishing operations. Some fur seals also are killed in Japanese drift gill nets set for squid in the northern North Pacific and in the declining Japanese high-seas salmon gill net fishery.

Attempts to establish a new international fur seal treaty have failed, thus leaving the Pribilof herd vulnerable to renewed killing at sea, particularly by fishing nations who view fur seals as competitors for valuable food resources.

FURTHER READING: Bartholomew and Hoel (1953), Eberhardt (1981), Fowler (1990), Gentry and Johnson (1981), Gentry and Kooyman (1986), Goebel et al. (1991), Lloyd et al. (1981), Perez and Bigg (1986), Scheffer et al. (1984), Veltre and Veltre (1987), Vladimirov (1987), York (1983), York and Kozloff (1987).

Guadalupe Fur Seal

Arctocephalus townsendi
Merriam, 1897

NOMENCLATURE: *Arctocephalus* comes from the Greek *arktos* for "bear" and *kephalē* for "head." Fur seals were called sea bears by early explorers, naturalists, and whalers because of their bearlike appearance. The Guadalupe fur seal was named after C. H. Townsend, who collected the original specimens, four weather-worn skulls, from a beach on Guadalupe Island in Mexican waters in 1892, at a time when the species was already feared by some to be extinct.

The first published observations of morphological differences (suggesting taxonomic distinction) among fur seals in the North Pacific were made by the German naturalist F. P. von Wrangel in 1839. Wrangel observed that the fur seals farther south, in California waters, were smaller and less silvery than those in Alaskan waters. Little additional attention was given to this distinction until 1892 when Townsend was sent to investigate the abundance of fur seals at Guadalupe Island off Baja California. By then, nearly all fur seals inhabiting rookeries on the coasts of Alta and Baja California had been killed by sealers, whalers, or sea otter hunters. Townsend observed only seven fur seals in the water near Guadalupe Island, but he collected some skulls from an abandoned rookery there. In 1897, C. H. Merriam used those skulls in describing the Guadalupe fur seal as a new species, distinct from the fur seal of the Galápagos Islands. The skulls used to represent the Galápagos fur seal were, however, actually those of Galápagos sea lions that

A Guadalupe fur seal bull (right center, with light head and dark body) amid California sea lion females and pups. The Guadalupe fur seal was driven nearly to extinction by overhunting. As it slowly recovers, pioneering individuals have begun to appear on islands other than Guadalupe. (San Nicolas Island, California, 10 June 1984: Brent S. Stewart.)

had been collected earlier by Townsend and mislabeled as *Arctocephalus*. The fur seal of the Galápagos finally was described correctly in 1904, although the first actual comparisons of skeletal material of Guadalupe fur seals with fur seals from other areas were made only in 1954, independently by J. King and E. Sivertsen. King concluded that the Guadalupe fur seals and those of the Juan Fernández Islands were the same species, while Sivertsen considered them to be distinct. In 1971, after examining cranial material from all known fur seal forms, with an adequate sample for each, C. A. Repenning and colleagues confirmed the specific distinction of the Guadalupe fur seal from all others.

DESCRIPTION: Guadalupe fur seals are sexually dimorphic in size, although few specimens have been measured or weighed. Two adult males were 193 and 182 cm in length, two adult females 137 and 119 cm. One recently dead adult male weighed about 100 kg.

Only the adult male of this species is readily distinguishable from other North Pacific otariids. The shape of the muzzle, size of the head, and appearance of the pelage are generally sufficient for distinguishing these fur seals from northern sea lions, northern fur seals, and California sea lions. The head of the Guadalupe fur seal bull is large, with a long, pointed muzzle. Juveniles, on the other hand, are hard to distinguish from northern fur seals and California sea lions when they occur together. Guadalupe fur seal flippers are larger and shaped differently from those of California and northern sea lions. The terminal flaps of the hind flippers of Guadalupe fur seals are of about equal length, whereas in California and northern sea lions the hallux (first toe) is longer and thicker than the other digits. The fore flippers are notably large and are partly haired on their dorsal surface, like California sea lions but unlike northern fur seals. Except for adult males, it is difficult to determine the sex of animals in the field. The testes of sexually immature males are small and often undetectable, and the penile opening is often difficult to confirm because of its small size and concealment by the thick fur coat.

Individuals of both sexes are dark brown or dusky black; the guard hairs on the back of the neck (mane) of older males are yellowish or light tan. Their coloration blends well with the black blocks of lava on the haul-out beaches and coastlines. Pups are born with a black coat colored much like that of adults.

The vocalizations are unique and provide another means of distinguishing these seals from other North Pacific otariids. Males bark quietly while patrolling their territories and herding females, and either growl, roar, or cough when threatening other males. Adult females make long bawling sounds when interacting with their pups. The vocal behavior of females is otherwise poorly known.

DISTRIBUTION: Guadalupe fur seals now breed only along the eastern coast of Guadalupe Island, approximately 200 km west of Baja California, where they prefer rocky habitat and volcanic caves. However, individuals have been seen with increasing frequency in other locations. From 1969 through 1989, 48 sightings of Guadalupe fur seals were made on the southern California Channel Islands, including one adult male that returned to San Nicolas Island and maintained a territory among breeding California sea lions from 1981 through 1990. In 1988, a second bull established a territory at San Nicolas and returned there in 1989, 1990, and 1991. Few records have been reported from Mexican islands other than Guadalupe.

The distribution and abundance of Guadalupe fur seals prior to the arrival of sealers, sea otter hunters, and whalers, who nearly exterminated them in the late 1700s and early 1800s, are not known, although there has been much speculation. From sealers' narrative accounts and archaeological investigations of Indian kitchen middens, it appears that Guadalupe fur seals may have ranged from the Revillagigedo Islands northward to Point Conception and may have bred on San Miguel and San Nicolas islands. The species evidently was exterminated from southern California waters by 1825. Commercial sealing continued, although with declining returns, in Mexican waters through 1894. Incomplete sealing records suggest that perhaps as many as 52,000 fur seals were killed on Mexican islands between 1806 and 1890, mostly before 1848; from 1877 to 1894, only some 6600 fur seals were harvested.

The superlative pelts of Guadalupe fur seals, such as this adult female's, commanded high prices and encouraged excessive harvesting. (Discovery Point, Guadalupe Island, Mexico, December 1971: Stephen Leatherwood.)

NATURAL HISTORY: Observations of a few adult males at Guadalupe Island and two at San Nicolas Island suggest that reproductive males are faithful to particular sites over a number of years. During any breeding season, the tenure of territorial males is 35 to 122 days. Ambient temperatures may reach 21°C at midday, so shelter from direct sunlight and access to water for cooling seem to be important to females in choosing where to give birth and to males in defining territorial boundaries.

Although births occur from mid-June through July, most females give birth in mid-June, 3 to 6 days after hauling out. Females are mated 7 to 10 days after giving birth. They then begin a series of trips to sea (lasting from 2 to 6 days) to feed, with briefer visits ashore to suckle their pups, who remain ashore and may gather to play in small groups. Adult males are absent from rookeries in winter; females may continue to haul out and nurse their pups through the following spring.

No information is available on age at sexual maturity or longevity.

Little is known of behavior or demography. Available data suggest that there are seasonal changes in the age and sex composition of the population of fur seals ashore at Guadalupe Island. Adult males, juveniles, and nonparous females may live at sea for all or part of some seasons, but their distribution at sea is unknown. Individuals, particularly males during summer, may linger near shore and groom themselves for several hours; this grooming evidently helps them stay cool on hot days.

Although their diet has not been well studied, Guadalupe fur seals apparently eat squid and lanternfish.

CONSERVATION STATUS: By 1897, the Guadalupe fur seal was believed to be extinct. None was seen until a fisherman found slightly more than two dozen at Guadalupe Island in 1926. He returned in 1928 and collected two for the San Diego Zoo. He is suspected of having gone back to the island later and, unhappy about payment for his services, killed all the seals that he could find. At any rate, no seals were seen again until 1949, when G. A. Bartholomew of the University of California at Los Angeles observed a lone young male at San Nicolas Island. In 1954, the late C. L. Hubbs of Scripps Institution of Oceanography headed an expedition to Guadalupe Island; the expedition found a group of 14 seals there. The colony has increased about 10 percent per year since then, but it remains relatively small. The population numbered at least 1073 by 1977 and approximately 1600 by 1984, including some 650 newborn pups. By 1977, seals were breeding discontinuously along 25 km of the island's eastern shoreline. In 1987, Mexican scientists counted 3259 Guadalupe fur seals at Guadalupe Island, including 468 adult males, 78 subadult males, 1134 females, 472 juveniles, and 998 pups.

The Guadalupe fur seal is protected under the MMPA, the Convention on International Trade in Endangered Species of Wild Fauna and Flora (CITES), and the California State Fish and Game Code. It is also fully protected by Mexican law, and its sole breeding habitat, Guadalupe Island, was declared a pinniped sanctuary by the Mexican government in 1975.

In summer 1988, an emaciated 1-year-old male Guadalupe fur seal hauled out near Oceanside in southern California. It was captured and taken to Sea World in San Diego. After several months of being nurtured, medicated, and trained to catch live fish, the seal recovered sufficiently to be returned to the wild at San Nicolas Island, near the site where Bartholomew had rediscovered the species in 1949.

FURTHER READING: Fleischer (1987), Peterson et al. (1968), Stewart, B. S. (1981), Stewart, B. S., et al. (1987).

Juan Fernández Fur Seal

Arctocephalus philippii
(Peters, 1866)

NOMENCLATURE: The species was named after R. A. Philippi who, in 1864 while director of the natural history museum in Santiago, Chile, collected the skin and skull of a young male from Más a Tierra Island. Peters described the species from this specimen.

DESCRIPTION: The snout is quite long and pointed, similar to that of the Guadalupe fur seal. Newborn pups weigh about 6.2 kg (females) to 6.9 kg (males) and are about 65 cm (females) to 68 cm (males) long. At 2 months of age, pups weigh 10 to 16 kg and are an average 76 cm long. Adult males are about 2.1 m and females about 1.5 m long. The adult male's heavy mane has silver-tipped guard hairs. Other body parts are a shiny blackish brown. Normally, a male's neck and fore flippers have many scars from fighting.

DISTRIBUTION: The species is restricted to the Juan Fernández island group (Más a Tierra or Robinson Crusoe, Santa Clara, and Más Afuera or Alejandro Selkirk) and the San Félix island group (San Ambrosio and San Félix), both off the coast of northcentral Chile. Breeding occurs on Más a Tierra, Más Afuera, and Santa Clara islands. Seals haul out in small numbers, but do not breed, on San Ambrosio Island.

NATURAL HISTORY: Breeding takes place from mid-November through January, although most pups are born in late November and

Juan Fernández fur seals, like this adult male (left) and smaller adult female (right), inhabit rocky shores and ledges of two groups of remote Chilean islands. (Juan Fernández Island, November 1988: John Francis.)

early December. Pup mortality during the breeding season is 4.5 to 8.2 percent. Lactation lasts 7 to 10 months; early lactation milk is 43 percent fat. The first foraging trip of lactating females lasts about 9 days.

Like Guadalupe fur seals, Juan Fernández fur seals often haul out on lava rocks at the base of cliffs. In the water, they frequently rest head down, with their hind flippers swaying gently above the sea surface.

The diet of these fur seals is poorly known. The stomachs of a few individuals contained mouth parts of five cephalopod genera (*Dosidicus, Octopoteuthis, Tremoctopus, Todarodes,* and *Moroteuthis*). Predators include local sharks (*Prionace glauca* and, perhaps, *Carcharodon*) and killer whales. Nothing is known about their diving behavior, patterns of seasonal dispersal and migration, survival, or reproduction.

History of Exploitation: Juan Fernández discovered the islands of San Félix and San Ambrosio in 1554. In 1683, the explorer William Dampier (1729) described the abundance of fur seals there: "Here are always thousands, I might say possibly millions of them, either sitting in the Bays or going or coming in the Sea around the Island, which is covered with them (as they lie at the top of the water playing and sunning themselves) for a mile or two from the shore." He also commented that "a blow on the nose soon kills them." Commercial sealing began at Más a Tierra in 1687 and at Más Afuera soon after. Perhaps as many as 3 million seals were killed during one 7-year period prior to 1824, by which time few remained alive. As an example of the magnitude of the slaughter as well as the former size of the seal population, the captain of the ship *Betsy* sold 100,000 seal skins in Canton in 1798. Nearly all of them were from Más Afuera. During the previous year, the crews of 14 ships were said to have been killing seals on Más Afuera at the same time. In 1801, it was reported that a single ship had carried a million skins to London from Más a Tierra in one cargo. Sealing continued on some islands until 1898, and by about 1900 the species was believed to be extinct.

CONSERVATION STATUS: Approximately 200 seals were discovered in 1965 at Más Afuera. Under protection from the Chilean government, the population has increased at annual rates as high as 21 percent since then, and by 1983 it was estimated at 6300 seals. Local fishermen, however, kill fur seals illegally for food and for meat to be traded for other goods.

FURTHER READING: Hubbs and Norris (1971), Norris and Watkins (1971), Torres N. (1987), Torres N. et al. (1985).

Galápagos Fur Seal

Arctocephalus galapagoensis
Heller, 1904

NOMENCLATURE: The specific name refers to the Galápagos archipelago, where this species is endemic.

DESCRIPTION: This is the smallest of the otariids. Adult females are an average of 120 cm long and weigh 27 kg; adult males, 152 cm and 64 kg. Pups weigh about 4 kg at birth, 9 kg when 6 months old, and 11 kg when 1 year old.

The snout is short and pointed. It and the belly are light tan; the sides and back grayish brown. The skull of the adult Galápagos fur seal is substantially smaller than that of any other fur seal. Skulls of adult males and females differ little in size, and the sagittal crest (see California sea lion account) is not developed in males.

DISTRIBUTION: The Galápagos fur seal is confined to the Galápagos Island group, centered at about 1° south of the equator and about 1100 km west of Ecuador. Most rookeries are along the western coasts of the islands, where nearshore upwelling is strongest. Breeding occurs on seven of the islands, and seals haul out on two others. Of the approximately 27,000 seals in the population in 1978, nearly a third hauled out on Isabela Island. There is no evidence that the historic distribution of the species extended outside the Galápagos.

NATURAL HISTORY: Pups are born from mid-August through mid-November, although most births are from late September through early October. The pups are nursed for 2 years or longer; 1- and 2-year-old pups continue to be suckled even after their mothers

In the equatorial sun, Galápagos fur seals gravitate to water, even small fetid tidal pools, to keep cool. (An adult female and two pups on Fernandina Island, June 1989: Frank S. Todd.)

have given birth again. However, the prospects for a newborn's survival are seriously compromised by the simultaneous suckling of a 2-year-old.

Pup mortality averages 9 to 15 percent but may be higher in some years, particularly those of El Niño events. For example, all pups born in 1982 died when prey available to nursing females was severely reduced by the intense 1982–83 El Niño.

Females are ashore for 2 to 3 days before giving birth. They remain with their pups for 5 to 10 days after giving birth, then begin alternating periods of 1 to 3 days at sea feeding with periods of a half day to a day and a half ashore nursing their pups. Parous females come into estrus about 8 days postpartum and then are mated by a local territorial male. While at sea, females forage mostly at night at depths of about 26 m; dives last 3 to 7 minutes. The longest recorded dive lasted 7.7 minutes; the greatest depth recorded was 115 m.

Males are territorial. However, because of the extended breeding season (nearly 3 months), individual males maintain territories for only approximately 27 days at a time. They may return later in the season to reclaim their territories.

The diet of Galápagos fur seals is poorly known. About three-quarters of the identified cephalopod beaks regurgitated by females have been from the squid *Onycoteuthis banksi*.

No data are available on survival, reproductive characteristics, or longevity.

Sharks apparently prey on these fur seals, judging by the wounds seen on some individuals. However, the effects of such mortality on the population are not known. Anecdotal information suggests that predation by sharks is relatively unimportant. On one occasion, a group of killer whales was seen stalking and killing a fur seal near shore; such attacks are not common. Feral dogs kill pups and attack adults on Isabela Island. They reportedly eliminated the subcolony on the southern end of the island.

There is no evidence for seasonal dispersal or migration.

History of Exploitation: Historically, Galápagos fur seals were killed primarily by whalers who stopped at the islands en route to whaling grounds or ports. Commercial sealers also operated in the Galápagos. Records from one voyage in 1816 indicate that approximately 8000 seals were killed. At least 22,500 were killed by U.S., British, and Spanish sealers between 1816 and 1933. Few seals remained by 1898, although small numbers continued to be killed on scientific expeditions through 1906. Eight seals were collected alive for display at the San Diego Zoo in 1932 and 1933.

Conservation Status: The Ecuadorian government officially prohibited the hunting of Galápagos fur seals in 1934, but the ban was little enforced among local residents. In 1959, most of the Galápagos archipelago was declared a national park, and conservation measures apparently have been fairly successful since then. Feral dogs have been eliminated from coastal areas of Isabela Island, but they have not been eradicated completely in the interior.

Further Reading: Bonner (1984), Clark (1975), Clarke and Trillmich (1980), Townsend (1934), Trillmich (1981, 1984, 1987).

South American Fur Seal

Arctocephalus australis
(Zimmerman, 1783)

Nomenclature: The specific name is from the Latin *australis,* meaning "southern." This seal is also known as the southern fur seal and Falkland fur seal. The Spanish names *lobo de dos pelos* and *lobo fino* are also used for other species of *Arctocephalus*.

Description: Adult males weigh 150 to 200 kg and are generally about 1.9 m long. Adult females are 1.4 m long and weigh 30 to 60 kg. Pups weigh 3.5 to 5.5 kg at birth; newborn males are 60 to 65 cm long, whereas newborn females are 57 to 60 cm. The newborn's coat is soft and black. Adult males are blackish gray; females and subadults, grayish black on the back and lighter on the belly. An adult male's hair is noticeably longer on the neck and shoulders than on the rest of the body.

Distribution: South American fur seals occur on the Falkland Islands (Volunteer Rocks, Elephant Jason Island, and New Island) and along the coasts of South America as far north as southern Brazil in the Atlantic and near Paracas, southern Peru, in the Pacific.

Their distribution and abundance along the Chilean coast are poorly known, but biologists counted approximately 40,000 there in 1976, mostly in southern Chile; in 1982, 228 seals were counted along the northern Chilean coast (22° to 23° N). In 1973, the Falkland Islands population was estimated at 14,000 to 16,000. In 1954, there were about 2700 along the Argentine coast, primarily in colonies at Staten and Escondida islands. Recently, two new breeding colonies were located on Staten Island, and numbers have increased on a small island near Ushuaia (50 pups). Nearly 20,000 fur seals inhabited the Peruvian coast in 1979, primarily at Point San Fernando, San Fernando Islet, and Point San Juan. The largest population is centered in Uruguayan waters, where an estimated 280,000 seals are present; more than 14,000 pups were counted at Lobos Island alone in 1981. Breeding occurs on five offshore islands: Raza, Encantada, and Islote in the Torres Island group, Lobos in the Lobos group, and Marco in the Castillo group.

NATURAL HISTORY: The breeding range of the South American fur seal overlaps that of the South American sea lion, but the two can be distinguished by the smaller size, thick underfur, and less robust muzzle of the fur seal. Differences in the breeding seasons of the two species apparently limit their competition for space.

Most females remain near rookeries year-round, but nothing is known about the seasonal movements of adult males and young seals. While migrations are not known, these seals are seasonally dispersed up to 200 km offshore.

Information on feeding is limited. South American fur seals are known to eat anchovies, sardines, and mackerel, as well as cephalopods, crustaceans, lamellibranchs, and gastropods. The cold, productive upwelling systems along the Peruvian and Chilean coasts usually support large prey populations. However, in some years food availability may be reduced dramatically as waters warm and upwelling is reduced due to the effects of El Niño.

On Uruguayan islands, births occur throughout November and December, although most pups are born in late November and early December.

South American fur seals are polygynous. Adult males defend territories but do not maintain harems. Losers in territorial disputes generally gather on the seaward periphery of the rookery. Younger males who have not attained full sexual status congregate around the rookeries and engage in mock combat after the breeding season is finished.

Along the Peruvian coast, pupping and breeding occur from mid-October through mid-December, with most births in November. During the breeding season, parous females come into estrus 5 to 8 days after giving birth. Some females reach sexual maturity at 3 years of age and give birth for the first time when 4 years old. Some nurse pups for more than a year, even at the expense of a newborn pup.

Postpartum females alternate between average periods of 4.6 days at sea feeding and 1.3 days ashore nursing their pups. While at sea, females dive and feed mostly at night. They reach depths

A tranquil moment in the life of a South American fur seal and pup. (New Island, Falklands, 22 December 1989: Frank S. Todd.)

of 40 m and remain submerged for nearly 3 minutes. Occasionally, they may dive for up to 7 minutes and to depths of 170 m. Evidently, a pup relies on its mother's milk rather than feeding at sea for most of its first year of life.

Pup mortality during the first month of life ranges from 10 to 47 percent, although it may be as great as 80 percent when intense storms occur during the breeding season. The high density of breeding seals on some beaches also results in high pup mortality.

Pups are weaned when 7 to 36 months old, depending on local environmental conditions. Recent studies have shown that in cold years, half of the survivors of the previous cohort are weaned when 9 to 10 months old. In warm years weaning is delayed, and pups may be suckled for 20 months or more.

No data are available on vital rates or life span, although one male tagged as a pup was resighted at age 21.

Several species of shark prey on young pups, and killer whales evidently eat some fur seals near the Uruguayan island rookeries. No important predators have been documented in Peruvian waters, but male South American sea lions may kill as much as 13 percent of the fur seal pups on some rookeries. Shark attacks on young and adult fur seals feeding offshore are rare.

HISTORY OF EXPLOITATION: South American fur seals contributed substantially to the diet of Fuego-Magellanic canoe peoples who occupied the Beagle Channel and Chilean fiords by at least 6000 years ago, and to that of the Charrua Indians along the Uruguayan coast until these people were exterminated by the Spaniards in the 1700s and 1800s. The fur seals were an important source of calories, protein, and fat, and their pelts were used for clothing.

The earliest record of exploitation by Europeans dates to 1515, when Juan Díaz de Solis explored the coast of Uruguay. The skins collected by his expedition were sold later in Seville, Spain. Extensive commercial exploitation, however, did not begin in Uruguay until 1724 and was not formally organized until 1792, when the king of Spain directed the Real Compania Maritima to supervise harvests. The main purpose of fur sealing at this time was to provide lighting oil for nearby towns. Since 1808, sealing has been regulated by the local government. Early in the twentieth century, the oil of fur seals and sea lions was used in hospitals throughout Uruguay to treat tuberculosis. More than 21,000 kg of this oil was produced in 1921 alone. The Uruguayan fur sealing industry is the longest sustained operation of its kind in the world. Between 4500 and 14,000 seals were killed each year until 1982. Since then, fewer have been killed because of a decline in the demand for furs. Relatively good records have been kept. It appears that sealing was within sustainable limits for at least the past century,with at least 750,000 seals killed from 1873 through 1983. Since 1950, mainly young males have been taken, largely for their pelts and oil. More than 23,000 pups were born at the Uruguayan rookery on Lobos Island in 1988, and more than 44,000 in 1989. The current annual catch in Uruguay is 7000 males.

United States whalers killed South American fur seals during the late eighteenth century. Also, beginning soon after the U.S. Revolutionary War, vessels from the United States sailed specifically to seal in the Southern Hemisphere. For example, the 1000-ton ship *States* (most of these ships were 150 to 200 tons) harvested 13,000 fur seal skins from the Falklands (these were sold in New York for 50 cents apiece and then resold in China for 5 dollars each).

In Peru, Native peoples hunted fur seals perhaps as early as 4000 years ago. Between 1925 and 1946, more than 800,000 seals were killed commercially, although it is not clear whether all were South American fur seals since South American sea lions also occur in the area. By 1943, few fur seals remained along the Peruvian coast. The killing of seals was prohibited between January and April yearly from 1946 through 1959, and entirely thereafter. With this protection, colonies increased from a remnant population of perhaps 40 seals in 1951 to nearly 22,000 by the early 1980s.

Today, South American fur seals are harvested commercially only in Uruguay.

CONSERVATION STATUS: Seal poaching, particularly by local fishermen, is common in Peruvian waters. Depletion of anchovy and overfishing of other prey organisms along the Chilean and Peruvian coasts during the past two decades may have prevented further increases in seal populations. It seems unlikely that the South American fur seal in these areas ever will become as abundant as it is thought to have been prior to commercial sealing. In the Chilean fiords some fur seals are taken for use as bait in crab fisheries.

The total population of South American fur seals is well over 300,000, perhaps close to 350,000. Although many local populations have been reduced substantially from historic levels, the species as a whole is not in immediate danger. However, South American fur seals as well as South American sea lions and some smaller cetaceans (Peale's dolphins and Burmeister's porpoises) that inhabit the fiords of the Chiloe Island region of southern South America face a new threat: the lucrative and expanding salmon culture industry. South American fur seals seasonally use river mouths, bays, and inlets fed by freshwater streams. The reasons for their dependence on these areas are not well understood. The salmon culture industry has occupied the same bays and inlets with large fish-rearing pens, thus excluding the fur seals. In addition, individual fur seals are killed when they become entangled and drown in the nets placed around the pens to protect the salmon.

FURTHER READING: Guerra C. and Torres N. (1987), Majluf (1987), Majluf and Trillmich (1981), Trillmich and Majluf (1981), Vaz-Ferreira and Ponce de Leon (1987).

South African and Australian Fur Seals

Arctocephalus pusillus
(Schreber, 1776)

NOMENCLATURE: The two subspecies, South African (*A. pusillus pusillus*) and Australian (*A. pusillus doriferus*) fur seals, are indistinguishable in the field, but slight cranial differences have been noted. Their subspecific status is based primarily on their disjunct geographic distributions. The South African or Cape fur seal was described in 1776 from a young specimen collected near the Cape of Good Hope. The Australian fur seal is also sometimes called the Tasman fur seal.

The specific name *pusillus* means "little" in Latin. The original description of the species was based on a picture of a pup. Ironically, these are the largest fur seals. The subspecific name *doriferus* is a composite of the Greek *dora* for "hide" or "skin" and the Latin *fero* for "I bear."

DESCRIPTION: Australian fur seals grow to lengths of 1.7 m (females) to 2.2 m (males) and weigh 110 kg (females) to 360 kg (males). Adult males are dark grayish to brown, with a dark mane of coarse guard hairs. Mature females are lighter silver gray with a yellowish throat and chest. Pups average 80 cm long and weigh 7.1 kg (females) to 8.1 kg (males) at birth; they attain weights of 10 kg (females) to 12 kg (males) within a month.

South African fur seal males may weigh up to 353 kg (the average is 250 kg), while females may grow to 122 kg (average 58 kg).

Birth length is 60 to 70 cm; birth weight is 4.5 to 6.4 kg. The pup's black coat is molted between late February and April.

DISTRIBUTION: The Australian fur seal is confined to waters off southeastern Australia, primarily along the coasts of Tasmania and Victoria. The two largest colonies are at Lady Julia Percy Islands and at Seal Rocks off Victoria. Smaller colonies occur on the small rocky islands in Bass Strait between Victoria and Tasmania. Australian fur seals haul out on Maatsuyker Island and nearby islets off southern Tasmania and along the east coast of Tasmania, but evidently do not breed there. Small numbers of nonbreeding seals haul out along the coasts of New South Wales.

The range of the Australian fur seal overlaps that of the New Zealand fur seal, but the much greater size of the former makes it relatively easy to tell the two species apart. The two also differ in their vocal behavior and pelage color and in the biochemical composition of some of their blood proteins.

The South African fur seal is believed to be the parent stock of the Australian fur seal although the two are now separated by 115° of longitude. The South African fur seal occurs along the coasts of South Africa and Namibia. There are 24 breeding colonies and 10 haul-out sites on sandy beaches from Port Elizabeth westward to Cape Town and then northward to Cape Frio, near the southern border of Angola. Occasionally, seals range as far north as 11° S, evidently assisted by the cold, northward flowing, coastal Benguela Current. South African fur seals often select small rocky offshore islands for pupping and breeding, but six colonies, including the four largest, are on the mainland.

A crèche of jet-black South African fur seal pups backed by part of the huge mainland colony at a nature reserve on Cape Cross, Namibia. (November 1988: Fred Bruemmer.)

NATURAL HISTORY: Adult male Australian fur seals arrive on the rookeries in late October and establish territories. Some females arrive in October, but most arrive in late November and give birth about 2 days later. Females remain ashore for about 6 days after giving birth and are then mated. After this, they go to sea to feed for a few days before returning for 2 to 3 days to nurse their pups. This pattern continues for about a year although feeding trips gradually increase in duration to a week. A few pups are suckled for more than 1 or, rarely, 2 years.

Nearly 15 percent of pups die during their first 2 months. Males live for about 18 years and females for 21 years. Females first ovulate when 3 to 4 years old, and a few give birth when 4 years old. Males are sexually mature when 4 to 5 years old but generally do not breed until they are 8 to 13 years old. Nearly 10,000 pups are born in Australian and Tasmanian waters each year.

This seal's diet consists of fish, cephalopods, and crustaceans. Snook is the most important fish, and the squids *Nototodarus* and *Sepioteuthis*, cuttlefish, and octopuses are the most important cephalopods. Captive females eat about 7 percent of their body mass each day.

Small numbers of seals are killed by large sharks, especially the white pointer.

A breeding colony of six Australian fur seals was established at the Taronga Park Zoo in New South Wales in 1977; ten pups were born there from 1981 through 1986.

South African fur seal bulls arrive at the rookeries in mid-October to early November, in advance of the females. The bulls defend territories for about 6 weeks. These territories vary in size according to the degree of competition in the particular area. An average territory contains 28 females. Births occur from late October through late December; the median birth date is around 1 December. After giving birth, females remain ashore for about 6 days nursing the newborn pup before coming into estrus and mating. They then depart for their first postpartum feeding trip of approximately 3 days, leaving the pups ashore. Attachment of the embryo to the uterine wall is delayed for 3.5 to 4 months. During the first 3 months of life, each pup's mother spends an average of 2.5 days ashore nursing, alternating with 3 to 4 days at sea feeding. While at sea, lactating females dive to an average depth of 40 to 50 m for 1.5 to 2.5 minutes. Maximum recorded dive depths and durations for two females were 204 m and 7.5 minutes, but dives deeper than 150 m were not common. Daytime dives are somewhat shallower than those at night.

Overall, the number of South African fur seal births increased by just under 4 percent per year from 1971 through 1983, despite large commercial harvests of pups. Mainland colonies have grown about 7.5 percent annually, while the smaller island colonies have been declining at about 3.5 percent per year. One of the largest

Female South African fur seals carry their young pups to the water's edge and leave them there while they make feeding trips to sea. (Cape Cross, Namibia, November 1988: Fred Bruemmer.)

rookeries is on the mainland coast near the diamond mining town of Kleinsee, which is also near the edge of the Namaqualand Desert. Daytime ambient temperatures during the summer breeding season often exceed 30°C on generally cloudless days; consequently, many seals crowd near the splash zone and in the near-shore surf at midday to cool themselves. Up to 17,000 fur seals, most of them juveniles, haul out during some seasons at Cape Frio.

These fur seals are believed to be nonmigratory. They feed within about 5 km of the mainland coast on more than 20 species of fish and cephalopods. The most important fish prey are mackerel, pilchard, Cape hakes, and anchovies. Besides these commercially valuable species, fur seals, particularly off Namibia, consume large quantities of noncommercial bearded gobies. Occasionally, the seals eat jackass penguins, Cape gannets, and cormorants, with which they compete for rookery space. The entire South African fur seal population consumes an estimated 1.4 million tons of food each year, about a million tons of which is fish. In comparison, commercial fishing operations catch about 700,000 tons yearly.

At the mainland colonies, fur seal pups may be killed by black-backed jackals and brown hyenas, which also scavenge placentas and seal carcasses, while sharks and killer whales eat a few seals at sea. A zoo animal lived for 20 years, but the oldest tagged seal recovered so far was only 13 years old.

South African fur seals have bred in captivity. Some have bred with California sea lions and produced hybrid pups.

History of Exploitation: Commercial sealers may have killed 200,000 Australian fur seals along the southern Australian coast, particularly in Bass Strait, between 1798 and 1825. However, sealing records did not distinguish between fur seal species that occurred there or between fur seals and the sympatric Australian sea lion. Residents of Bass Strait killed seals there indiscriminately from 1825 until 1889, when the Fisheries Act of 1889 restricted harvests. In Tasmanian waters, residents of Cape Barren Island were permitted to kill seals in summer until 1923, and in winter thereafter until 1970, when harvesting was prohibited by the National Parks and Wildlife Act. Australian colonies along the Victorian coast were legally protected by the Game Act of 1890 and the Wildlife Act of 1975.

Although Portuguese explorers discovered large numbers of South African fur seals on islands off the southern coast of Africa in the late fifteenth century, commercial sealing did not begin there until the early 1600s, when Dutch settlers established a sealing industry. About 45,000 seals were taken near the Cape of Good Hope in 1610, and most of the colonies near Cape Town were destroyed soon afterward. French and British sealers and merchants continued sealing through the 1600s. In the late 1700s, British and U.S. sealers began extensive operations at the large colonies on the west coast. With no legal restrictions on the numbers, ages, or sex of seals killed, the fur seal stocks were depleted rapidly. By the late 1800s, at least 23 colonies had been destroyed.

The first legal restrictions on sealing in southern Africa came in 1893 under the Cape Fish Protection Act that required government licenses for all sealers. Commercial harvests during the breeding season were prohibited beginning in 1909. Sealing along the Namibian coast was not regulated until 1922, with the introduction of the Sealing and Fisheries Proclamation. Additional restrictions came in 1949 under the Sealing and Fisheries Ordinance. However, at some colonies, commercial killing of "pups" for pelts has continued with little interruption. Pups are defined by law as males up to 3 years old and females up to 2.

More than 2.5 million South African fur seals were taken between 1900 and 1983. The South African government conducted the harvests until 1979, when private concessioners took over the operations at 9 of 23 rookeries. Pups were killed by a blow to the head with a heavy club; bulls were shot in the head. Between 2600 and 9300 seals were killed each year from 1900 through 1910. The harvest fluctuated between none and about 170,000 per year during the 1930s. Annual harvests were between 27,000 and 45,000 during the 1950s and increased to between 62,000 and 81,000 during the 1970s. Adult males, usually fewer than 5000, were also taken in summer. The harvest of bulls increased during the 1980s to as high as more than 20,000 in 1984, largely to supply a market for the genitalia, dried and sold in powdered form as an aphrodisiac in the Far East. At most colonies, only the genitalia were collected; the rest of the animal, including the blubber and skin, was wasted.

Conservation Status: The killing of Australian fur seals is prohibited in Tasmanian waters under the National Parks and Wildlife Act of 1970 and in Victorian waters under the Wildlife Act of 1975. The seals breed at 11 colonies and haul out seasonally at another 21 sites around southeastern Australia and Tasmania today. Numbers have remained fairly constant since about 1945, with approximately 10,000 pups being born each year. Increased commercial harvesting of fish and increased mortality of seals in fishing gear may prevent the Australian fur seal population from expanding.

The Sea Birds and Seals Protection Act has protected South African fur seals on island colonies since 1973. Landing on the islands and disturbing, capturing, or killing seals requires a government permit specifying the site, season, age, and sex of any seals expected to be disturbed or taken. Commercial sealing operations on the South African coast were suspended in 1990, and this suspension was extended through at least mid-1992.

In 1983, the number of South African fur seals was estimated at 1.1 million, of which about 271,000 were pups. The current population may be close to 1.4 million. The declines and recoveries of fur seal herds have affected the breeding success of seabirds. As an example, the seal colony on Mercury Island, off Namibia, had been extirpated by the end of the nineteenth century. Numbers of breeding seabirds increased dramatically. Seals began recolonizing Mercury Island in 1979 or 1980, and 3600 pups were born there in 1985. With the return of the seals, bank cormorants and Cape gannets were displaced from the island, and the numbers of jackass penguins and Cape cormorants breeding there declined precipitously.

Fishermen complain bitterly about the way fur seals rob them of their catches, even taking fish off baited hand lines. Fur seals also interfere with purse-seining operations by frightening shoals of anchovy and pilchard and causing them to sound before the nets can be pursed, or by entering the nets in large numbers and raising havoc there. Attempts to frighten the seals away from purse-seine nets by using firecrackers and other loud explosive devices or by broadcasting the sounds of killer whales have been ineffective. The seals move a short distance away, but most return soon after the sounds stop.

Further Reading: Bruemmer (1988), David (1989), Lipinski and David (1990), Shaughnessy (1985), Shaughnessy and Warneke (1987), Warneke and Shaughnessy (1985).

New Zealand Fur Seal

Arctocephalus forsteri
(Lesson, 1828)

NOMENCLATURE: The species is named after J. G. A. Forster, a naturalist on Captain James Cook's circumglobal voyage of 1772–1775. Forster described fur seals encountered at the South Island, New Zealand.

DESCRIPTION: Pups are about 55 cm long at birth. Males may weigh 3.9 kg at birth, gain 45 to 74 g per day for the next 60 days, and weigh 14 kg when 10 months old. Adult males weigh 180 to 200 kg and are about 2 m long. Females may weigh 3.3 kg at birth, gain 46 to 61 g per day until 2 months old, and weigh 12.6 kg when 10 months old. Adult females are substantially smaller than adult males, and most weigh 30 to 35 kg. An exceptionally large 12-year-old female weighed 50 kg, and the largest male weighed was 154 kg.

At birth, pups have a coat of long black hair. At 2 to 3 months, the natal coat is replaced by shorter hair, similar to that of adults. Adult coloration is similar in both sexes: dark gray-brown dorsally and lighter ventrally. However, the pelage of New Zealand fur seals is lighter brown than that of Australian fur seals, an important distinction for recognizing the two species where they occur together.

DISTRIBUTION: The New Zealand fur seal is distributed discontinuously around the Southern Ocean. It is present on the coasts of the South Island and the southern part of the North Island, New Zealand; certain New Zealand subantarctic islands; Macquarie Island; the southern coast of Australia eastward from Eclipse

A male New Zealand fur seal rests on the rugged New Zealand coast at Kaikoura. (February 1981: Stephen Leatherwood.)

Island and the Recherche Archipelago to Kangaroo Island; and the Maatsuyker Island group off southern Tasmania.

In New Zealand, most colonies are on the rocky exposed western coast of the South Island. Archaeological evidence suggests that some New Zealand fur seals hauled out on the North Island until the fourteenth century. There have been few records of their occurrence on the North Island during the past 200 years until recently. Now New Zealand fur seals haul out regularly around the southern coast of the North Island in the vicinity of Cook Strait and Wellington. Smaller haul-outs have been established recently as far north as Auckland. Breeding colonies inhabit the Auckland, Stewart, Campbell, Chatham, Bounty, and Snares islands south of the South Island. An estimated 50,000 fur seals were present in New Zealand waters in 1982, mostly on the New Zealand mainland; 16,000 were counted at the Bounty Islands. New Zealand fur seals appear to be increasing in number, and new colonies have been established at Kaikoura, Banks Peninsula, and Otago along the Nelson coast and in a number of locations on the west coast of the North Island.

The breeding colonies on the rocky islands off South and Western Australia are mainly in caves and crevices. The main colonies in South Australia are at Cape du Couedic and Cape Gantheaume on Kangaroo Island, nearby Casuarina Islets, the Neptune Islands, and the Whidbey Island group. In the mid-1980s, the Australian population numbered over 2500, with 600 to 700 in Western Australia and about 2000 breeding on South Australian rookeries. Most seals that haul out at Cape du Couedic are immature nonbreeders, but a few pups were born there in 1983. At least 100 New Zealand fur seals were present on Maatsuyker Island in the late 1980s, producing at least 15 pups per year. Fur seals also occur along the southeastern coast of Australia, but these are Australian fur seals.

New Zealand fur seals also haul out, but do not breed, at Macquarie and the Antipodes islands. Fur seals are present on rookeries and haul-out sites year-round, although the proportions of different sex and age classes vary seasonally. In New Zealand waters, bulls and adolescent males move north after the breeding season and return south the following spring. At Macquarie Island, young males are present year-round, particularly at Northeast Peninsula. Nearly 1200 seals were counted in 1982, with most hauled out from November through March and a few present from July through November. These New Zealand fur seals share space with resident breeding antarctic fur seals and subantarctic fur seals, although they outnumber both.

Natural History: Adult males arrive on New Zealand and Australian rookeries and establish territories from mid-October through November. Males fast for up to 2 months while defending their territories and herding or mating with females. They are inactive about 90 percent of the time ashore, allowing them to conserve energy and thus maintain territorial tenure for such a long time without eating. Nearly all adult males have returned to sea by mid-January, and few are ashore in winter. Females give birth 2 to 3 days after hauling out from late November through early January, with most births occurring in early to mid-December. Females remain with their newborn pups for approximately 9 days before mating and going to sea to feed. They then alternate feeding trips of 1 to 5 days, while pups remain ashore, with 2 to 7 days on shore nursing. Feeding trips lengthen as pups age. Females and pups can be found at the rookeries year-round, although the number of seals on shore is lowest during the winter (March to October).

Nearly 80 percent of the pups survive until 50 days old, and 60 percent survive from birth until weaning at about 300 days old. There are no data on postweaning survival. Males are able to maintain territories when 10 years old, although they presumably mature sexually several years earlier. Some females give birth for the first time when 5 years old, but there are no detailed data on their rates of survival or reproduction. Maximum longevity is greater than 15 years for males and females.

New Zealand fur seals have no natural terrestrial predators, but sharks and killer whales occasionally kill them at sea. Adult male New Zealand sea lions occasionally may kill and eat fur seal pups on some rookeries where they occur together. Small crater wounds on a seal stranded on the Australian coast were believed to be from a cookie cutter shark.

These fur seals eat mostly cephalopods (the squids *Nototodarus* spp. and octopus) and fish, especially barracouta. Some have been seen eating rockhopper penguins at subantarctic Campbell Island south of New Zealand. Captive males eat up to 10 kg of food per day, females up to 7.5 kg per day, and juveniles up to 3 kg per day.

History of Exploitation: In Australian waters, New Zealand fur seals were hunted by Aborigines for meat and skins long before Europeans ventured to this region. Commercial sealing began soon after Matthew Flinders discovered the fur seals in 1798. The enterprise continued along the South Australian coast through 1840. At least 70,000 seals were killed at Kangaroo Island and nearby islands from 1804 through 1834, and more than 143,000, although not necessarily all New Zealand fur seals, were taken in southeastern Australia.

Approximately 4500 fur seals were killed in New Zealand waters in 1792–93, although large-scale commercial sealing did not begin until sealing in South Australia began declining in the early 1800s. After the New Zealand subantarctic island colonies of fur seals were discovered, sealing expanded quickly, and seal populations soon were exterminated at most sites. At the Antipodes islands, more than 60,000 seals were killed in 1804–05. Australia's Macquarie Island was discovered in 1810, and almost 200,000 seals were killed there before 1820. Between 62,000 and 100,000 seals were killed in 1811 alone, and harvests declined dramatically afterward as few seals remained. By 1820, no seals survived at Macquarie Island, and by 1830, none were left at the Auckland and Snares islands. It has been impossible to determine from sealing records the proportions of New Zealand, Australian, subantarctic, and antarctic fur seals killed during this intensive sealing era.

Conservation Status: Fur seals are protected by national legislation in Australia, although there is little management by the Australian National Parks and Wildlife Service because of the remoteness and inaccessibility of most colonies. An exception is the colony at Cape du Couedic on Kangaroo Island, where many thousands of tourists per year, in addition to local residents, have relatively easy access to the approximately 1000 seals, mostly nonbreeders, which haul out there.

Sealing was prohibited by the New Zealand government in 1894, although some licenses were granted between 1913 and 1916 and in 1922 at Campbell Island. In the early 1900s, local fishermen began complaining that fur seals were depleting fish stocks, and in 1946 permits were granted to kill approximately 6000 fur seals at the South Island, Stewart Island, and nearby islands. In 1978, the New Zealand Marine Mammal Protection Act was passed, prohibiting the killing of fur seals and other marine mammals along the New Zealand coast and out to 200 nautical miles from the coast.

Some fur seals become entangled and drown in commercial set nets or in plastic shipping bands and discarded netting from fishing vessels. Also, a recently developed and expanding fishery for hoki off the west coast of the South Island killed several hundred fur seals in 1989 and 1990. The effect of this mortality on the local populations is not clear. In 1989, the combined population of fur seals at Westland and Fiordland was estimated at 20,000 to 22,000. These sites were the likely sources of most of the seals killed by the hoki fishery.

FURTHER READING: Abbott (1979), Beentjes (1990), Brothers and Pemberton (1990), Cawthorn et al. (1985), Ling (1987), Mattlin (1987), Robinson and Dennis (1988), Shaughnessy and Fletcher (1987).

Antarctic Fur Seal

Arctocephalus gazella
(Peters, 1875)

NOMENCLATURE: The specimen from which this species was described was carried from Kerguelen Island on the German vessel SMS *Gazelle;* the antarctic fur seal was named in the ship's honor. It is also commonly known as the Kerguelen fur seal.

DESCRIPTION: Pups weigh 4.8 to 5.2 kg (females) or 5.4 to 5.9 kg (males) at birth and about 14 kg (females) to 17 kg (males) when weaned 110 to 115 days later. Adult males are 1.9 m long and average 188 kg in weight (230 kg maximum) at the beginning of the breeding season. Females are smaller, averaging only about 1.2 m and 35 kg (maximum 50 kg).

Both sexes are gray-brown, although females have a lighter belly and adult males have a dark mane. A small percentage of pups are leucinistic.

DISTRIBUTION: The present breeding range is from about 61° S north to the Antarctic Convergence. Small but increasing numbers also breed on two colony sites at subantarctic Marion Island, where they hybridize to some extent with subantarctic fur seals. The overwhelming majority of antarctic fur seal pups (about 200,000) are born at South Georgia and nearby Bird (47,000 in 1989 there alone) and Willis islands. In recent years, colonies have been established at the South Orkney, South Shetland, and South

Sandwich islands and at Bouvet, Heard, McDonald, Kerguelen, and Macquarie islands. Growth at many of those colonies has been rapid, primarily fueled by immigration from the large colonies at South Georgia.

In the austral summer of 1986–87, almost 30 years after the first births were observed at Cape Shirreff on Livingston Island, nearly 4000 pups were born at various small rookeries throughout the South Shetland islands; most of the births were on Elephant and surrounding islands (approximately 780), King George Island (approximately 160), and Nelson and Livingston islands (approximately 3050).

In late summer, some adult males can be seen hauled out on ice floes along the ice-edge zone in the Weddell Sea. Many young males haul out at or near Signy Island in summer, and similar, but smaller, nonbreeding aggregations occur in many parts of the northern Antarctic Peninsula. It has been suggested that females migrate north of the Antarctic Convergence. A few antarctic fur seals have been seen in recent years at the Juan Fernández Islands, hauled out among Juan Fernández fur seals.

NATURAL HISTORY: Adult males arrive at rookeries from October through early December. At Bird Island, the mean arrival date for all males is between 24 November and 1 December. Older, experienced individuals often arrive a week or two earlier. Males maintain breeding territories for 20 to 40 days, using ritualized boundary displays and vocal threats. Fights occur frequently. Territorial bulls weigh up to 205 kg early in the breeding season but may lose 1.4

A large antarctic fur seal bull presides over his harem at Seal Island, a small rocky islet near Elephant Island, off the Antarctic Peninsula. Chinstrap penguins, occasional prey of these aggressive seals, are in the background. (December 1987: Brent S. Stewart.)

to 1.5 kg per day during their tenure, which averages 29 to 31 days. Of this mass loss, 54 percent is fat, 10 percent is protein, and 36 percent is water.

Females give birth from late November through late December, about 2 days after arriving at the rookeries. The median birth date is between 4 and 8 December at South Georgia. Females remain ashore nursing their pups for 5 to 8 days postpartum. After mating at the end of the perinatal period, they begin feeding at sea for 3 to 6 days at a time while the pups remain ashore. The females make shore visits of 1 to 2 days to nurse their pups. At birth, pups weigh an average of 4.2 kg. They gain 70 to 100 g of body mass per day, so that by the time they are weaned, approximately 117 days after birth, they weigh about 14.7 kg (females) or 17.8 kg (males).

Pup survival varies with the density of breeding seals, the weather, and the availability of krill in the vicinity of nursing rookeries. Many pups die from starvation when their mothers have difficulty finding food during foraging trips offshore.

Annual mortality of adult females is about 8 percent. Males and females die at similar rates until they are 7 years of age, when male mortality increases to about 30 percent per year. Females are sexually mature when they are 3 years old, and 57 percent of them give birth for the first time at 4. Males are sexually mature when they are 3 or 4 years old, generally do not breed until 7 years of age, and may live for 14 years or more. Approximately 75 percent of the females return to breed at the same site for two consecutive years; about 29 percent return to the same site every year that they breed.

In summer, lactating females feed primarily on krill in most areas. Some scientists have suggested that large reductions in krill-eating competitors, particularly whales, have encouraged the rapid growth of fur seal populations. Thus, antarctic fur seals may be more numerous now than they were before sealers depleted them. During the nonbreeding season at South Georgia and at Macquarie Island, some fur seals eat fish as well as krill.

While at sea, lactating females dive and feed on krill mostly at night, usually to depths of 30 to 40 m. Some dives may exceed 250 m. Average dive duration is about 2 minutes.

At Bird Island, fur seals (mostly certain subadult males) occasionally kill and eat macaroni penguins and, perhaps, king and gentoo penguins. Penguins do not appear to be an important part of their diet, however, as only a few individuals engage in this behavior. Moreover, these seals eat only small portions of the birds they kill. The partially eaten penguin carcasses are taken by giant petrels that feed them to their chicks; this additional food may be important to Bird Island's increasing petrel population.

Although killer whales may eat some fur seals, they are not considered significant causes of mortality as they are not common

Antarctic fur seals haul out on tussock grass, such as this clump in a fresh-water melt stream at Gold Bay, South Georgia. As populations of fur seals increase, they damage the vegetation, increasing erosion. (18 February 1990: Stephen Leatherwood.)

around the larger rookeries at South Georgia. Leopard seals kill and eat fur seals near some rookeries. At Seal Island, near Elephant Island, leopard seals also have been observed to haul out regularly among breeding antarctic fur seals and to stalk and kill young fur seal pups on shore.

Because of the rapid increase of fur seals at South Georgia, where 95 percent of the world population of antarctic fur seals now breed, more than 60 percent of the tussock grass habitat in certain areas (mainly Bird Island and the northwestern extremity of the main island) has been destroyed by females and pups that wander inland to rest in summer. Antarctic hair grass also has suffered along the stream banks and meadows due to trampling by fur seals as they move between the coast and inland haul-out areas. This destruction of plants by fur seals also has had detrimental effects on resident endemic birds, such as the pintail and pipit, which depend on tussock grass habitat for nesting sites.

A "golden" antarctic fur seal contrasts with the normally colored individuals in the background; this blonde color phase occurs in a small percentage of births. (Gold Bay, South Georgia, 18 February 1990: Hans Reinhard.)

HISTORY OF EXPLOITATION: Commercial sealing for antarctic fur seals by U.S. sealers began at South Georgia in the early 1790s. The peak of that industry was in the early 1800s; in 1801, nearly 112,000 seals were killed. Few seals remained alive by 1822, by which time perhaps as many as 1.2 million had been killed. Sealing was resumed at South Georgia between 1870 and 1907, and all seals that could be found at the recovering but small colony were killed. No sealing for fur seals has occurred there since 1907. Sealing began at the South Shetland Islands in 1820, a year after their discovery. Nearly 250,000 seals were killed in 1821, depleting the colonies there, and smaller colonies at the South Orkney (probably mainly bachelor bulls) and South Sandwich islands also were rapidly depleted. When their exploitation stopped, only a few hundred antarctic fur seals may have been left alive throughout their range.

CONSERVATION STATUS: From 1958 to 1972, the colony at South Georgia increased by about 17 percent per year. Although the rate of increase has slowed considerably since the early 1970s, the population at South Georgia now numbers approximately 1.8 million. All other extant colonies on the South Shetland, South Sandwich,

Kerguelen, McDonald, Heard, Bouvet, and Marion islands evidently were established by pregnant females immigrating from South Georgia. Most of those colonies also have been growing quickly in recent years. At least 0.1 percent of fur seals at South Georgia, and perhaps as many as 1 percent, have neck collars of debris such as packing bands, bits of nylon string, and fragments of fishing net. Approximately 71 percent of the entangled fur seals observed have been males, and 88 percent of these males were 4 years old or younger. The mortality from entanglement apparently has not yet begun to influence the rate of population growth significantly, but scientists are concerned since similar rates of entanglement of northern fur seals are thought to be partly responsible for their substantial decline in the Bering Sea.

FURTHER READING: Aguayo L. (1978), Bonner (1968, 1985), Boyd and McCann (1989), Condy (1978), Croxall et al. (1990), Kerley (1983), McCann (1980b), Payne M. R. (1977, 1978, 1979), Smith, R.I.L. (1988).

Subantarctic Fur Seal

Arctocephalus tropicalis
(Gray, 1872)

NOMENCLATURE: The specific name is from the Greek *tropikos,* "tropical," referring to the location along the north coast of Australia where the first specimen was mistakenly believed to have been collected. As most rookeries are on subantarctic islands north of the Antarctic Convergence, the species is appropriately called the subantarctic fur seal. It is also sometimes known as the Amsterdam Island fur seal.

DESCRIPTION: Males grow to a length of about 1.8 m and attain an average body mass of about 131 kg (maximum 165 kg). Females are smaller, growing to about 1.4 m long and an average body mass of about 35 kg (55 kg maximum).

At birth, pups are black, weigh between 4 kg (females) and 4.4 kg (males), and are about 63 cm long. When 1 year old, they weigh about 9.5 kg (females) to 12.9 kg (males). Adults are dark brown dorsally, with a lighter belly and a whitish-orange chest, nose, and face. Males have a tuft of white-tipped hair on top of the head; females lack this tuft.

The species was considered conspecific with the antarctic fur seal until recently, when the two were separated based on differences in cranial, dental, and biochemical features as well as geographical distributions.

DISTRIBUTION: Subantarctic fur seals haul out and breed north of the Antarctic Convergence in the South Atlantic and Indian oceans, mostly on the subantarctic islands of Amsterdam, Saint Paul, Crozet, Gough, Marion, Prince Edward, and Macquarie. The largest colony is at Gough Island in the Tristan da Cunha island group, where nearly 20,000 seals breed. A mother and pup observed at Heard Island in the late 1980s provided the first evidence of breeding south of the Antarctic Convergence. Breeding on Macquarie Island was confirmed only recently, and 15 births were recorded there in 1982 and 20 in 1983. In 1986, 37 fur seal pups were counted there, although that included some pups of antarctic fur seals that also breed there in very small numbers. New Zealand fur seals also haul out, but do not breed, on Macquarie Island. A subantarctic fur seal was born at Heard Island, in the midst of antarctic fur seals, in 1988.

Adult males occasionally appear on South Georgia among breeding antarctic fur seals, but interspecific matings have not yet been observed. A few seals occasionally wander to Brazil, Cape of Good Hope (South Africa), and, rarely, Australia. In October 1983, two fur seals were observed near the Kwanza River on the coast of Angola; one of them was positively identified from photographs as an adult male subantarctic fur seal. In recent years, a few subantarctic fur seals have been seen among breeding Juan Fernández fur seals on the Juan Fernández Islands off western South America. The species clearly is capable of dispersing over great distances.

At Gough Island, most rookeries are on the western, windward coasts, where cooling from ocean spray and wind reduces heat stress. Few males are ashore in winter, but their numbers swell rapidly in November as they arrive to establish territories. Peak numbers occur in mid-December. The males begin leaving the island in January, and few remain by the end of the month. Females and pups come ashore year-round, although fewer are hauled out at any one time from April through October than in the summer.

Natural History: Pups are born from late November through early January. Most births take place between 9 and 13 December; at Prince Edward Island, the median birth date is 17 December. The sex ratio at birth is 1:1. Mating occurs about a week postpartum, but attachment of the embryo to the uterine wall is delayed for 128 days. In January and February, lactating females at Marion Island alternate between periods of about 5 days at sea feeding and 2.5 days on shore nursing their pups. Pups are nursed for about 300 days. Nearly 90 percent of the newborns survive for 6 weeks. Females are sexually mature at 4 to 6 years of age and live for 23 years or more; by age 5 nearly 80 percent of females have ovulated at least once. Males are sexually mature when 4 to 8 years old, although few breed before they are 8. Full adulthood is usually not reached until 10 or 11 years of age. At Gough Island, all males reach puberty at 4 years of age, but they generally are not socially mature for another 4 years. Bulls are physically mature when 8 to 11 years old, and they may live for 18 years or more. Adult males and females molt once each year between March and May.

Giant petrels and skuas scavenge dead pups. Killer whales, which commonly occur around some rookery islands, may eat some seals, although such predation has not been observed directly.

The diet of these fur seals consists mostly of cephalopods of several families, some fish, and, occasionally, rockhopper penguins.

Subantarctic fur seals share space with small numbers of antarctic fur seals on the Prince Edward Islands where they hybridize to some extent; approximately 0.1 percent of seals there in 1982 were believed to be hybrids.

History of Exploitation: Some 300 years after Gough Island was discovered in 1505, U.S. sealers began killing seals there. Approximately 5600 seals were taken in 1790, and by 1820 few remained. Some sealing resumed between 1860 and 1890, but so few seals were left alive that sealing completely stopped by 1892.

Sealing began at the Prince Edward Islands soon after they were discovered in 1772, and few seals remained by the early 1800s. Some small-scale sealing occurred in the early 1900s, but it has been prohibited since 1948.

Sealing at the Crozet Islands began in 1803 and continued through 1850, although seal numbers were low by 1825 and few if any remained by 1900.

At St. Paul and Amsterdam islands, sealing began in 1789. By 1835, seals no longer bred on St. Paul; by 1850, only a few remained on Amsterdam. Although by 1900 the species was believed to be locally extinct, a small colony was found on Amsterdam Island in 1950.

Conservation Status: Births increased on St. Paul and Amsterdam islands at annual rates of 16 to 17 percent between 1971 and 1985. Approximately 11,000 pups were born at Amsterdam Island in 1982, and 66 were born at St. Paul Island in 1985.

The colony at Gough Island has increased by 13 percent annually since 1955; nearly 59,000 pups were born there in 1978. Births increased by 11 percent annually at the Prince Edward Islands from 1951 until 1975 and by 15 percent a year from 1974 through 1982; in 1982 approximately 3800 pups were born at Marion Island and approximately 3000 on Prince Edward Island.

From 1979 through 1985, births at the Crozet Islands increased by 18 percent per year; nearly 100 pups were born there in 1984.

The world population was estimated at 270,000 in 1983.

Further Reading: Bester (1981, 1982, 1987), Roux (1987), Roux and Hes (1984).

Northern (Steller) Sea Lion

Eumetopias jubatus
(Schreber, 1776)

NOMENCLATURE: The generic name is from the Greek *eu,* meaning "typical" or "well," and *metopion,* "having a broad forehead," referring to the prominent forehead. The specific name is from the Latin *jubatus,* "having a mane," referring to the well-developed mane of adult males. These sea lions are also commonly known as Steller or Steller's sea lions, after the German surgeon and naturalist aboard the Russian ship *Vitus Bering.* While shipwrecked at Bering Island in 1742, Steller described the sea lions that he saw there. Aleuts call them *seevitchie;* Russians, *sivuch,* meaning "sea wolf."

DESCRIPTION: Newborns are dark brown to black and weigh 16 to 23 kg; they reach 40 kg when 6 to 10 weeks old. At 4 to 6 months of age, they shed their natal pelage and grow new hair that is lighter brown. Through successive molts, the pelage becomes lighter, until by age 2 it is similar to the adult color.

Adults are sexually dimorphic in size. Males grow to 3.25 m (on average, 2.8 m) and 1120 kg (on average, 566 kg); females to 2.9 m (average, 2.3 m) and 350 kg (average, 263 kg). Both sexes are light tan to reddish brown, slightly darker on the chest and belly, but females are generally lighter than males. The snout is short and straight. There is a space between the upper fourth and fifth postcanine teeth.

DISTRIBUTION: Northern sea lions occur throughout the North Pacific rim from central California to Japan. They formerly bred in small numbers as far south as the southern California Channel Islands. In 1982, only two pups were born there (at San Miguel Island); no northern sea lions have been seen on the Channel Islands since 1984. Small rookeries now exist in California only on Año Nuevo Island, Southeast Farallon Island (where some 30 pups are born annually), and Cape St. George in northern California. Several hundred pups are born each year at Orford and Rogue reefs off the Oregon coast. Small numbers haul out along the Washington coast seasonally, mostly in October and November, but none breed there.

Small breeding rookeries and haul-out sites exist along much of the British Columbia coast. The main rookeries are on small islands near the northern tip of Vancouver Island, at Cape St. James, and North Danger Rocks. The total British Columbia population is 4000 to 5000.

Most reproduction occurs at scattered rookeries along the central coast of the Gulf of Alaska and in the central Aleutian Islands. Few of the rookeries now produce more than 1000 pups each year. In 1989, there were approximately 2200 northern sea lion births at Marmot Island, 700 on Chirikof Island, 360 on Bogoslof Island, 530 on Seguam Island, and 300 at Lier Cove on Kiska Island. The northernmost rookery is at Seal Rocks in Prince William Sound. Although large numbers of pups were born on the Pribilof Islands in the Bering Sea historically, none are known to have been born at St. Paul Island or St. George Island since about 1916. Only about 300 births now occur each year at nearby Walrus Island, down from 3000 in 1960. Small numbers of sea lions haul out seasonally, but no breeding occurs, on St. Matthew, St. Lawrence, Diomede, Round, Walrus, and Amak islands and along the Alaskan coast in Bristol Bay at Cape Newenham. There are small rookeries in the western North Pacific on the Commander Islands and in the Okhotsk Sea on Robben, Iony, and Yamskiye islands. The largest western Pacific rookeries are on the Kuril Islands.

The rangewide population was estimated to be approximately 290,000 in 1985, nearly 68 percent of which were in Alaskan waters, but numbers have continued declining throughout the range from unknown causes. Only 25,000 sea lions were counted in 1989 from the central Gulf of Alaska to the central Aleutian Islands, down from more than 67,000 in 1985.

NATURAL HISTORY: Breeding is polygynous. Males arrive on the rookeries in May and establish territories through ritualized displays and vocal threats. Vigorous fights occur occasionally but are usually brief. Males fast for 1 to 2 months while they maintain breeding territories. Females arrive in late May and June and give birth to single pups about 3 days later; most pups are born in late June. Mothers remain ashore and nurse their pups for 9 days before going to sea to feed. They then alternate 1 to 3 day feeding trips (36 hours on average) with a few hours to 2 days ashore nursing (21 hours on average). The foraging trips lengthen as the pups age. The pups ingest about 1.8 liters of milk each

An unsettled atmosphere prevails on this northern sea lion rookery, as two pairs of animals exchange aggressive displays. (Marmot Island, Alaska, August 1972: Fred Bruemmer.)

day and gain 0.4 kg per day. Estrus and mating take place 11 to 14 days postpartum. The embryo continues to develop for a few weeks but then stops until it attaches to the uterine wall in late September or early October, at which time growth resumes.

While the mothers are at sea feeding, the pups remain ashore, often joining one another to form small groups near the water's edge. They gain swimming experience by playing in tide pools. When a month old, some pups begin accompanying their mothers on short trips to sea. Most are weaned before the end of their first year, although a few may continue getting milk until 2 or, rarely, 3 years of age. Nonreproductive females and juveniles molt in late

A northern sea lion bull grooms himself, the picture of contentment. (Marmot Island, Alaska, August 1972: Fred Bruemmer.)

July and August, followed by lactating females, and then by adult males later in autumn.

Individuals are believed to range widely during the nonbreeding season. Males in particular evidently move north at the end of the breeding season and do not return south to the rookeries until the following spring. The pelagic movements and diving behavior of northern sea lions are being studied by scientists at the U.S. National Marine Mammal Laboratory in Seattle. They have found that some females routinely make 6-week trips to locations more than 550 km south of their haul-out sites. One female ranged to just south of Sitka in southeastern Alaska, making a round-trip of more than 2200 km. Northern sea lions evidently feed in the water column at relatively shallow depths. The deepest dive yet recorded was to 277 m.

Females reach sexual maturity when 3 to 8 years old (average, 6 years); they are physically mature when 10 and may live for

30 years. Males become sexually mature when 3 to 7 years old, although few breed before they are physically mature at 10 years of age. Males are not known to live for more than 18 years. Although the sex ratio is 1:1 at birth, about 74 percent of males, compared with 53 percent of females, die before they are 3 years old. Afterward, males continue to die at slightly higher rates than females, so that the adult sex ratio is about 3:1. Approximately 87 percent of females between the ages of 8 and 20 are pregnant each year.

Northern sea lions forage mostly near shore and over the continental shelf. Seasonal shifts in diet in various areas reflect local changes in the distribution of prey. In the Bering Sea and Gulf of Alaska, walleye pollock, important prey for northern sea lions, are also consumed by harbor seals and northern fur seals and harvested by a rapidly expanding commercial fishery. In Canadian waters, northern sea lions primarily eat herring, rockfish, cod, squid, and octopus. Some sea lions travel up freshwater rivers, particularly the Rogue River in Oregon, where they prey on lampreys and, to a much lesser extent, salmon parasitized by lampreys.

At St. George Island, in the Pribilofs, young male northern sea lions kill and eat northern fur seal pups as they play near shore between August and November. Nearly 5000 fur seal pups were estimated to have been killed by male sea lions in 1975. In other areas, male northern sea lions occasionally eat harbor seals, ringed seals, and, perhaps, sea otters.

HISTORY OF EXPLOITATION: The stocks of sea lions in British Columbia may have been depleted prior to the 1800s by Indians, who killed them for meat, hides, and oil. However, when the Native

A confrontation between two distantly related otariids—a northern fur seal on the left and a northern sea lion on the right. The two species were abundant in the cold temperate and subarctic North Pacific until recently; populations of both have declined alarmingly over the past several decades. (St. Paul Island, Pribilofs, July 1972: Fred Bruemmer.)

population declined precipitously because of introduced diseases in the 1800s, the number of sea lions evidently began increasing. Because of complaints by local fishermen, the Canadian government encouraged the killing of northern sea lions, with bounties, organized culls, and commercial harvests between 1913 and 1968. Numbers in Canadian waters were considerably reduced by the early 1970s, and the population that breeds in British Columbia has not fully recovered.

Small numbers of pups were taken by Russian sealers on rookeries in the Okhotsk Sea in the 1930s and 1940s. During the mid-1970s, about 50 sea lions, mainly adult males, were killed by Japanese hunters for a bounty in the area of Rebun Island, northwest of Hokkaido, Japan.

In U.S. waters, northern sea lions were killed in California in the early 1900s because of complaints by fishermen that the sea lions were competing with them for prey. Pups were harvested commercially from 1959 to 1972 in the eastern Aleutians, particularly at Ugamak and Akutan islands. More than 45,000 pups were killed at Sugarloaf and Marmot islands and a few other small rookeries in the Gulf of Alaska between 1963 and 1972. This commercial harvest was primarily for skins used to make clothing.

Northern sea lions occurred at San Miguel and, perhaps, Santa Rosa islands through the 1950s, although their abundance declined during the early and mid-1900s. Archaeological evidence from kitchen middens of Chumash Indians suggests that northern sea lions were never very abundant in southern California waters. In the 1800s and 1900s, commercial sealers killed an unknown number of northern sea lions for hides and oil on the Channel Islands, particularly San Miguel. Adult males also were killed for their testes and bacula (penis bones). These were sold in the Orient to be dried and ground to a powder that was considered an aphrodisiac.

CONSERVATION STATUS: The species is protected in U.S. waters under the MMPA, and no northern sea lions may be killed without a permit issued by the U.S. government. Nonetheless, in 1988 a multiyear exemption was granted to commercial fisheries operating in U.S. territorial waters of the North Pacific and Bering Sea. Fishermen have been allowed to kill up to 1350 sea lions incidentally during fishing operations, without penalty. On 5 April 1990 the northern sea lion was classified as a "threatened species" under the U.S. Endangered Species Act. In Canadian waters, northern sea lions have been protected from commercial and intentional killing since 1970 under the federal Fisheries Act.

Despite the end of commercial hunting in the 1970s, northern sea lion abundance has not increased. In fact, numbers declined somewhat in California, Oregon, and Washington, and by nearly 52 percent in the central Gulf of Alaska and the eastern Aleutian Islands from 1960 through 1985 . Declines, though not as great, also have occurred in the central and western Aleutians. At rookeries in central and northern California, births fell from more than 650 a year in the early 1970s to approximately 450 in 1981. In the Gulf of Alaska, there was an overall decline in births of approximately 25 percent from 1979 to 1984. Rangewide surveys in 1989 documented further substantial declines: nearly 63 percent fewer animals were counted in 1989 than in 1985 in the Aleutian Islands. At Alaska's Marmot Island, one of the largest northern sea lion rookeries, there were approximately 35 percent fewer births in 1989 than in 1988, and at one beach where 800 pups were born in 1988, none were born in 1989.

The reasons for these declines are not known. However, researchers believe that incidental mortality of sea lions in commercial fishing gear, shooting by fishermen, and reduction of important prey species, such as walleye pollock, by commercial fishing operations have contributed significantly to the observed declines. The total population in 1991 was estimated at about 40,000.

Further Reading: Bigg (1988b), Braham et al. (1980), Harestad and Fisher (1975), Kenyon (1962), Loughlin et al. (1987), Mathisen et al. (1962), Merrick et al. (1987), Orr and Poulter (1967), Pitcher (1981), Pitcher and Calkins (1981), Thorsteinson and Lensink (1962).

California, Galápagos, and Japanese Sea Lions

Zalophus californianus
(Lesson, 1828)

NOMENCLATURE: The generic name is from the Greek *za,* an intensive prefix, and *lophos,* meaning "crest" and referring to the large sagittal crest on the adult male's skull. Three subspecies have been recognized. The names of two, *Z. californianus californianus* and *Z. californianus japonicus,* refer to obvious geographic locations. The name of the third subspecies, *Z. californianus wollebaeki,* was assigned to the Galápagos sea lion for A. Wollebaek, who in 1925 collected from the Galápagos the sea lion skull from which the subspecies was described. In California, Mexico, and the Galápagos, sea lions are also commonly called *lobos marinos.*

DESCRIPTION: The California sea lion is perhaps the most familiar of all pinnipeds because of its appearance in zoos, aquaria, and circuses. Young individuals are trained easily and become versatile performers.

Males grow to 2.4 m and 390 kg; females to 2.0 m and 110 kg. Newborn pups average 0.8 m and 6.0 to 7.7 kg (females) or 6.7 to 9.0 kg (males). When 3 months old, pups normally weigh 16 to 18 kg (females) or 20 to 22 kg (males).

A young male California sea lion dives from the rocks at Los Frailes, Baja California, Mexico. (22 March 1985: Stephen Leatherwood.)

Pups are dark brown to black at birth, molt into a lighter brown pelage within the first month, and molt again when 5 to 6 months old. After this second molt, they grow the blond to light brown hair of adult females. Adult males are generally dark brown, but there are some blonds. Both sexes are lighter dorsally than ventrally.

Males begin developing a pronounced forehead and a broader chest when they enter puberty. The distinctive forehead is formed by the sagittal crest, a bony flange on the skull, combined with the soft tissue, mainly musculature, lying on each side of the crest. In everyday language, researchers often speak of the entire forehead of the male California sea lion as its "sagittal crest." The hair on the forehead is much lighter than that on the rest of the body. The "sagittal crest" distinguishes adult and subadult males from adult females and immatures, both of which have a relatively straight muzzle-forehead profile. The fore flippers are relatively broad and long and are haired distal to the wrist on the dorsal surface. Skulls of adult sea lions from California measure up to 309 mm in condylobasal length. Those of Galápagos sea lions are smaller (to 276 mm) and more slender.

DISTRIBUTION: The writings of early explorers, whalers, and natural historians indicate that California sea lions once hauled out and bred at many sites along the coasts of Baja California and southern California, including several of the southern California Channel

Islands. Locations and sizes of rookeries prior to the twentieth century are difficult to verify as estimates often are based on anecdotal accounts written in retrospect. Even the writings of the nineteenth-century whaling captain and naturalist Charles Scammon, which contain most of the information on early sea lion distribution and natural history, become confusing when inspected closely. Thus, in the final analysis, they are of little use in assessing the historic distribution and abundance of this species. When interpreting Scammon's accounts in the light of those by other nineteenth-century authors and present knowledge, it is clear that he was not aware of the distinctions between the two North Pacific species of sea lions in certain geographic areas. Scammon often blended the natural histories of different species, often from other areas, into his accounts of abundance, distribution, commercial harvests, and so on, taking secondhand information at face value, with no critical evaluation. Paul Bonnot (1951), who studied pinnipeds in California, stated that, "Before 1860, sea lions were extremely numerous along the California coast." The basis for this statement, and others concerning breeding range and location of rookeries, cannot be verified. It was apparently derived from Scammon's earlier, general, qualitative statements. Bonnot's additional conclusions that numbers had declined steadily until the late 1870s were also apparently based on Scammon's limited observations. In fact, statements about the magnitude of historic abundance and commercial and noncommercial killing of sea lions, originally based on anecdotal, or at least highly subjective, reports, often appear in later literature as authoritative statements. It is virtually impossible to evaluate the present abundance of California sea lions relative to historic abundance.

Today, California sea lions breed on San Miguel, San Nicolas, Santa Barbara, and San Clemente islands in southern California waters; small numbers also haul out seasonally at Santa Rosa, Anacapa, and Santa Catalina islands. The largest California sea lion colony is on San Miguel Island. A few pups are born occasionally at South Farallon and Año Nuevo islands off the central California coast. In Mexico, California sea lions haul out and breed on the Coronados, Guadalupe, San Martin, Cedros, and the San Benito islands off the Pacific coast of Baja California; and there are many smaller colonies on islands in the Gulf of California. An adult male sea lion (California or Galápagos) was recently taken in a shark gill net near the Mexico–Guatemala border (14°42′ N).

In autumn and winter, adult and subadult males from Channel Islands rookeries range as far north as the northern tip of Vancouver Island. Along the way, some haul out in small numbers along the central California coast; larger numbers haul out on Año Nuevo and South Farallon islands in November.

A race of California sea lions evidently lived off the coasts of Japan, and possibly Korea, through the early to mid-1900s. They

Galápagos sea lions inhabit equatorial waters of the Galápagos Islands, off northwestern South America. Although the islands are hot and dry, the surrounding waters are cooler than expected for these latitudes. (7 January 1990: Hans Reinhard.)

reportedly were widely distributed–on the Pacific coast near Tokyo, off Shikoku, in the Seto Inland Sea, off Kyushu, and at various locations in the Sea of Japan. By the 1950s, however, they were only reported to be present in small numbers (50 to 60) on Takeshima Island (37°15′ N, 131°52′ E) and speculated to exist in larger numbers on Korean islands. For more than 30 years, there has been no confirmation that California sea lions survive in the northwest Pacific. Those that did exist there were 8500 km from the nearest colonies in California. Since only eight specimens from Japan have been available for study, it has been impossible to determine whether there were significant morphological differences between the sea lions in the eastern and western North Pacific.

Galápagos sea lions are restricted to the Galápagos archipelago where they haul out and breed on flat sandy and rocky beaches on all of the larger islands and many smaller islets.

NATURAL HISTORY: Adult males establish breeding territories along the water's edge from May through July in Baja California and southern California and from May through January on the Galápagos Islands. They defend these beach areas against neighboring males; some males also patrol waters immediately adjacent to shore and defend access to small stretches of beach where females give birth and nurse their pups. Most interactions among territorial males are ritualized visual and vocal displays, but occasionally these

Copulating California sea lions. The much larger, darker male is also distinguishable from the female by his thick shoulders and mane and by his domed forehead. (San Miguel Island, California, July 1989: Brent S. Stewart.)

displays escalate to physical combat. Then males lunge at each other, attempting to bite the opponent's vulnerable fore flippers rather than the thick, well-protected neck and chest. The males' boundary-display vocalization, which sounds like a dog's bark, may be heard at any season but is incessant during the breeding season. Physiographic features such as small reefs, rocks, or tide pools often are adopted by bulls as territorial boundaries.

Females give birth 4 to 5 days after coming ashore in May and June; most births on California rookeries are in late June. Births on the Galápagos Islands are from late May through January, but they peak at different times on various rookeries. The sex ratio at birth is 1:1. Pups are nursed for about 8 days before their mothers go to sea to feed for about 2 days. Lactating females then alternate between feeding trips of 60 to 100 hours and nursing periods ashore of 30 to 70 hours. These cycles continue until the pup is weaned, generally after 4 to 8 months on California rookeries but frequently after more than a year and sometimes 3 years on the Galápagos Islands. Pup mortality during the first month of life is 10 to 15 percent. Pups on the southern California Channel Islands weighed 25 to 35 percent less in 1983 than in 1982. This difference was attributed to the poorer nutrition of pregnant and lactating females in 1983 because of the 1982–83 El Niño.

Estrus occurs 3 to 4 weeks postpartum, and most females are bred in the water or at the water's edge. The frequent movements of females between land and sea, and the wanderings of pups when left on the rookeries during this time, prevent males from effectively defending territories for extended periods of time. The breeding system is therefore not strictly polygynous but rather is more promiscuous; estrous females are mated as they depart for or return from feeding trips or rest near shore cooling themselves, by males who are successful in defending territories temporarily.

The life history parameters of California and Galápagos sea lions are poorly known. Approximately 71 to 79 percent of sexually mature female California sea lions at San Nicolas Island gave birth in 1984. The ages at sexual maturity and first birth are not known, but California sea lion females younger than 5 years old have not been observed to give birth. Females may live for 25 years.

Lactating California sea lion females at San Miguel Island forage within 100 km of the rookeries in highly productive areas of upwelling near Point Conception and south of San Miguel and Santa Rosa islands.

In southern California waters, California sea lions eat more than 50 species of fish and cephalopod, but mainly northern anchovy, Pacific whiting, juvenile rockfish, Pacific and jack mackerel, and market squid. There are seasonal and annual variations in their diet, depending on what prey species are most abundant locally. They feed at relatively shallow depths (26 to 74 m), although Stewart and colleague R. L. DeLong recorded a 376 m dive near San Miguel Island. Most dives are brief, lasting about 2 minutes on average. The longest dive recorded by Stewart and DeLong lasted 8 minutes.

Immature and resting adult female California sea lions molt in August and September, lactating females and subadult males molt in September and October, and adult males molt from November through February.

Adult male California sea lions leave the rookeries in August and September and migrate north during the autumn and winter; they return south to the rookeries from March through May. Males from Baja California rookeries arrive at the Channel Islands in December and January; those from southern California travel as far north as Puget Sound and British Columbia. The seasonal movements of females are unknown, but some researchers have proposed that they remain near the rookeries year-round. Some immature sea lions make northward migrations, shorter than those of adult males, and may remain north of the breeding rookeries for a year or more.

California sea lions are eaten occasionally by great white, hammerhead, and blue sharks, and by killer whales, but such mortality has insignificant effects on the population.

Galápagos sea lions do not seem to suffer any significant mor-

In most places, otariids have learned to be wary of people. Not so in the Galápagos. (Isla Plaza, 23 January 1980: William T. Everett.)

tality from aquatic or terrestrial predators. Large sharks that occasionally swim near the territories of breeding males usually are chased away by one or more of the bulls.

HISTORY OF EXPLOITATION: California sea lions and other pinnipeds apparently were important to the subsistence of Chumash and Nicoleño Indians on the California Channel Islands from as long ago as 7000 years throughout prehistoric time.

California sea lions were hunted in the 1800s and early 1900s on islands off Baja California and southern California. Until the late 1800s, they were mainly killed for their whiskers, testes, and bacula (penis bones), primarily to be marketed as aphrodisiacs in China. The blubber often was processed for oil, although the sea lions were not killed exclusively for the oil. Thousands of barrels of sea lion oil may have been procured along the coast of California in some years, but by the beginning of the twentieth century comparatively little was produced. Although females and pups occasionally were killed, it appears that most of the sea lions harvested were adult males. The direct effects of these harvests on the populations are not clear, although several early authors believed that it was the primary explanation for the low numbers of California sea lions on the Channel Islands in the early 1900s. Northern sea lions apparently also were harvested on the Channel Islands, but the relative proportions of the two species in the catch are obscured in the accounts of "sea lion" harvesting. In the late 1800s, sea lions also were killed for their hides, which had become valuable as glue stock. Similar harvests for meat, oil, and skins may have been made on Mexican rookeries by Indian and European inhabitants of Baja California in the 1800s, but the intensity and duration of such harvests are unknown. Some of the sea lions killed in predator control programs in British Columbia in the early to mid-1900s may have been male California sea lions, which are now known to migrate there in autumn and winter.

The trends in abundance of sea lions in southern California and Baja California during the 1800s are not well known, although some early twentieth-century biologists claimed that California sea lions steadily increased in numbers from the 1860s through the 1870s along the California coast. In 1899, the U.S. Fish and Game Commission concluded that sea lions (presumably both California and northern) were too numerous and that they should be reduced in numbers at rookeries on federal lighthouse reservations. This conclusion apparently was based only on the complaints of fishermen, as no systematic surveys of sea lions had been made. As a result of this decision, several thousand sea lions (probably all northerns) were killed intentionally at Año Nuevo Island in May 1899 before permission for continued killing was revoked on 31 May. Nonetheless, the California Fish and Game Commission directed deputies to kill sea lions in areas other than federal reservations in the spring of 1899 and 1900. It is not clear from available documentation whether California sea lions were hunted, how many were killed, or in what geographic areas the killing took place. At least 12,105 pinnipeds were killed for bounties along the Oregon coast between 1921 and 1926, but it is not clear whether any of them were California sea lions.

In 1909, legislation was passed forbidding "the killing, maiming, or capturing of sea lions in the waters of Santa Barbara Channel (excluding Santa Catalina Island) and on land adjacent thereto . . . to prevent the extermination of the sea lion." This legislation remained in force through 1927 despite attempts by the expanding southern California fishing industry to repeal it. Professional hunters, who wanted sport hunting of sea lions legalized, killed some sea lion bulls, cows, and pups at San Miguel Island. Although few sea lions remained on southern California rookeries by 1920, unauthorized killing of bulls apparently continued through the 1930s. Indiscriminate killing continued to some extent through the 1960s. More than 3000 sea lions were collected alive for display from Anacapa, Santa Cruz, San Nicolas, and San Miguel islands between 1878 and 1955, and some collecting continued at the last two through 1973.

At the San Benitos and Cedros Island in Mexican waters off western Baja California, relatively small numbers of California sea lions were harvested in 1937, and perhaps 1938, for use as pet food in the United States.

Conservation Status: Killing and harassment of California sea lions have been prohibited in U.S. waters, except under federal permit, since the passage of the MMPA. Nonetheless, as many as 1500 may have died each year in the early 1980s after becoming entangled in monofilament gill nets and drowning or from being shot by commercial fishermen. Despite recent attempts to regulate and limit gill net fisheries in California waters, some California sea lions still become entangled and drown in them. Some animals survive when they are removed or escape from gill nets, but many of these animals have small pieces of net remaining around their necks. Such fragments and pieces of other plastic marine debris eventually kill the sea lions by creating deep wounds around their heads and necks as they grow.

Killing of California sea lions has been prohibited in Canadian waters since 1970 under the federal Fisheries Act.

Diseases and pollution apparently affected California sea lions in the late 1960s and early 1970s. Large numbers of premature pups were observed at San Nicolas and San Miguel islands, and DDT and PCB contamination and bacterial and viral infections were believed to be the causes.

During the mid-1980s, adult male California sea lions began frequenting the Chittendon Locks in the Lake Washington ship canal near downtown Seattle, Washington, in autumn and winter. A few of them have been consuming large numbers of steelhead trout that spend several days equilibrating to fresh water at the base of the lock before traveling up river to spawn. The sea lions increased from 2 to 4 in 1985–86 to 30 to 40 in 1988–89 and ate as much as 62 percent of the returning steelhead trout each year. This fish population already was considered endangered due to overfishing and habitat destruction. Recent attempts to capture and translocate the sea lions have been successful, but most returned to the area within days or weeks after being transported and released several hundred miles down the Washington coast or in southern California waters.

California sea lions have become especially familiar to residents of Monterey and San Francisco, in central California. Portions of marinas in both cities have become autumn and winter haulouts for sea lions. The animals that congregate on piers are smelly, noisy, and aggressive. Such "invasions" of coastal sites that have been transformed by humans are one consequence of efforts to protect wild populations, allowing them to grow and reoccupy their traditional ranges.

Adult and subadult California sea lion males have become relatively abundant in Canadian waters since the early 1970s. In 1984, approximately 4500 were counted along the eastern and western coasts of Vancouver Island. More than 1500 congregate on log booms near Nanaimo, interfering with operations of pulp and paper mills but also attracting the attention of tourists. The sharp increase in numbers hauled out in the early 1980s may have been due to the intrusion of warm water from the south during El Niño of 1982–83. California sea lions are seasonal visitors to Canadian waters and are present primarily from November through April. While there, the sea lions feed mostly on herring, squid, and cod, all targets of commercial fisheries. In recent years, fishermen frequently have claimed that California sea lions interfere with their operations by damaging fishing gear, scaring fish away, and eating large numbers of valuable fish.

The total population of California sea lions numbered approximately 160,000 in 1989, about half in the United States and half in Mexico. Nearly 19,000 pups were born on rookeries on the Channel Islands in 1989, 25,000 in 1990, and 31,000 in 1991. The present abundance of Galápagos sea lions is not known, but there were an estimated 20,000 to 50,000 in 1963. The Japanese/Korean race of California sea lions is extinct.

FURTHER READING: Antonelis et al. (1990), Barlow (1972), Bigg (1988a), Bonner (1984), DeLong et al. (in press), Feldkamp et al. (1989), Lowry, M.S., et al. (1991), Miller (1974), Odell (1975), Peterson and Bartholomew (1967), Rowley (1929), Stewart, B. S., et al. (1987, in press).

South American Sea Lion

Otaria byronia
(de Blainville, 1820)

NOMENCLATURE: The name *Otaria* comes from the Greek *otarion*, meaning "little ear." It refers to this sea lion's small external ear. There is uncertainty about the valid specific name of the species. The name *byronia*, honoring Commodore J. Byron of HMS *Dolphin*, who brought home the skull of the type specimen after a voyage to the South Seas, is widely used. Some researchers continue to use *flavescens*, from the Latin *flavus*, for "yellowish."

The name South American sea lion is preferable to southern sea lion, as the species is sometimes called, since this prevents confusion with the Australian and New Zealand sea lions, both also confined to the Southern Hemisphere. South Americans call the sea lion inhabiting their coasts *lobo-marinho-de-um-pelo* or *leão marinho* in Brazil; *león marino sudamericano*, *lobo de un pelo*, or *lobo ordinario* in Uruguay and Argentina; *lobo común*, *lobo chusco*, or *león marino austral* in Chile; and *lobo marino de un pelo* or *lobo chusco* in Peru.

DESCRIPTION: Males grow to a maximum length of more than 2.8 m and weights of 300 to 340 kg; females to 2.2 m and 144 kg. Bulls lose nearly 50 kg of weight by the end of their breeding-season fast. There is extreme sexual dimorphism in body form as well as size. Adult males have a massive neck and broad head, with a characteristically upturned muzzle. Their thick mane of long guard hairs begins on the forehead and around the eyes and chin and extends onto the nape and chest. The foreparts are so bulky that the hindquarters appear peculiarly feeble by comparison. Females are also powerfully built but are much more evenly proportioned than males, with a smaller head and no mane.

Newborns are shiny black dorsally and dark grayish orange ventrally. The hair on the back is curly. After 1 month of age, the pups become dark chocolate brown or gray, often acquiring a reddish tinge and becoming paler during the first year. Subadults of both sexes vary in color from dark or reddish brown to orange, with whitish areas on the face. The color of males tends to deepen with age, although some bulls become pale gold. The mane is often a lighter color than the rest of the body. Adult females vary from brownish orange to light orange or yellow; they often have a generally variegated coloration.

Male pups are 79 to 85 cm long and weigh 13 to 15 kg at birth; females are 73 to 82 cm and 10 to 14 kg.

Distribution: South American sea lions are distributed along the coast of South America from southern Brazil, clockwise, to northern Peru, including the Falkland Islands, Tierra del Fuego, and Staten Island. They are not known to inhabit the Juan Fernández Islands off Chile. The northern limit for wanderers along the east coast is Rio de Janeiro (approximately 23° S) or São Paulo (there is even one record from Salvador, Bahia State, at 13° S); along the west coast, Zorritos, Peru (4° S) (there is a record of a dead adult male at the Galápagos near the equator). The northernmost breeding sites are at Recife das Torres, Uruguay (approximately 29° S), in the east and Lobos de Tierra Island, Peru (6°30′ S), in the west. Sea lions occupy some islands well south of Cape Horn, including Chile's Diego Ramírez Islands and Argentina's San Martin de

South American sea lions on Isla Marta, Chile—an adult male on the left, an adult female on the right, and two pups in the foreground. (5 February 1981: Stephen Leatherwood.)

Tours. Individuals appear occasionally on South Pacific atolls as far west as Tahiti.

There are no breeding colonies on the Brazilian and Uruguayan mainlands today. The Uruguayan breeding areas are on the Coronilla, Castillos, and Torres island groups and on Lobos Island. In Argentina, there are more than 70 haul-out sites, of which more than half are on the mainland. In the Falklands, some 64 rookeries have been identified, most of them remote islets or rocks. Sea lions occupy dozens of mainland and island sites in Chile and Peru.

NATURAL HISTORY: Individual sea lions, especially males, may wander widely. For example, pups tagged on Uruguayan breeding rookeries have been recovered as far as 835 km southwest of their birthplace, and many sea lions follow the coast of Brazil northward to Rio de Janeiro, at least 1930 km away from the nearest regular breeding site. Groups of sea lions have been observed more than 200 km north of the Falklands in late December, when the breeding season has begun at most rookeries. Some sandy beaches where pupping occurs during the summer are deserted during the winter. Nevertheless, most populations of South American sea lions are essentially sedentary, and most of the larger colonies are located near zones of upwelling where productivity is high.

It is not unusual for South American sea lions to wander into estuaries, even into freshwater systems. In some portions of their range they live sympatrically with southern elephant seals and South American fur seals. The territorial sea lion bulls are tolerant of fur seals trespassing onto their territories, perhaps because they mistake them for females or juveniles of their own species.

The kinds of areas used for pupping and breeding may depend partly on the nature and extent of human disturbance. In Chile, sea lions breed and give birth in rocky areas, sometimes in caves, generally inaccessible from land. In Argentina, where there is generally less persecution, they come ashore on open sandy or pebbly beaches.

At Punta Norte, on the Valdés Peninsula, adult males and females arrive at the rookeries during the first half of December. The males initially defend territories along the high-water mark. Later, they devote most of their energy to defending a breeding center containing 1 to 10 (average about three) females and to sequestering the females. Once a female has been incorporated into a harem, the male prevents her from defecting by blocking her with his body or by grasping her in his jaws and holding her, shaking her, and hurling her into place.

About half of the aggressive encounters between males amount to nothing more than an exchange of vocal threats. Fighting involves biting of the opponent's neck and chest, violent shaking and twisting of the neck, and pushing. Most of the territorial males

A pug-nosed bull South American sea lion, fast asleep. (Seal Island, Falklands, January 1988: Frank S. Todd.)

get wounded at some point during the breeding season, but they are rarely killed outright. Some lose and break teeth.

The type of polygyny practiced by South American sea lions depends on the topography and substrate and on the thermal environment. On open, uniform beaches where there are no tide pools or shade to provide relief from the summer heat, males sequester females and do not defend rigid territories. In areas with a rockier, uneven substrate, where tide pools form and rocks or cliffs provide patches of shade, males defend preferred territories that attract females.

Although pupping begins at Valdés Peninsula in mid-December and does not end until early February, most pups are born in January, usually 3 days after the mother's arrival at the rookery. Copulation occurs on land about 6 days after parturition, and by the end of January, breeding activity declines and the harem structure begins to break down. Subadult males, who were peripheral to the central breeding area during the peak season, now attempt to mate as the females begin entering the water in groups. These younger, nonterritorial males pose a threat to the pups, sometimes tossing them considerable distances through the air.

Mothers begin leaving their pups temporarily and foraging offshore about a week after giving birth. The foraging trips generally last 3 days and are followed by 2 day nursing bouts on land.

Once they are mobile, the pups wander somewhat and play together in large groups. They enter the water for the first time at 3 to 4 weeks of age and are good swimmers by 2 months of age.

The period of pup dependence is prolonged. Some pups are suckled until the mother's next pup is born.

The timing of reproductive events varies somewhat among rookeries, presumably due to climatic and energetic factors. Although pupping generally begins in mid-December in Argentina and Uruguay, it peaks shortly before 15 January at the Falklands, and later at Valdés Peninsula. In Chile, males begin establishing territories in early September, and the females arrive toward the end of September. Pups are born between September and March, the peak of pupping varying by latitude.

Pup mortality varies greatly according to conditions at a given rookery. On crowded rookeries and at sites where many aggressive young males are present, half the pups may die or be killed. At uncrowded rookeries, very few pups die. Causes of pup mortality (apart from predation) include starvation, disease, trampling by bulls or cows, drowning, and biting, shaking, or tossing by mothers other than the pup's own or by males. Subadult males sometimes injure or kill unweaned pups. The pups may function as female substitutes for the frustrated pubertal males. The unsettled atmosphere of sites where large numbers of young itinerant males are present can itself exacerbate other threats to pup survival.

Ages at sexual maturity have not been verified but are thought to be about 4 years for females and 5 for males.

South American sea lions eat a large variety of organisms. In some areas they eat mainly squid, "lobster krill" (pelagic larval stages of the rock lobster), and various kinds of fish, particularly demersal and mesopelagic species. They sometimes eat large medusae (jellyfish). They forage mainly in shallow water (less than 300 m deep) near coasts or on productive offshore fishing banks, often in multispecies assemblages with large numbers of seabirds and cetaceans. Fur seals (both adult females and pups), ducks, and penguins (including rockhopper, gentoo, and Magellanic) also are included in their diet. On Bird Island in the Falklands, gangs of sea lions sometimes surround and trap rockhopper penguins on land. They have been seen to flay the penguins before eating them, leaving the skin with head, flippers, and feet still attached.

Like other sea lions, South American sea lions swallow stones. It is not unusual to find more than a kilogram of pebbles and sharp-angled stones in the stomach.

These sea lions are themselves preyed on by large sharks, killer whales, and leopard seals. Detailed studies of killer whale predation have been done at Punta Norte. There, killer whales hunt sea lions (and southern elephant seals) either by cooperatively corralling them at sea or by individually rushing directly into the surf zone and capturing them at the land-sea interface. Frequently the rushes into the surf zone result in the whale's beaching itself temporarily. Pups and small subadults are the most vulnerable, but the whales are known to kill adults as well. Land predators, such as mountain lions, probably attack sea lion pups in some areas. Vampire bats prey on some South American sea lions at coastal rookeries in Peru and Chile, and Andean condors often visit coastal Peruvian rookeries to scavenge sea lion carcasses.

South American sea lions are held in a few South American zoos and aquaria, and live animals have occasionally been exported to North America. Several births have taken place at the Montevideo Zoo in Uruguay. Individual sea lions have lived in captivity for more than 15 years.

History of Exploitation: There is a long history of exploitation throughout most of their range. Prehistoric peoples living near the southern tip of South America hunted sea lions for food and pelts, as did explorers and adventurers beginning at least as early as the sixteenth century. Fuegian Indians killed adult males in winter and females and pups in summer, storing the oil in sea lion stomachs and using the hides to make boats or rafts.

More recently, the hides of adults have been used to make leather and suede items. Those of pups 3 to 5 weeks old are commercially more valuable as they are used to make ladies' coats. The oil has been used in the tanning process and the meat for animal food. A 260 kg sea lion produces approximately 34 kg of oil and 40 kg of meat.

The government-run commercial sealing industry in Uruguay is one of the oldest operations of its kind in the world. Before the sea lion hunt ended in 1977, males were killed selectively, by clubbing on the nose with a stout wooden club and then stabbing in the heart with a knife, after being driven toward the interior of the rookery island. The annual catch did not exceed 3260 from 1963 to 1976. During the last 2 years of the fishery (1976 and 1977), 2500 sea lions were taken at Lobos Island and Cabo Polonia, combined.

Sea lions have been heavily exploited in Chile. Between 1821 and 1822, at least 52,000 sea lion pelts were obtained in an area between Mocha and Santa María islands. British and U.S. sealers killed large numbers of fur seals and sea lions in western Patagonia between 1825 and 1865. By the late 1860s the sealing industry in southern Chile had been taken over by Chilean nationals, who continued sealing until 1907, when a ban was decreed by the government. An officially sanctioned seal hunt resumed in Chile in 1976. This hunt is mainly for pups, called "poppies." Quotas are set by region and vary from year to year. During the late 1970s the total catch in some years exceeded 11,000 pups. Quotas for adults and subadults are lower and apply mainly to areas where fishery conflicts are considered acute.

Commercial sealing for sea lions in Argentina continued at high levels through the 1950s but declined in the 1960s for economic reasons. Argentine scientist Enrique Crespo has estimated that commercial sealing reduced the sea lion population in northern Argentina by 80 to 90 percent; signs of recovery have been detected only recently. Sea lions became protected in Argentina in 1974 under a 1928 decree that protected whales.

At the Falkland Islands, sea lions were not extensively harvested before 1928 because their pelts were not as valuable as those of South American fur seals and their oil yield was much less than that of southern elephant seals. In 1928, the newly formed Falkland Islands and Dependencies Sealing Company began operations with the South American sea lion as its primary focus. Nearly 40,000 sea lions were killed for oil between 1928 and 1939, when there was a temporary halt to operations. Sealing resumed in 1949 and continued for 4 years; approximately 3050 animals were killed. Interest in sealing was renewed in 1962, and licenses were granted to permit the killing of 1500 sea lions for pelts.

CONSERVATION STATUS: South American sea lions, although widely persecuted, remain abundant. The world population was estimated in the early 1980s as well over 300,000. Peru has approximately 34,000; Chile, 90,000; Argentina, more than 170,000; and Uruguay, at least 30,000. In 1937, some 300,000 sea lions, including 80,000 pups, were estimated to inhabit the Falklands archipelago. Surveys in 1965 indicated that only approximately 30,000 (approximately 19,000 counted, including 5500 pups) remained, and the decline has continued. Recent counts indicate that perhaps 3000 sea lions are present in the Falklands today. Although the commercial harvests in the 1930s and 1940s reduced the stock somewhat, they alone cannot account for such a major population decline. The principal cause is still not known. One suggested cause is commercial fishing by foreign trawlers, which may have substantially reduced fish populations in the Falklands region. However, since the major decline in the sea lion population was under way before there was any fishing off the Falklands, and when there was still very little trawling in the western South Atlantic generally, trawling cannot be the sole cause of the sea lion decline.

Virtually throughout their range along the South American mainland, sea lions are viewed by fishermen as competitors and pests. They steal and mutilate catches from hooks and nets, often damaging gear in the process. This has meant that in spite of protective laws, sea lions have continued to be killed in fairly large numbers. For example, in Peru, although fishermen use concentrations of sea lions to indicate the presence of fish schools, they readily shoot or dynamite any of the sea lions that get near their boats or wrapped inside their nets. The sea lions' tendency to follow fishing boats makes them all the more vulnerable. They also get entangled in gill nets. In Chile, sea lions are valued as crab bait and in some areas as human food. A lucrative and expanding salmon culture industry in the Chiloe Island region is in conflict with sea lions and other marine mammals using the same inlets and embayments. Apart from the disturbance and appropriation of space by the industry, individual mammals die in the nets placed around the pens to protect the salmon.

The South American sea lion's wide geographic distribution, with many essentially sedentary local populations, has been, on balance, favorable to the survival of the species. Some local groups have been extirpated, and the aggregate population probably has declined considerably over the past several centuries. However, groups in remote, inaccessible places have served as reservoirs for repopulating adjacent areas where overharvesting has occurred. The pelt's lack of commercial appeal has been a saving feature of these pinnipeds.

Some rookeries have special protection. There are many faunal reserves along the Patagonian coast of Argentina, and a small haul-out area used by subadult males near Mar del Plata, northern Argentina, is protected by a local conservation group. Recent field studies have suggested that colonies in Tierra del Fuego have recovered to near presealing numbers; there are approximately 1500 sea lions in the Argentine sector of the archipelago.

In Brazil, the colony at Recife das Torres, which numbered 200 to 300 individuals in 1953, had been extirpated by the 1970s. It is now being reestablished, and since 1983 the area has been an ecological reserve.

In Uruguay, the commercial killing of sea lions ended after the 1977 season, but clandestine and accidental killing by fishermen continues. The Uruguayan sealing industry maintained a stable sea lion population at least from 1954 to the early 1970s. Sealing in La Plata estuary appears to have been managed on a monopolistic, sustained basis for more than four centuries. Today, guided tours leave Punta del Este daily during the summer to visit the sea lion rookery at Lobos Island.

Further Reading: Campagna (1985), Campagna and Le Boeuf (1988a, 1988b), Campagna et al. (1988a, 1988b), Hamilton (1934, 1939), Sielfeld K. (1983), Vaz-Ferreira (1981).

Australian Sea Lion

Neophoca cinerea
(Péron, 1816)

NOMENCLATURE: This single-species genus has undergone many revisions since the name *Neophoca,* meaning literally "new seal," was introduced by J. E. Gray in 1866. The specific name comes from the Latin *cinereus,* meaning "ash colored." It refers to the grayish color of females and juvenile males.

DESCRIPTION: Adult males are 2.0 to 2.5 m long and may weigh up to 300 kg; adult females are 1.3 to 1.8 m and weigh up to 110 kg. In addition to the sexual dimorphism in body size, there are color differences between the sexes. Although males and females are both silvery gray to fawn above and creamy below during the first 2 years of life, maturing males acquire some spotting on the chest and their muzzle darkens. Later, the male's body darkens to a rich chocolate brown, and the top of the head and the nape become whitish, which in earlier times led to its being called the counselor seal. Males that have not quite reached their prime can be recognized by the dark coat, white nape, and white ring around the eye. Females' coloring remains essentially unchanged throughout life.

The adult male is also much bulkier than the female. Apart from size and color, adult males are readily distinguished from adult females by their much thicker, more powerful neck, covered by somewhat longer, rougher hair, which gives the impression of a mane.

Newborn pups are approximately 70 cm long and weigh about

A male Australian sea lion sits on the beach at North Fisherman Island, Western Australia, while 13 females cool themselves in the nearby shallows. (September 1979: John Ling.)

6 to 8 kg. Their soft, thick natal fur begins to molt at about 2 months of age. Pups are chocolate brown at birth and become gingery by the time they begin to molt.

DISTRIBUTION: These sea lions are confined entirely to Australia, where they can be found today on many offshore islands and at a few mainland sites along the west and south coasts, from Houtmans Abrolhos in the west to Kangaroo Island in the east. In South Australia, sea lions have been observed on 69 islands and reefs and at 3 mainland sites. Individuals have appeared as far north on the west coast as Shark Bay (25° S) and as far east on the south coast as Portland, Victoria (142° E). The largest breeding colonies are in the eastern half of South Australia, particularly at Seal Bay on

An Australian sea lion in a coral garden. (Neptune Islands, South Australia, 2 February 1988: Howard Hall.)

the south side of Kangaroo Island; Dangerous Reef, a rocky islet off Port Lincoln; and The Pages, off the eastern tip of Kangaroo Island. Although scarce at present east of Kangaroo Island, these sea lions used to occur in Bass Strait between the Australian mainland and Tasmania. They may have been particularly abundant in the Furneaux islands in the strait, where they were hunted along with the more abundant and valuable fur seals.

NATURAL HISTORY: Australian sea lions form relatively small breeding colonies, thinly distributed along 3000 km of the western and southern coastline of Australia. There are few records of more than 200 animals hauled out together at one time. This may reflect a natural tendency, since accounts by Europeans who visited southern Australia during the early nineteenth century suggest that they found the animals in similarly modest aggregations even then. The largest breeding colony, at Seal Bay, contains only some 330 adults.

Australian sea lions prefer sandy beaches and smooth rocks as breeding substrates. They are well known for their habit of straying inland, sometimes appearing several kilometers from the seashore.

Recent studies indicate that there are several asynchronously breeding populations, each with a breeding cycle of 17.5 months. The breeding season of each population spans 5 months. At Dangerous Reef, the breeding season lasts from October to January. On Kangaroo Island and in Western Australia, the peak of pupping is often in June, but it can also be in October. In fact, there is now good evidence for an 18-month breeding cycle at Kangaroo Island. The adaptive value of this strategy remains unclear.

The territories defended by bulls are fluid in the sense that they are readily changed or abandoned in response to environmental conditions and the availability of receptive females. A normal harem contains four to six females. The bulls are aggressive toward one another and ruthlessly herd the females. Pups are subject to savage attacks by bulls and cows alike. Young bulls sometimes sequester older pups and bite them, mount them, or hold them underwater.

Copulation occurs within 10 days after the pup is born. The gestation period, estimated from three captive births, is 14 to 15 months. However, pupping trends of wild females suggest a total gestation period of 17.5 months. Delayed implantation has been demonstrated to occur, but the timing is uncertain.

The pup's dependence is prolonged. For the first 10 to 14 days, the mother generally remains on the beach, aggressively defending her pup. Thereafter, she makes feeding forays offshore, but she and her pup are quite sedentary until the pup is 3 months old. Suckling continues for at least a year, although pups likely take solids well before this time. Mothers about to give birth are sometimes still nursing the pup from the previous season, which may

Like other otariids, male Australian sea lions engage in serious combat for the right to mate with females. (Kangaroo Island, South Australia, January 1988: Hans Reinhard.)

be almost as large as its mother. In these circumstances, some newborn pups are rejected and die of starvation.

Assorted fish bones and squid beaks (mouth parts) have been found in scats, and stones are often present in the stomach. Adult males often prey on little penguins that nest in burrows on many of the islands occupied by these sea lions. There was an observation of a sea lion at the surface shaking and then apparently consuming a large fiddler ray; these rays are common in the shallow waters off southern Australia. Cuttlefish are also eaten. No detailed study of the diet of Australian sea lions has been completed, but they probably feed mainly on shallow-water benthic prey. They are thought not to be deep divers, and apparently they do not travel far from land. In one study, the deepest dive by a female sea lion was to 92 m.

Large sharks, particularly white pointers, which are common near the rookeries in South Australia, are probably these sea lions' most serious enemies.

Australian sea lions adapt well to captivity, and they are fairly common in zoological parks in Australia, where some have bred successfully. Many of those obtained from the wild in recent years had been orphaned or debilitated.

History of Exploitation: Sea lions probably were hunted to some extent by Aborigines; their bones have been found in kitchen middens in northern Tasmania. Early European visitors killed sea lions for food. Intensive commercial sealing began in southern Australia at the beginning of the nineteenth century, and within

Until recently, most of what was known about seals came from observations on land or ice. We are learning, however, that many species can be observed and filmed underwater. (Australian sea lion, Neptune Islands, South Australia, 2 February 1988: Howard Hall.)

scarcely 25 years the more accessible colonies were eradicated. By 1825, the large-scale commercial sealing industry in Australia was finished. Further exploitation of the residual stocks continued as human settlement and use of the land proceeded.

Sea lions were of less commercial value than fur seals. They were taken mainly for their oil rather than their pelts.

CONSERVATION STATUS: All pinnipeds have been protected legally in Western Australia since 1892 and in Gulf St. Vincent and Spencer Gulf, South Australia, and adjacent waters north of Kangaroo Island since 1919. In 1954, the main haul-out ground on the south coast of Kangaroo Island became a sea lion sanctuary. However, full legal protection throughout South Australia did not come until 1964. During the 1950s and 1960s, sea lions were still being taken in considerable numbers for shark bait, and some were destroyed when interfering with fishing nets. Since 1987, all pinnipeds have been protected by national legislation.

In spite of the long period of legal protection in much of their range, the population of Australian sea lions remains low. They are among the rarest of pinnipeds. An estimated 700 were present in Western Australia, and there were 2300 (minimum, including pups) in South Australia during the 1970s. It is now known that the estimate for Western Australia was too low; there are considerably more than 700 sea lions there. The current total population is probably between 2000 and 5000. Most importantly, there is no evidence that the population is increasing or expanding its range. Many of the islands that had sea lion rookeries before the sealing days remain abandoned.

Fishermen kill sea lions at least occasionally when they interfere with fishing operations. Perhaps because their population is sparsely distributed relative to other otariids, these sea lions have not evoked the usual widespread hostility of fishermen. However, at Cheynes Beach, Western Australia, people fishing for Australian salmon have complained that sea lions attack their nets and that individual sea lions have learned to rob rock lobster pots set far offshore. Australian sea lions are said to like eating shark livers; they swim along 3000-m nets set for schooling sharks, systematically removing the catch. Particularly as juveniles, Australian sea lions often become entangled in monofilament netting. In addition to the direct mortality that results, some individuals that escape become burdened with a permanent line necklace.

The breeding colony in Seal Bay Conservation Park on Kangaroo Island is economically important because it helps to support a rapidly expanding tourist industry. This group of sea lions is accustomed to visitors, and people can mingle with the animals on the beach in designated areas. Human activity there is closely monitored and intensively managed. Point Labatt Conservation Park, approximately 50 km south of Streaky Bay, South Australia, is the only easily accessible mainland haul-out site. About 30 to 80 sea lions of mixed ages come ashore there regularly; the reserve was visited annually by some 25,000 to 30,000 people during the late 1980s.

FURTHER READING: Marlow (1968, 1975), Marlow and King (1974), Robinson and Dennis (1988), Walker and Ling (1981a).

New Zealand Sea Lion

Phocarctos hookeri
(Gray, 1844)

NOMENCLATURE: This sea lion has been variously assigned to the genera *Arctocephalus, Otaria,* and *Neophoca.* Taxonomists now generally place it in the monotypic genus *Phocarctos,* derived from the Greek *phōcē,* for "seal," and *arktos,* for "bear." The skull is slightly reminiscent of a bear's. Sir Joseph D. Hooker, a British botanist, secured the first specimens during the British Antarctic Expedition of 1839–1843.

DESCRIPTION: Adult males are 2.0 to 3.25 m long and weigh up to at least 400 kg; adult females are 1.6 to 2.0 m and can weigh 160 kg. Bulls have a well developed neck and chest, covered with a mane of hair that is rougher and longer than that on the rest of the body. Their profile differs from that of Australian sea lion bulls; the face is blunter and more rounded, with a shorter muzzle. They appear snub-nosed by comparison with their Australian counterparts. Females are superficially indistinguishable from Australian sea lion females. They are buff to creamy white, with darker areas around the muzzle and flippers. Young males (9 to 18 months old) have the same silvery gray dorsal and cream ventral coloration as females, but they darken and become black or dark brown overall. Bulls do not have the pale region on the head and nape that is characteristic of Australian sea lion bulls.

Pups, 75 to 80 cm long at birth, are sexually dimorphic. Males

On Enderby Island, New Zealand sea lions sometimes wander inland from the beach, entering the thick rata forest. (February 1981: Frank S. Todd.)

are larger than females at birth (7.9 kg versus 7.2 kg) and at 20 days of age (13 kg versus 11 kg). They also are usually darker colored. Most male pups are chocolate brown except for a lighter area on top of the head extending back onto the nape and forward, as a stripe, to the nose. Female pups are predominantly light colored.

DISTRIBUTION: This species has an extremely limited distribution, centered at the Auckland Islands, approximately 400 km south of Stewart Island. Although some may be encountered at virtually any suitable site on the main Auckland Island and its associated smaller islands, Sandy Bay, on the south side of Enderby Island, and Dundas Island are the principal rookery areas. There are small breeding colonies on Campbell Island and the Snares Islands to the south and north of the Aucklands, respectively. A few, mainly bulls, haul out regularly in the Port Pegasus area of southern Stewart Island. Individual stragglers also appear sporadically on the New Zealand mainland. A small colony of males is present year-round at Papanui Beach on the Otago Peninsula. Some breeding occurred on the west coast of the South Island during the nineteenth century. The remains of New Zealand sea lions have been found in Maori middens on the North Island, suggesting that the prehistoric distribution of the species may have extended north to Cape Kidnappers and the Coromandel Peninsula. Subadult and socially immature males regularly travel to and from Macquarie Island (54°40′ S), more than 600 km southwest of the Aucklands. Several young males have been known to come ashore there in consecutive years.

NATURAL HISTORY: The distribution of these sea lions during the nonbreeding season is diffuse; no regular migration has been documented. Those animals not at sea can be found, at least on Enderby Island, resting deep in the forest or on the grass-covered cliffs. Their proclivity for wandering on land can lead to sojourns of several kilometers from the sea. There is a marked preference for readily accessible beaches with soft sand. While hauled out, the sea lions, especially females in advanced pregnancy, help to cool themselves by flipping damp sand over their bodies.

Most females become sexually mature at age 3 and produce their first pup when 4 years old. Males mature later, becoming sexually mature when 5 but not reaching social maturity until at least 8 years of age.

The breeding behavior of New Zealand sea lions differs from that of Australian sea lions. Bulls begin hauling out on the rookeries from late October to early November. Progressively from late November they establish and defend territories, each amounting to a 2 m circle of sandy beach, their "personal space" that can, however, be relocated as circumstances require. Combat among adult males, however fierce it may appear, is largely ritualistic. Heavier animals usually prevail in excluding lighter individuals from disputed territory.

Pregnant females begin arriving at the rookeries during the fourth week of November and reach peak numbers in Sandy Bay by the third week of December. They are gregarious and crowd together on the beach. Bulls do not often herd the cows, and they continue to defend their spot of sand even if no females choose to join them. Harems contain 8 to 25 females (the average is 12). During the entire breeding season the males remain ashore and do not eat. They are unaggressive toward cows and pups alike. Mating occurs on land 7 to 10 days after the pups are born. After mating, a bull's devotion to the task of keeping a territory wanes. By late February the harem structure has dissolved, and the bulls take to the sea, some having fasted for as much as 4 or 5 months.

Pups are born mainly from early December to early January, or up to 10 days after the females arrive. Southern skuas are attracted by the afterbirth and can be relied upon to indicate a pup's arrival. The pups are sedentary and lie quietly beside their mothers for 2 or 3 days, after which they become quite active. Mothers go to sea for feeding and return every 24 to 48 hours to nurse their pups. While the mothers are away, the pups form groups that move higher up the beach or into the tussock and scrub, sleeping or play fighting. They sometimes play in freshwater pools. On their return from the sea, the mothers call their pups with a mooing sound. Pups recognize distinctive features of their own mother's call and respond by bleating. This vocal system of mutual recognition can function over distances of at least 100 m. By 6 weeks of age the pups become noticeably bolder about approaching the water, and

A mother New Zealand sea lion and her pup on the boardwalk at Beeman Hill, above Perseverance Harbour, Campbell Island, New Zealand. (January 1990: Roger Moffat.)

soon afterward they begin swimming, closely attended by their mothers. Lactation continues for at least 8 months, and a pup may stay with its mother, and be suckled, for a year or more.

The molt begins in late February, soon after the breeding season.

The behavior of New Zealand sea lions at sea has been little studied. Like other sea lions, they can be seen just offshore swimming at high speed and porpoising.

Their diet includes small fish, such as flounders, octopuses and other cephalopods, crustaceans (including crabs, crayfish, and prawns), seabirds, and penguins. The visiting sea lions at Macquarie Island have been seen preying on gentoo penguins. Some individuals develop a habit of eating fur seal and elephant seal pups. Like some other otariids, these sea lions have learned to follow fishing vessels, taking advantage of discarded or escaped fish. They also remove fish and squid directly from trawl gear at the surface. Observations of haul-out patterns of males at Papanui Beach suggest that they are nocturnal feeders.

At the granitic Snares Islands, geologists were perplexed at finding patches of basaltic pebbles on the beach. However, it finally was determined that these were disgorged from sea lions, who sometimes vomit as many as 20 pebbles, along with squid tentacles and small fish.

Predation by sharks occurs, and some sea lions have missing appendages and wounds that bear witness to encounters with sharks. Some predation by killer whales also may occur. Pup mortality is caused mainly by starvation, often after the pup has crawled into a rabbit burrow and found itself unable to back out. Rabbits were introduced to Enderby and Rose islands in the middle of the nineteenth century and have proliferated there.

Maximum ages of 23 (male) and 18 (female) years have been documented.

We are unaware of any New Zealand sea lions in captivity at present.

History of Exploitation: The Maori exploited New Zealand sea lions to some extent in prehistoric times, and shipwrecked seamen and settlers certainly used them for food.

Sea lions were abundant on the Auckland Islands at the time of the islands' discovery in 1806, but their numbers were quickly reduced by the intensive seal fishery that ensued. Fur seals were vastly more important commercially, and it was the market for their pelts, particularly in the Far East, which made the sealing industry lucrative. Sea lions were essentially incidental victims taken for hides and oil. Enderby Island was a focal point of commercial sealing, and its populations of fur seals and sea lions were much reduced by 1830. Although commercial sealing on a large scale was no longer commercially viable in New Zealand after the middle of the nineteenth century, and fur seals and sea lions were nominally protected by law beginning in 1894, attempts to reestablish a sealing industry continued sporadically until the end of World War II. Commercial sealing has been forbidden in New Zealand since 1946.

Conservation Status: Like the Australian sea lion, the New Zealand sea lion exists in relatively small numbers and has a restricted distribution. The total population is not more than 6000, of which more than three-quarters live on and near the Auckland Islands. More than 1000 sea lions—perhaps 200 bulls, 450 cows, and 450 pups—use Enderby Island during the breeding season. This number has remained essentially unchanged for the past 40 years.

A tourist meets sea lion pups on Sandy Bay, Enderby Island, New Zealand. (10 February 1991: Stephen Leatherwood.)

The colony on Dundas Island had approximately 3550 animals, including some 1700 pups, in 1978.

The remoteness and harsh living conditions of New Zealand's subantarctic islands have discouraged human settlement, and this has worked to the advantage of the sea lions. However, exotic fauna, notably rabbits, introduced by people, have had some impact on the sea lions. Also, a large trawl fishery for squid recently started near the Auckland Islands has caused some mortality (particularly of mature females) and is a continuing cause for concern.

FURTHER READING: Beentjes (1989, 1990), Bruemmer (1983), Cawthorn et al. (1985), Gaskin (1972), Marlow (1975), Marlow and King (1974), Walker and Ling (1981b).

True Seals

Family: Phocidae

Harbor Seal

Phoca vitulina
Linnaeus, 1758

NOMENCLATURE: The generic name is from the Greek *phōcē* for "seal," a term said to be derived from the Sanskrit root *sphâ,* meaning "to swell up," referring to the animal's plumpness. The name *vitulina* is from the Latin *vitula,* for "calf," and the suffix *inus,* meaning "like." In Norway, the vernacular name is *steinkobbe* or *kystsael;* in France, *phoque común;* in Germany, *seehund;* in the Netherlands, *zeehond;* in Denmark, *sael;* in Sweden, *knubb säl;* in Iceland, *landselir.* In Great Britain and English-speaking Europe, the harbor seal is usually called the common seal. In the fur trade of Canada and Europe, harbor seal pelts were called rangers, and in subarctic regions of Canada this name is still sometimes used.

In the vast range of this species, along the continental and island coasts of the temperate Northern Hemisphere, there is much local and clinal variation in morphology, physiology, and behavior. The spotted or largha seal was recognized as a separate species only in the 1970s. The matter is still not completely settled. Soviet scientists in particular continue to treat the spotted seal as a subspecies of *P. vitulina.* The taxonomy of harbor seals in the North Pacific and Bering Sea has been controversial, although two subspecies generally are recognized by most workers: *P. vitulina richardsi,* the coastal seal of the eastern Pacific; and *P. vitulina stejnegeri,* the coastal seal of the western Pacific. The small populations on the Kuril Islands are considered by Soviet scientists to be an endangered race, previously known as a subspecies, *P. vitulina kurilensis.* As the distribution of harbor seals is continuous across the Aleutian and

Commander islands, the putative subspecies may be regarded more realistically as a single trans-Pacific cline. Japanese scientists refer to the harbor seals in the far western North Pacific as Kuril seals. Harbor seals of the western North Atlantic are called *P. vitulina concolor;* those of the eastern Atlantic, including Iceland, *P. vitulina vitulina.* Taxonomic separation of Atlantic from Pacific harbor seals is based on disjunct distributions rather than on morphological evidence. Although the harbor seals inhabiting the Seal Lakes region of Québec's Ungava Peninsula were assigned initially the status of a subspecies, *P. vitulina mellonae,* this assignment has not gained wide support because of the small sample size used.

DESCRIPTION: Harbor seals in Alaska and the western Pacific are significantly larger than those in the Atlantic. We give the size ranges for the species as a whole. Adult males are 1.4 to 1.9 m long and weigh up to 140 kg (a 150 kg specimen has been reported). Adult females are 1.2 to 1.7 m long and weigh close to 80 kg. Males are consistently larger than females. The range of adult weights for European harbor seals is 65 to 110 kg. Birth length is from 75 to 100 cm (average 82 cm); weight, 8 to 12 kg (average 9 kg).

Most pups are born with the short, stiff hair coat typical of

Geographic variation in harbor seal pelage patterns is demonstrated here by the extreme light and dark color phases. In the Pacific, the light phase dominates in the north, clinally changing to a greater percentage of the dark phase in the south. (San Miguel Island, California, July 1980: Brent S. Stewart.)

adults. Some pups in the Pacific are born mostly covered with a longer, softer white or gray coat (lanugo) that is shed within 10 days. The closely related spotted seals of the North Pacific are normally born in lanugo. There are two basic color morphs of adult harbor seals. Some are white or light gray to silver with dark spots; others are black or dark gray to brown with white rings. Various combinations of color, spots, and types of rings are intermediate between these two types. The presence and extent of an orange, rust, or green cast on the pelage varies among populations. In some areas, such as San Francisco Bay, this color occurs on as many as one-fifth of the seals, while at Sable Island off eastern Canada it has not been observed at all. Such coloring may be due to algal growth (orange and rust in some areas, green in others).

Harbor seals hauled out on ledges or rocks often assume this arched posture. (Medny Island, Commander Islands, western Bering Sea, July 1979: Sasha Zorin.)

Harbor seals hauled out on rocks and ledges often assume a characteristic banana-shaped profile. At such times they are on their side, with the head raised and the hind flippers elevated and pressed tightly together. Harbor seals and juvenile gray seals are sometimes difficult to tell apart while in the water. In addition to their small, round head, slightly upturned tip of the nose, and white mottling, harbor seals can be distinguished by their V-shaped nostrils. The nostrils of gray seals are more nearly parallel.

DISTRIBUTION: The harbor seal is confined to temperate and subarctic regions of the Northern Hemisphere. In the eastern North Atlantic, it is distributed from Iceland, Svalbard, and northern

Norway south to Brittany, France. A few individuals have reached the northwestern coast of Portugal. Historically, harbor seals bred in small numbers along the English Channel coast of France, but no breeding has occurred there since about 1930. By 1980, only a small number of seals, mostly young ones that had strayed from the British coast, hauled out seasonally in the Bay of Somme. The group of at least 500 seals at Svalbard is centered along the west coast of Prins Karls Forland. As it is some 900 km away from the nearest population in northern Norway, this population is considered insular. The largest concentrations of harbor seals are around the British Isles (especially along the west coast of Scotland, throughout the Hebrides and northern isles, and in certain estuaries and on sandbanks along the North Sea coast) and in parts of Scandinavia (for example, the Kattegat-Skagerrak) and the Wadden Sea. Harbor seals also inhabit rocky coasts on Scotland's Orkney Islands as well as mud flats and sandy beaches of The Wash (a large embayment on the North Sea coast of England). A large population of harbor seals is present in County Down, northeastern Ireland. They are scarce in the west of Ireland but common in the east. Vagrant harbor seals occasionally enter Scotland's famous Loch Ness, possibly contributing to monster lore there. Four small colonies exist on the Swedish Baltic coast.

In the western North Atlantic, harbor seals breed at least from Cape Dorset on southwestern Baffin Island to Massachusetts. They occur along the west coast of Greenland as far north as Disko Island. Even though some harbor seals are found in southwestern Greenland just a few kilometers north of Cape Farewell, they are absent from the east coast. Small numbers are present in parts of Hudson Strait, Ungava Bay, and Hudson Bay. In these areas, harbor seals are encountered most often at river mouths. Small, local populations inhabit some rivers and lakes of western Hudson Bay, moving as far as 240 km inland. A small population lives in the Seal Lakes (Lacs des Loups Marins) at the headwaters of the Nastapoka River in northern Québec. Harbor seals have been seen as far north as Ellesmere Island (79° N) and as far south as Daytona Beach, Florida (29° 05′ N). A large population breeds on Sable Island, some 125 km off Nova Scotia.

In the eastern Pacific, harbor seals breed from San Quintín Bay, Baja California, to Nome, Alaska. The northernmost portions of their range in the Bering Sea include Bristol Bay, the Pribilof Islands, and the Aleutian and Commander islands. The southern limits are Hokkaido, at about 43° N, in the west and Cedros, Natividad, and the San Roque islands, Baja California (about 28° N), in the east. Harbor seals are abundant in protected inlets, bays, and fiords but are generally less abundant along simple, exposed coasts and around small islands in the Commander, Aleutian, and Pribilof islands. They breed at nine rocky haul-out sites along the east coast of Hokkaido, Japan, including Erimo Cape.

While there is much site fidelity at haul-out grounds from year to year, harbor seals are also capable of long-distance movements. They often cross the English Channel. One seal tagged at Netarts Bay, northern Oregon, was later found dead in a gill net in Humboldt Bay, California, 550 km to the south. A pup tagged on the south coast of Iceland was recovered 9 months later on the north coast more than 550 km from the tagging site. A young seal tagged at Sable Island off eastern Canada was recovered in New Jersey, a straight-line distance of 1475 km.

Some short movements may be associated with seasonal availability of prey and with breeding. However, in many areas harbor seals are present year-round.

Natural History: Harbor seals haul out on intertidal ledges, rocky islets, reefs, mud flats, log rafts, piers, and isolated sand or cobble beaches. In some areas they also haul out on glacial and sea ice. Some harbor seals live mainly in fresh water, either far up rivers or in lakes. The factors influencing haul-out behavior probably include season, time of day, tide, wave height or intensity, wind chill, and disturbance. Disturbance by humans, whether approaching overland or from the water, and by dogs strongly affects haul-out behavior. In many developed coastal areas, intentional or inadvertent harassment excludes harbor seals. Often, young harbor seals are left behind on disturbed beaches. Well-meaning but poorly informed people pick them up to "rescue" them and take them to oceanaria or other rehabilitation facilities, even though the mothers may be waiting just offshore. In most instances, it would undoubtedly be best to clear the area of humans immediately and give the mothers a chance to come ashore and retrieve their pups. This may take some time, since the females sometimes wait and watch for several hours before venturing back onto the beach.

Harbor seals are gregarious on land, although most of their aggregations are smaller than those of other gregarious pinnipeds. Thousands may gather on coasts with a suitable haul-out substrate. Seals hauled out alone spend significantly more time scanning for predators than seals hauled out in groups. Thus, members of groups can rest more than lone individuals. Harbor seals are moderately aggressive toward one another while hauled out. However, their biting and clawing, accompanied by barks and growls, rarely result in contact. More typically, hauled-out seals sleep, rub, scratch, yawn, and occasionally scan for predators.

Spotted seals and harbor seals coexist off eastern Hokkaido, the Kuril Islands, eastern Kamchatka, and the Commander Islands and in the southeastern Bering Sea. The two species generally live in different habitats and differ in the timing of their annual reproductive and molting cycles. There is some confusion of the two in the older literature from the North Pacific, where spotted seals (formerly not distinguished from harbor seals) pup mainly in April, at least a month before harbor seals. In the North Atlantic, harbor seals are sympatric with gray seals; many islands and coastal sites have populations of both species. There is no overlap in the timing of the pupping season, however: harbor seals give birth mainly in spring and summer, gray seals in autumn and winter.

An interesting feature of reproduction in harbor seals is that regional populations have widely different pupping seasons, extending at least from late January to October. Harbor seals give birth in the western Pacific and Gulf of Alaska in May and June; in Bristol Bay and along the Aleutian and Pribilof islands in June and July; and progressively later from there to British Columbia. In British Columbia and Washington, pupping occurs from June to September. However, in Willapa Bay, on the outer coast of Washington, and in Oregon most pups are born in May and June. Pupping is progressively earlier from Washington and Oregon southward to Baja California, where it takes place in February and March. Pupping in central California begins in late March and peaks in the first half of May.

In the eastern Atlantic, most pups are born in June or July. In Great Britain, the season is late May to the end of June. The peak of the 3 week pupping season in the Wadden Sea is in late June or the beginning of July. The latter half of June is the pupping season at Svalbard and Norway. In the western Atlantic, the main pupping season is May and June in Nova Scotia and Newfoundland, June and July farther north. At Sable Island, harbor seals pup from early May to mid-June, with a 2 week peak in late May. Twinning is rare, and it is unlikely that twins could survive to weaning.

Because they often are born in the intertidal zone, pups must be ready to enter the water almost immediately after birth. Within 5 minutes, the precocious pup can follow its mother on land and, if necessary, in the water. The female recognizes her pup by its

Harbor seal pups are precocious. They begin swimming within hours after birth, remaining in close company with their mothers. (Sable Island, Nova Scotia, Canada, May 1972: Fred Bruemmer.)

vocalizations that allow the two to maintain contact at sea as well as on land. During its first week of life, the pup often rides on its mother's back as she dives and surfaces. During the 1 to 2 weeks immediately prior to weaning, most pups spend as much time in the water as out. Most nursing takes place on land, but some is done in the water. Mothers are solicitous, occasionally defending their pups aggressively. On Sable Island some have been seen to grab their pup's flipper with their mouth and tow it beyond the surf out of the reach of men stalking the seals from land.

The lactation period is usually 4 weeks, but some pups are weaned after slightly more than 2 weeks and some are nursed for as long as 6 weeks. Pups have a thin blubber layer at birth (about 1.3 cm), but it thickens rapidly as they gain about 0.7 kg per day while they are being nursed, and they weigh 25 to 30 kg when weaned. Diving records for lactating females in California and northwest Atlantic waters indicate that they continue to forage while nursing their pups, although perhaps at a lesser rate and at shallower depths than before giving birth. Lactating females at Sable Island lose about 37 percent of their initial postpartum mass by the time they wean their pups.

For the first few weeks after weaning, young harbor seals eat benthic crustaceans, including shrimp (*Crangon* sp.). This tendency has been confirmed for at least the east and west sides of the Atlantic and for the eastern Pacific.

Mating takes place within a few days after the pups are weaned. Implantation is delayed for 1.5 to 3 months. Harbor seals generally have been considered serially monogamous or promiscuous. However, since courtship and copulation occur mainly in the water, these activities are difficult to observe. Adult males display during the breeding season by slapping the water with their fore and hind flippers. They also sustain many wounds and bear scars from fighting during this period, and they have a higher natural mortality rate than adult females. These characteristics are typical of polygynous species. There is some evidence that nursing mothers form subclusters within groups of hauled-out harbor seals. They are aggressive and threaten nearby animals.

Females become sexually mature at 3 to 6 years of age, males at 3 to 7 years.

Harbor seals molt in the Bering Sea and Gulf of Alaska from late June to early October, at the southern California Channel Islands from May through August, in the eastern Atlantic in August and September, at Scotland's Orkney Islands from early June through mid-September, and in the western Atlantic in July. Individual differences in the timing of molt are related to sex, age, and reproductive status. Impregnated adult females molt during the period of delayed implantation; immature and mature males molt after females do.

Captive harbor seals eat 3 to 10 percent of their body weight in fish per day; growth requires about 909 kilocalories for each 100 g increase in body mass. Wild harbor seals probably consume 6 to 8 percent of their body weight in food per day. They are significantly fatter (that is, their blubber is thicker) in winter than during the reproductive and molting periods. In Alaskan waters during winter, blubber accounts for 27 percent of the total body mass of males and 30 percent of females. Their diet varies regionally, seasonally, and geographically according to local prey availability. Since harbor seals have such a wide range and occupy such diverse habitats, they eat an extremely large variety of organisms, including pelagic and benthic fishes, cephalopods, and crustaceans. Sand lance and herring are important prey near Cape Cod, with the herring eaten mostly in March and April. In Iceland in late winter, they eat mainly capelin and sand lance. In Netarts Bay, Oregon, they prey on chum salmon. The annual winter spawning run of eulachon in the Columbia River attracts some 2000 harbor seals. Around the southern California Channel Islands, harbor seals eat mainly plainfin midshipmen, spotted cusk eels, octopus, and rockfish. In Danish waters, where harbor seals feed mostly on the sandy bottom, pleuronectids (flatfish) constitute 75 percent of the diet. Prey up to 30 cm long are consumed whole or in pieces.

In a recent study, researchers recorded some 35,000 dives of 10 harbor seals in southern California. Maximum dive depths and durations varied from 54 to 446 m and from 7 to 27 minutes. Seals dived, on average, for 3 to 7 minutes to depths of 17 to 87 m. Satellite telemetry has shown that adult seals remain close to the Channel Islands year-round, feeding in epibenthic habitats near shore.

Rates of mortality vary by area. In the Kattegat-Skagerrak, mortality was relatively low during a period of rapid increase in the seal population. During the 1980s, the estimated annual mortality rate for adult males in this population was 9 percent. The estimated rate of survival from birth to sexual maturity was 20 to 45 percent. In one study in eastern Canada, an estimated 12 percent of pups died within their first month, and about a quarter died before the end of their first year. Abandonment or orphaning, leading to starvation, is a major cause of pup mortality.

Large sharks may be important predators of harbor seals, especially pups, in some areas. For example, in 1980 during the pupping season at Sable Island, at least 23 pups were killed by sharks in 3 days along an 8 km stretch of beach. In waters bordering Vancouver Island, British Columbia, killer whales prey regularly on harbor seals. Northern sea lions are confirmed predators in the North Pacific, and eagles are suspected predators of pups in various regions. Coyotes prey on pups in parts of Alaska and Washington. Bears and canids may prey on harbor seals in certain other areas, but we are unaware of confirmed observations.

Pups believed to be orphaned or abandoned commonly are

rescued and rehabilitated along the east and west coasts of the United States and in Great Britain. In at least a few cases, seals nourished with infant formula and nursed to health with vitamins and antibiotics have been returned to the wild. Young seals also can be reared directly on a diet of raw oily fish such as herring or mackerel.

Harbor seals may live for at least 30 years, although few live longer than 25. Males have a shorter life span than females, on average, probably due at least partly to the stress of fighting during the breeding season.

Outbreaks of disease play a role in controlling harbor seal populations in some regions. A well-documented example occurred in 1979–80 when more than 400 seals, mostly juveniles, died in New England from acute pneumonia associated with an influenza virus. In spring of 1988 a devastating epizootic began in northwestern Europe, spreading rapidly from the Kattegat, between Denmark and Sweden, to the North Sea and eventually throughout much of Great Britain. Before it had run its course, more than 18,000 harbor seals had died. The primary cause of the die-off was a virus closely resembling that of canine distemper. This newly discovered virus was named phocine distemper virus (PDV). Once their immune response systems had been weakened by PDV, the seals became susceptible to a variety of other debilitating viral and bacterial agents.

Gut nematodes are common in most harbor seal populations. In Washington, heartworm was found in approximately 47 percent of the seals; there was also a high incidence (approximately 45 percent) of seal louse, which may be an intermediate host of the heartworm. Heartworm also has been identified in seals from European waters. Mites (Halarachnidae) are common parasites in the nasal passages of harbor seals in the Bering Sea. These mites cause inflammation and swelling of nasal tissues.

History of Exploitation: Harbor seals have been hunted and netted for centuries over most of their range. Traditional hunting for subsistence continues in Alaska, Greenland, and northern regions of Canada and the U.S.S.R.

Bounties were paid on harbor seals during various periods. The government of Alaska spent more than a million dollars on seal bounties from 1927 to 1967. Also in Alaska, depth bombs were used to kill some 27,000 seals near important fishing grounds from 1951 to 1957. Phocid pelts had high value during the 1960s, and licensed hunters in Alaska sold 170,000 harbor seal pelts during 1964–1970. The annual commercial catch in Alaska was 40,000 to 60,000, mostly pups, in the mid-1960s. During 1964 and 1965, almost the entire pup production on Tugidak Island in the Gulf of Alaska was taken. Most of the commercial catch during the 1960s was to supply a market for sealskin sports clothing in Europe. The annual commercial catch in Alaska declined to between 8000 and 10,000 by the early 1970s. Since the 1972 ban on commercial harvesting, subsistence hunters in Alaska have accounted for 2500 or fewer harbor seals each year.

An estimated 200,000 to 240,000 harbor seals were killed in British Columbia between 1913 and 1969, including both bounty and commercial harvests. Oregon paid $25 per pelt to professional hunters from 1954 to 1964, and Washington had a bounty until 1970. There was a bounty on harbor seals in Massachusetts from 1888 to 1962. Hundreds of bounties were paid on harbor seals in the Canadian Maritimes each year from the late 1920s through 1976, although the number of bounty claims declined through time. The Canadian bounty was not only supposed to prevent competition and damage to fishing gear but also to control codworm. Although gray seals are more significant hosts of this parasite, harbor seals are also infected with codworm larvae to some extent.

During 1911 to 1953, a bounty was paid in England for harbor seals killed in The Wash. Mainly juveniles and adults were taken, with the goal of reducing their impact on local fisheries. A large-scale commercial hunt for the fur trade began in The Wash during the early 1960s, concentrating on pups. As many as 850 pups were taken in a single year (1966), and the average annual catch was more than 700 during 1965 to 1970. After the Conservation of Seals Act came into force in 1970, giving protection to the seals during the pupping season, the quota was 400 pups per year. The commercial harvest at the Shetland Islands during the late 1960s and early 1970s removed nearly 90 percent of the annual pup production, and this situation led to a complete ban on seal hunting there in 1973.

In Norway, more than 1000 harbor seals were killed in government-sponsored culls from 1980 to 1986. An average of more than 5600 pups were killed each year in the Icelandic commercial seal fishery from 1962 through 1976. Since 1976, the pup harvest has declined, but the annual harvest of adults has increased in recent years to more than 1000. Between 1962 and 1987, approximately 117,000 harbor seal pups and 7300 adults were taken in Iceland.

Large numbers of harbor seals die from entanglement in gill nets in some areas. Perhaps as many as 5 percent of the harbor seals in California waters become entangled and drown in coastal gill nets annually. Off eastern Hokkaido, substantial numbers die in salmon trap nets, and fishermen deliberately kill them.

CONSERVATION STATUS: The world population of harbor seals probably is still at least half a million. They remain widely distributed across the cool middle latitudes of the Northern Hemisphere. However, in many parts of their range, the populations have been greatly reduced or eliminated by exploitation and habit modification. For example, the harbor seals inhabiting Lake Champlain and Lake Ontario during the nineteenth century have been extirpated. Also, during the early twentieth century there were approximately 300 harbor seals in the Dutch Wadden Sea (north Netherlands) and approximately 1300 in the Delta area (southwest Netherlands). They are now virtually absent from the Delta area and much reduced in the Dutch Wadden Sea. Overhunting, encouraged by a bounty, occurred between 1930 and the 1950s. Seal hunting was banned in The Netherlands in 1961. However, dams constructed since 1953 to protect estuaries in the southwestern part of the country limited the amount of available habitat. Disturbance from increased shipping and yachting probably has also contributed to the seals' decline. Pollution had caused a reduction in either pup production or survival, or both, in the Dutch Wadden Sea before 1988.

Although greatly reduced by hunting, from perhaps 17,000 in 1890 to 2000 in the 1960s, the population in the Kattegat-Skagerrak was protected beginning in 1965. It then grew rapidly and had made an impressive recovery before crashing as a result of the viral outbreak in 1988.

In the Baltic Sea, there were two geographically distinct groups totaling nearly 150 seals along the Swedish coast in the mid-1980s. Three Baltic haul-out sites in Sweden have been protected as seal sanctuaries since 1978, and several other sites fall within bird sanctuaries. Sealing has been prohibited in the Swedish sector of the Baltic since 1974.

Great Britain's harbor seal population was estimated at approximately 25,000 in the mid-1980s, but the die-off in 1988 reduced it somewhat. The population along the Scottish coast was little affected. Several breeding sites in northeastern Ireland are in nature reserves.

At least 3000 harbor seals live in scattered groups in western and northern Norway. The population in the Hvaler area (Oslofjorden) was reduced by the epizootic in 1988 to about a quarter of its previous size. The large population of approximately 30,000 around Iceland apparently was unaffected by the 1988 epizootic.

Since 1979, seals that have been found stranded along the coast of France have been rehabilitated and then released in attempts to reestablish breeding colonies.

The suspension of the bounty on harbor seals in eastern Canada in 1976 allowed the population that pups on Sable Island to increase rapidly. From 1978 to 1987, the number of pups born on Sable Island increased at an annual rate of more than 6 percent. An estimated 13,000 harbor seals were present in eastern Canada south of Labrador during the mid-1980s.

Protected since 1972, harbor seals are now abundant in parts of the United States. Still, there are no pupping rookeries on the east coast south of Maine (they pupped in much of coastal Massachusetts in the past). Nearly 13,000 harbor seals were counted at haul-out sites in Maine in June 1986. There are several hundred harbor seals in New York waters, where numbers reach a peak from

Harbor seals at Sable Island, off Nova Scotia, Canada, seem to have habituated to human activities near their breeding and haul-out areas. In most other areas, however, harbor seals have reacted to human intrusion by abandoning sites or altering their haul-out patterns. (May 1972: Fred Bruemmer.)

November through April. In recent years, they have increased in Massachusetts and New Hampshire and now number more than 4000; most of these are near Cape Cod and Nantucket. A large aggregation also can be found in winter at Monomoy Island and nearby shoals. Although this substantial seal population may be perceived as a threat to regional fisheries, the predominance of sand lance, a noncommercial fish species, in the diet of the seals in Cape Cod waters should moderate calls for predator control. In Maine, however, lobstermen complain that seals rob both bait and lobsters from traps. They also claim that seals eat shedding lobsters and consume "short" (legally undersized) ones before the fishermen can release them.

The total population in the North Pacific was estimated at more than 300,000 in 1979, with at least two-thirds of this number in Alaskan waters alone. Approximately 29,000 harbor seals were present along the Alaska Peninsula and eastern Aleutian Islands, mostly near Port Moller, Port Heiden, and Cinder River.

They are virtually unexploited in most of the Aleutian, Commander, and Pribilof islands. Nonetheless, their numbers declined by nearly 50 percent during the 1980s. At Tugidak Island in the Gulf of Alaska, formerly the site of one of the world's largest harbor seal colonies, the population dropped by 85 percent, from approximately 20,500, between 1976 and 1988. The causes of the decline have yet to be clearly identified, although it parallels the great declines in northern fur seals and northern sea lions in the region during the same period. The population of harbor seals in the Kuril Archipelago was estimated as 2000 to 2500 in the early 1960s and had declined to about 1500 by 1970. Only about 350 survive in eastern Hokkaido. Japanese salmon fishermen have resisted attempts at seal conservation because the seals remove or damage fish caught in trap nets.

Since becoming protected in 1970, harbor seal populations in British Columbia had increased almost tenfold by 1988, with an estimated post-pupping population in provincial waters of 75,000 to 88,000. There was no sign of a decline in the growth rate as of 1988.

FURTHER READING: Bigg (1969), Bonner (1972), Boulva and McLaren (1979), Corbet and Harris (1991), Dietz et al. (1989), Fay and Fedoseev (1984), Härkönen (1987), Härkönen and Heide-Jørgensen (1990a), Markussen et al. (1990), Olesiuk et al. (1990), Payne, P. M., and Selzer (1989), Shaughnessy and Fay (1977), Stewart, B. S., et al. (1989), Temte et al. (1991), Thompson (1988, 1989), Thompson and Rothery (1987), Yochem et al. (1987).

Spotted Seal

Phoca largha
Pallas, 1811

NOMENCLATURE: The name *largha* comes from the Tungus word for this species. The Tungus people live along the shores of the western Sea of Okhotsk, where spotted seals are abundant. For many years the spotted seal was regarded as a subspecies of the harbor seal. However, over the past 15 years the argument for its recognition as a distinct species has gained wide acceptance.

DESCRIPTION: Spotted seals are generally 77 to 92 cm long and weigh 7 to 12 kg at birth. By the time they are weaned at 3 to 6 weeks of age, they can weigh as much as 36 kg. However, much of this weight is lost during the difficult first few weeks on their own. Several weeks after weaning, juveniles may weigh as little as 20 to 25 kg. Adult males grow to 1.7 m long and females to 1.6 m; they weigh 82 to 123 kg.

The variable color pattern usually consists of a dark gray mantle on a silvery or brownish yellow background. There are many small, dark, and irregular spots, particularly on the back and sides. Light gray rings often are also present on the mantle of juveniles. Pups are born in a dense, woolly, grayish to off-white lanugo that is retained until weaning.

DISTRIBUTION: Spotted seals occur throughout the Sea of Okhotsk, in the northern Sea of Japan, and thence south to at least the Shantung Peninsula in the Yellow Sea. Three breeding concentrations are known in or adjacent to the Sea of Okhotsk: one in Shelikhova Gulf, one east of Sakhalin Island, and one in Tatar Strait. Other concentrations are in Peter the Great Bay in the Sea of Japan and in the Po Hai Sea, an appendage of the Yellow Sea. Spotted seals haul out on rocky reefs at Point Notoro and on grass- or sandbars off the east coast of Hokkaido.

In the Bering Sea, the greatest densities of spotted seals are found during winter and spring (February through May) within approximately 25 km of the irregular and shifting southern margin, or front, of the pack ice. They give birth, nurse their young,

Amid a tumble of spring ice, a spotted seal arches to survey its surroundings. (Bering Sea ice front, northeast of the Pribilof Islands, May 1978: Lloyd F. Lowry.)

and mate on the small, dispersed, moving floes characteristic of this broad band stretching from Alaska to Siberia. They rarely are encountered more than 100 km from the pack ice edge. In April, adults generally are found as isolated pairs with a white-coated pup. As the Bering Sea ice front breaks up and disintegrates in May and June, the seals stay with what remains of the ice, forming small herds to molt and bask. Their net movement in late spring and early summer is northward and coastward. In late summer and fall, spotted seals essentially replace ringed seals in estuaries and embayments, the latter having moved north with the receding pack ice. Spotted seals often haul out on sandbars and beaches during the open-water season. In summer, they haul out on St. Matthew and nearby islands in the northcentral Bering Sea.

Especially high densities of spotted seals have been found in April in outer Bristol Bay, the central Bering Sea between 175° W and 180°, and Karaginskii Gulf. Morphological evidence suggests considerable differentiation, and thus some degree of geographical isolation, between regional populations, so separate western, central, and eastern stocks have been proposed for management in the Bering Sea. The western stock winters mainly in Karaginskii Gulf and disperses toward the coasts of Koryak and Kamchatka. The central stock is distributed in winter and spring south of Cape Navarin to St. Matthew Island, dispersing into Anadyr Gulf. The eastern stock is centered in the southeastern Bering Sea from the Pribilofs to outer Bristol Bay and disperses northward through Bering Strait and along the shores of the Chukchi Sea. About 2000 spotted seals occupy Kasegaluk Lagoon near Point Lay, Alaska, from July to freeze-up each year.

NATURAL HISTORY: Spotted seals and harbor seals are widely sympatric in the southern Bering and Okhotsk seas. Members of the two species may haul out in the same areas, even during their respective breeding seasons, but they generally form separate groups and rarely (or never) interbreed. Differences in parasite faunas reinforce the view that these two superficially similar phocids are socially and nutritionally divergent. Their breeding seasons are about 2 months out of phase where their ranges overlap most extensively.

Spotted seals form "family" groups on the pack ice during the spring breeding season. These groups are separated from one another by at least 0.2 km. The male is thought to join the female approximately 10 days before parturition and to remain with her until copulation occurs immediately after the pup is weaned. It is unlikely that a given pup will have been sired by the adult male attending its mother during its own birth and lactation. Spotted seals are considered annually monogamous and seasonally territorial. Gestation lasts 10.5 months, including a period of 1.5 to 4 months of delayed implantation. Off Hokkaido, the pupping season extends from late January to mid-April, with a peak of births during the second half of March. Japanese investigators estimate that spotted seal pups are weaned 2 to 3 weeks after birth.

Female spotted seals give birth and suckle their pups on small floes in the 5 to 30 mile wide ice front. (Bering Sea, northeast of the Pribilof Islands, April 1977: Kathryn J. Frost.)

Mothers show devotion to their pups and often attempt to drive away intruders. Pups remain on the ice for the first few weeks and make their first awkward attempts at swimming shortly before being abandoned by their mothers. Their buoyancy makes diving difficult initially, and it takes several more weeks for them to become proficient swimmers. Within a month after weaning, they can dive to at least 80 m to feed. Adults dive to depths of at least 300 m.

Female spotted seals become sexually mature at 3 to 4 years of age, males at 4 to 5 years.

Spotted seals eat a broad spectrum of prey, including many kinds of fish and invertebrates, especially crustaceans and octopus. Their diet varies by area and season. Among the more important fish eaten are walleye pollock, capelin, arctic cod, saffron cod, sand lance, herring, and rainbow smelt. In northern Bristol Bay and the Yukon-Kuskokwim delta, major haul-out sites are near spawning areas of herring and capelin. Spotted seals have been estimated to consume 3 to 13 percent of their body weight in fish and other prey each day, but these values vary with age, sex, and season.

There is a progression in the spotted seal's diet according to age. During the first year, they eat mainly small crustaceans, including amphipods, shrimps, and euphausiids. They also often swallow algae, sticks, and other debris. At 1 to 4 years of age, they eat schooling fishes, larger shrimps, and some octopus; seals 5 years or older rely more on benthic pleuronectids (flounders, halibut) and cottids (sculpins), as well as on crabs and octopus.

Spotted seals reach a maximum age of 35 years. They do not appear to suffer heavy predation from polar bears or arctic foxes because few of these predators are present in the breeding grounds. Killer whales presumably take some, and walruses are known to eat pups. Northern sea lions and large sharks kill spotted seals at least occasionally, and avian predators, including eagles, ravens, and gulls, are a threat to young pups in some circumstances. In summer, when spotted seals haul out on land in the Sea of Okhotsk particularly, they are sometimes killed by brown bears and wolves. Other threats include crushing by ice in storms and becoming stranded out of the water when the surface freezes over suddenly.

History of Exploitation: At least a few hundred spotted seals have been taken annually by shore-based hunters in the U.S.S.R. since the 1930s, to feed animals on fur-farming collectives. A commercial ship-based harvest of seals was begun by the U.S.S.R. in 1932 in the Sea of Okhotsk, and relatively small numbers of spotted seals were taken. This remained true after the operation moved into the Bering Sea in 1961. Reported annual Soviet catches of spotted seals in the Bering Sea from 1969 to 1983, including both ship- and shore-based harvests, averaged 3500. Since 1970, quotas have been in force in the U.S.S.R.: 5000 spotted seals from ships and 2000 from shore in the Sea of Okhotsk, 6000 from ships and 2000 from shore in the Bering Sea. In 1988 the reported Soviet harvest was 2273; in 1989, 2315. Nearly all of the seals were from the ship-based hunt in the Bering Sea.

A Japanese commercial hunt for spotted seals and other ice seals (mainly ribbon seals) continued into the mid-1970s. The main catching grounds were in the Sea of Okhotsk along the east side of Sakhalin Island. The winter and spring catches were centered in three areas: the mouth of Terpeniya Bay off Sakhalin; Kitami-Yamato Bank, Lake Saroma, and Point Notoro; and between Shiretoko Peninsula and Nemuro Peninsula. Hunters in skiffs shot the seals with shotguns or rifles and delivered them to a mother ship for processing. Some also were caught in fishing nets off the Pacific coast of Nemuro Peninsula during fall. Close to 400 spotted seals were taken in the 1975 hunt. Since the mid-1970s, this Japanese commercial sealing operation has been closed because of the U.S.S.R.'s enforcement of a 200 mile limit. Ice seals are now hunted by the Japanese only in years when the pack ice approaches the coast of Hokkaido. Some spotted seals continue to die in the salmon trap nets along the Nemuro Peninsula.

The subsistence catch of spotted seals in Alaska was estimated at approximately 2400 per year in the 1960s and 1970s. Subsistence hunting has continued. In one study, it was learned that five villages had a combined catch of nearly 1000 spotted seals between September 1985 and June 1986.

CONSERVATION STATUS: There are probably at least 200,000 spotted seals in the Bering-Chukchi population and perhaps 130,000 or more in the Sea of Okhotsk. The world population is considered essentially stable. Although direct hunting does not appear to be excessive, the intensive commercial fisheries for pandalid shrimps, pollock, and herring in the southern Bering Sea may threaten the seals' food supply. Also, oil and gas exploration, extraction, and transport activities are under way or scheduled in portions of their winter range. These could have a significant impact on the seals themselves or on their food resources.

FURTHER READING: Fay and Fedoseev (1984), Naito and Konno (1979), Naito and Nishiwaki (1972, 1975), Quakenbush (1988).

Ringed Seal

Phoca hispida
Schreber, 1775

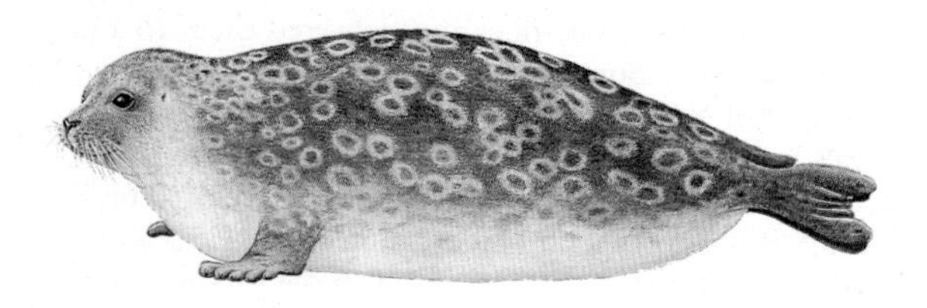

NOMENCLATURE: The specific name is from the Latin *hispidus,* meaning "hairy" or "bristly," in reference to the coarse pelage of the adult. An obsolete specific name, *foetida,* refers to the strong odor characeristic of adult males during the breeding season.

The common name is from the irregular rings of lighter pigmentation on the adult pelage. In the eastern Canadian Arctic, this seal is called the jar, with newly molted juveniles known as silver jars. Historically, some English-speaking visitors to the Arctic called ringed seals floe rats. The Norwegian name is *ringsel,* the Danish *ringsael,* the German *ringelrobbe,* and the Icelandic *hringanóri*. Finns refer to the Saimaa ringed seal as *norppa*. Eskimos call the ringed seal *netsiq, netsiak,* or *natchek.*

Various subspecies have been proposed, mainly on the basis of geographical isolation: the circumpolar *P. hispida hispida* of the Arctic Ocean, *P. hispida ochotensis* of the Sea of Okhotsk and northern Japan, *P. hispida botnica* of the Baltic Sea (including the gulfs of Bothnia and Finland), *P. hispida saimensis* of Lake Saimaa, and *P. hispida ladogensis* of Lake Ladoga, U.S.S.R.

DESCRIPTION: The average birth length is 60 to 65 cm; the weight, only 4.5 to 5.5 kg. Maximum size is somewhat more than 1.6 m and 110 kg, but most ringed seals do not grow larger than 1.1 to 1.5 m and 50 to 70 kg. Males are somewhat longer than females. There is considerable geographical variation in the sizes attained by ringed seals within and among areas. Also, their weight varies seasonally. Males, in particular, lose considerable weight during the molt in early summer. In the Svalbard area, the average length of sexually mature males and females is about 1.3 m. During the molting season, when seals are not feeding, males weigh an average of 53.5 kg and females 61.4 kg. Maximum lengths and masses in this area are 1.57 m and 90 kg (males) and 1.54 m and 92 kg (females). In Bothnian Bay, males weigh 80 kg, on average, and females 87 kg.

The pup's fine, white, woolly lanugo begins shedding at 2 to 3 weeks and is completely lost by 6 to 8 weeks of age. It is replaced

A ringed seal pup that has molted its woolly lanugo, or fetal coat, is now covered by the unspotted juvenile pelage. (Spitsbergen, Norway, July 1961: Fred Bruemmer.)

by a largely unspotted coat, silver on the belly and dark gray on the back. The background color of adults is generally the same, but gray or whitish rings are prominent on the back and sides.

The ringed seal's body is plump; in winter the girth can exceed 80 percent of the body length. The muzzle is short; the face more catlike than doglike. Pups are born with the full complement of permanent adult teeth.

Distribution: The ringed seal is sometimes described as ubiquitous in the Arctic and Subarctic. It is indeed very widespread and abundant. However, the density of its occurrence is extremely variable, depending principally on winter ice conditions. In Amundsen Gulf, for example, there is a consistently high density of seals in Prince Albert Sound on the west coast of Victoria Island. Outside this sound, there are large fluctuations in seal density from year to year. In Minto Inlet, the next embayment north of Prince Albert Sound, early summer aerial surveys indicated an increased density from 0.62 seals per km^2 in 1972 to 3.66 per km^2 in 1973. Scientists documented a large decrease in the number of ringed seals in the southeastern Beaufort Sea from 1974 to 1977. Between 1974 and 1975, they declined by 50 percent, and the number born in the prime breeding habitat declined by 90 percent. These dramatic changes are believed to have been caused by exceptionally heavy ice conditions in the winter of 1973–74, resulting in low pup survival and large-scale emigration. After 2 more years (1976 and 1977) of low abundance, the regional ringed seal population more than doubled in 1978 and appeared to have "recovered" by 1979, probably through a combination of increased productivity, high survival, and substantial immigration.

Elsewhere in Canada, the ringed seal occurs throughout the High Arctic, in Hudson Strait, Hudson and James bays, and along almost the entire coast of Labrador to Newfoundland. It also inhabits Nettilling Lake on Baffin Island.

Alaskan waters provide a tremendous amount of sea ice habitat used by ringed seals. Tens of thousands have been counted on the shore-fast ice during the spring molt. Highest densities occur in the land-fast ice zone, which extends 5 to 40 km seaward from the shore. This zone is the preferred habitat for adults, while juveniles are more common in the flaw zone and pack ice. During winter and spring, ringed seals occur in the Bering Sea as far south as the limit of sea ice, usually at least to Nunivak Island or Bristol Bay. As the seasonal sea ice cover melts in spring, many ringed seals move north to spend the summer in the pack ice of the northern Chukchi and Beaufort seas. In the western Pacific, ringed seals are present in the Okhotsk Sea and off northern Japan.

Ringed seals from the Eurasian Arctic occur sporadically off the north coast of Norway. There is a population in the northern and eastern Baltic Sea, although it has been considerably reduced from historic levels of abundance. Two relict lake populations exist in the Baltic region, one in Lake Saimaa in eastern Finland and one in Lake Ladoga near Leningrad. Both lakes are connected to the Gulf of Finland, Lake Saimaa by a canal and Lake Ladoga by the Neva River. However, neither of these freshwater seal populations appears to mix with the marine population. The Saimaa seal stock was cut off from the closest ringed seal populations in the White and Barents seas during early postglacial time, 9000 to 9500 years ago. The stock is itself now segmented into three or four islated groups by human activities and development. The town of Savonlinna occupies a narrow strait separating the southern and central groups, and the eastern group is probably isolated from the others by the long, narrow Hanhivirta Sound.

Like other seals, ringed seals occasionally wander far outside their normal range. For example, a young individual was caught alive on a beach south of Lisbon, Portugal, in summer 1968, and in the mid-1970s a young ringed seal was found on a beach in southern California.

NATURAL HISTORY: Ringed seals, particularly breeding adults, occupy land-fast ice in winter. They are able to maintain breathing holes in ice at least 2 m thick by abrading it with the heavy claws of their fore flippers. The preferred adult winter and spring habitat

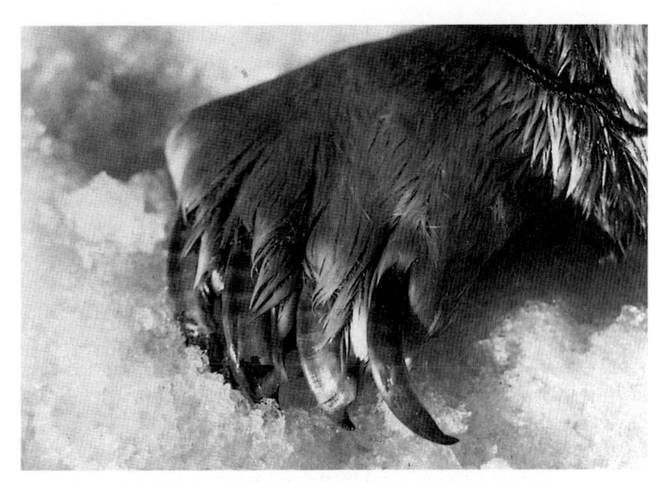

The robust claws on the ringed seal's fore flippers are useful for abrading ice to maintain breathing holes. (Off Moriussaq, northwestern Greenland, 3 June 1988: Stephen Leatherwood.)

is the shore-fast ice of bays, coastlines, and solidly frozen inter-island channels. Ringed seals, particularly juveniles, are also widely distributed among moving pack ice. The edges of cracks and refrozen leads and polynyas are often lined with their breathing holes. Although most ringed seals remain closely associated with pack ice and alongshore remnants of fast ice during the summer, some animals, again mainly juveniles, remain in ice-free areas through the summer.

Female ringed seals generally are not reproductively active until at least 4 years of age. In some populations, the mean age at sexual maturity is between 5 and 6 years. Males reach sexual maturity at 7 years of age, on average. Although the ovulation rate in Baltic Sea ringed seals has been estimated as more than 90 percent, the pregnancy rate has been extremely low (as low as 28 percent) since 1970, apparently related to high levels of PCBs and DDT in the Baltic ecosystem.

The fetal sex ratio and sex ratio at birth are 1:1. After giving birth in March or April (earlier in the Baltic), ringed seals nurse their pups for about 5 to 7 weeks. Nursing females forage at least occasionally but lose weight at a rate of about 450 g per day. Virtually all of the lost mass is body fat. Ovulation and mating occur near the time of weaning. Implantation is delayed for close to 3 months. It has been suggested that abnormally small, or stunted, ringed seals are the result of poor nourishment during the suckling or early postweaning period.

The use of subnivean (snow-covered) birth lairs is a unique behavioral feature of ringed seals. All other pinnipeds give birth on the exposed ice surface, on beaches, or occasionally in caves. Ringed seals in the Sea of Okhotsk apparently do not give birth in lairs, but those in all other areas do. Lairs are constructed principally around ice hummocks, along pressure ridges, and in the snow accumulated at the feet of active glaciers. Several different types of lair are used. The most common birth lair is an elongate chamber, about 2.5 m×3.5 m, excavated under overlying snow. The walls of the lair become more solid as ice forms on their inner surface. A hole approximately 30 cm in diameter at the surface of the ice is maintained by the mother to provide access to the water. The pup is born and nursed in the lair. The female leaves the pup unattended while she forages. The insulation provided by the snow cover can be critical, as it reduces heat loss and prevents the pup from freezing. A birth lair can be distinguished from a haul-out or resting lair by the tunnels dug off the main chamber by the pup and by the presence of lanugal hair on the walls of the lair. Some multichambered lairs are used by more than one adult seal. In the Beaufort and Chukchi seas, adult male and female ringed seals may use as many as four lairs from March through June. Lairs of sexually active adult males have a lingering, pungent odor apparently caused by an oily secretion from facial skin glands.

The ringed seal's productivity is closely attuned to the availability of stable, snow-covered fast ice, which in turn is governed largely by latitude (climate) and the configuration of the coastline. The older, more productive seals tend to occupy complex coasts, while younger seals are relegated to simpler coasts. In the Canadian High Arctic, the density of birth lairs is highest in the drifted pressure ridges across the mouths of bays and in interisland channels where the fast ice tends to become firmly locked in place.

Judging by the bite wounds on adolescent males, they apparently are excluded from breeding areas by the aggressive territorial adult males. Ringed seals establish territories at freeze-up and remain in the area for the winter, maintaining their individual breathing holes. A social structure of sorts is thought to be maintained by vocalizations.

Adult ringed seals tend to be much more sedentary than young seals; in the High Arctic, many occupy the same bay or fiord year-round. The dispersal pattern of subadults is directed from complex to simple coasts. In the western Canadian Arctic young ringed seals migrate west along the coasts of Amundsen Gulf and the southeastern Beaufort Sea during autumn. The movements of four ringed seals from northwest Greenland were monitored using satellite transmitters in summer and autumn 1988. One seal traveled north for more than 200 km, one west more than 200 km across Baffin Bay, one south, and one south and east along the Greenland coast. Tagged ringed seals have been caught more than 1000 km away from the tagging site.

Contrary to the situation in most areas where ringed seals have

A scientist releases a ringed seal with a satellite transmitter. (Wolstenholme Fiord, Greenland, 6 June 1988: Stephen Leatherwood.)

been studied, there are large breeding populations in the pack ice of the Sea of Okhotsk and Baffin Bay. Some ringed seals are also born on the drifting pack ice of the northern Chukchi Sea and Arctic Ocean. The seals born on the pack ice generally are smaller than their fast-ice counterparts.

Saimaa ringed seals live in a labyrinthine habitat that requires a somewhat different type of lair construction. Their lairs are usually situated within a meter of shore, along the northern or eastern sides of islands, rocky islets, or capes. These are the only areas where sufficient snow drifting occurs. The lairs are partly on water and partly on ground, and the water below the lair hole is less than a meter deep.

Ringed seals molt mainly in June and July, when they bask on the ice and stop (or almost stop) feeding. Considerable blubber loss occurs during the molt. The blubber "content" of breeding females in Svalbard decreases from 50 percent of body mass in March to 31 percent in June. In males, it declines from 41 percent of body mass in March to 29 percent in June. During this March through June interval, females lose body mass at 160 g per day, adult males at 100 g per day. Seasonal changes in the body mass of all ringed seals are due almost entirely to changes in the amount of body fat. Unlike other ringed seals, Saimaa seals molt on land.

The ringed seal is a versatile feeder, eating a wide variety of planktonic, nektonic, and benthic organisms. The longest dive time achieved by a captive ringed seal under experimental conditions was just under 18 minutes. In deep offshore waters, planktonic amphipods (for example, *Themisto libellula* and *Thysanoessa raschii*) are commonly eaten, whereas in bays and other inshore waters, arctic cod, saffron cod, and various crustaceans are taken. In some areas at certain times, squid and sculpins are important prey. In the northcentral Bering Sea during summer, ringed seals prey mainly on shrimp (*Pandalus* spp., *Eualus* spp., *Lebbeus polaris, Crangon septemspinosa*). The intensity of feeding activity varies seasonally. In Alaskan waters, ringed seals appear to concentrate during summer and early autumn in areas where euphausiids and hyperiid amphipods are abundant. Similarly, they congregate in Siberian waters to feed on local concentrations of arctic cod. Adults eat less and lose weight in spring and early summer, but by late summer and fall they are feasting and gaining weight. Because of the diversity of their prey and their own large numbers, ringed seals are viewed as competitors of many fishes, birds, and other marine mammals.

Thomas G. Smith (1987) described recurrent nearshore feeding complexes of adult ringed seals and gulls in Prince Albert Sound. Groups of up to 30 ringed seals were commonly seen feeding on schools of small arctic cod, attended by flocks of glaucous or Thayer's gulls. The gulls landed among the surfacing seals and lunged and pecked at their heads, forcing them to release fish from their mouths. Individual seals killed on such occasions have had as many as 400 tiny (3 to 5 cm long) cod in their stomachs. These feeding complexes are formed mainly during September and October. High-density, open-water concentrations of ringed seals are also seen in other areas, such as Eclipse Sound where at least 80 were seen in late August within a 1 km radius of a stationary boat.

Ringed seals eat whole fish at least 20 cm long. Much larger fish are also eaten. On one occasion, a 127 cm, 13.6 kg wolfish was found floating in the breathing hole of a medium-sized ringed seal. It had been bitten and clawed by the seal.

Although they are by no means loquacious, ringed seals do have a limited vocal repertoire. Their sounds, which are produced day and night, year-round, have been described as mainly low-pitched barks and high-pitched yelps or yowls. There is no suggestion that ringed seals echolocate. The intriguing question of how they locate prey and breathing holes under ice during winter darkness remains unanswered.

The maximum age attained by ringed seals in the Baltic may be close to 40 years; a 43-year-old specimen was reported from the eastern Canadian Arctic. Mortality during the first 10 years of life is nearly 84 percent in each sex in the northern Baltic Sea; afterward, males die at slightly greater rates than females. In the Okhotsk Sea, 35 percent of pups die during their first year; yearlings to 12-year-olds die at annual rates averaging 11 percent.

Ringed seals have several natural predators. In the birth lairs, pups are often killed and eaten by arctic foxes. However, polar bears are unquestionably the most important predators of ringed seals in most areas. The behavior of hauled-out animals in summer attests to their fear of polar bears. A ringed seal constantly alternates between lying flat and raising its head to scan for predators. In one study, looking time averaged 7 seconds, lying time 26.3 seconds. Eskimos stalking ringed seals on the ice behind white screens mimic the polar bear by moving forward and "freezing," alternately. Polar bear predation is believed to be most intense in the flaw zone and the broken pack ice, but polar bears also hunt ringed seals in the fast ice, and they dig into lairs. The depth, and thus thickness of the lair's roof, may influence the bear's ability to detect and prey on the seal inside. Most bear predation is on pups and subadults; newborn ringed seals are often killed but abandoned. Bears often eat mainly the blubber and leave the meat, although at least occasionally they eat an entire carcass. Predation rates vary from year to year and between areas, but for some areas and years it has been estimated that one-third or more of the pups born were killed by polar bears. Walruses also prey on ringed seals. Eskimos say that ringed seals avoid areas where walruses are plentiful, a belief supported by survey results.

Neither polar bears, arctic foxes, nor walruses inhabit the shores of lakes Saimaa and Ladoga. The ringed seals in these lakes probably experience little natural predation today. Red foxes occasionally show interest in the birth lairs, but actual predation on lake seals by these foxes has yet to be documented (a description of a red fox killing and eating a ringed seal pup in the Beaufort Sea was published recently). Ravens, wolves, dogs, and wolverines also have been mentioned as ringed seal predators. Glaucous gulls and ravens kill ringed seal pups when they are born outside birth lairs or when mild temperatures erode the lairs and expose young pups. Some researchers have suggested that avian predation on pups might be an important factor in limiting the southern breeding range of the ringed seal.

Occasionally a young ringed seal hauled out near a crack in the ice finds the crack closed when it attempts to return to the water. It must then crawl across the ice in search of another opening. Such individuals are vulnerable to predation or starvation.

Ringed seals are relatively easy to capture alive and to maintain in captivity.

History of Exploitation: People in the Far North have hunted ringed seals for millennia. Many capture methods have been devised. Christian Vibe, who studied marine mammal hunting in Greenland's Thule district during 1939–41, described five principal ways used by the Polar Eskimos to catch ringed seals: shooting and harpooning from kayaks in open water, shooting or harpooning the seals at their breathing holes while standing on the ice, shooting seals in ice cracks while standing at the ice edge, netting with set nets, and shooting hauled-out seals after stalking them behind "shooting screens." Vibe judged the ringed seal to be the most important game animal in the life of the Polar

Eskimos. As he put it, "One may get tired of all other kinds of meat, but never of Seal meat, either it is eaten boiled, dried, frozen or raw" (Vibe, 1950). The people also ate some of the blubber. The small intestine, boiled or frozen, was considered a delicacy, as was the raw or frozen liver. Some of the blubber was used for oil lamps and some was fed to dogs. The skins were used for making bootlegs, winter anoraks, woolen mittens, dog harnesses, tent and kayak coverings, "bladders" or floats for hunting, and lashes.

Accurate statistics on the number of ringed seals killed are almost impossible to collect. Skins traded represent only a part of the animals secured, as some are used by the hunters and their families. Estimates of the total annual catch in Canada during the 1970s and 1980s are on the order of 50,000. The people of three villages on eastern Baffin Island (Pond Inlet, Clyde, and Broughton Island) take 12,000 to 15,000 ringed seals per year. In Greenland, the estimated annual catch is 50,000 to 70,000 and growing. The estimated annual harvest in Alaska declined from between 7000 and 15,000 during the 1960s and early 1970s to between 2000 and 3000 by the late 1970s. The catch in recent years, however, is thought to have been more than 3000 per year.

Hunting loss is significant in all areas. Seals shot in open water or along ice edges often sink, especially at the end of the fasting season when their blubber reserves are low and their specific gravity is high. As many as half of the seals killed in midsummer can be lost by sinking. At river mouths where the water is less salty, sinking occurs with greater frequency.

In the Okhotsk Sea, Soviet commercial sealers killed approximately 415,000 ringed seals in 6 years during the 1950s and 1960s. In 1988, the Soviet Union reported harvesting 7321 ringed seals by ship- and shore-based hunting: 6871 in the Bering Sea, 213 in the Chukchi Sea, and 237 in the East Siberian Sea.

Of the three seal species inhabiting the Baltic Sea (harbor, gray, and ringed), the ringed seal has been the most important commercially. During the second decade of the twentieth century, more than 100,000 were killed by Swedish and Finnish hunters.

CONSERVATION STATUS: Ringed seals are widespread and abundant, with a world population in the millions. In some parts of their range, they have been the backbone of Native economies. Many of the maritime Eskimos' requirements for food, clothing, and fuel formerly were supplied by these animals. More recently, as trading companies have become permanently established in northern communities, the hunt for ringed seals has also been driven by the need for cash. Pelt prices have been strongly influenced by conservationist action against the commercial harp seal hunt in eastern Canada, and this has caused considerable hardship in communities where the sale of ringed seal pelts is a major source of money. While ringed seals are still harvested for food, the use of their pelts for clothing and their blubber for fuel has greatly declined.

In Canada, the exploitation of ringed seals is regulated under the Seal Protection Regulations (Fisheries Act). These allow residents to kill seals for food (including dog food) and to trade or sell pelts. In Alaska, coastal Natives are allowed to hunt ringed seals for subsistence.

Overexploitation has occurred in some areas, but the impact has been local or regional. The catch of ringed seals in the northern Baltic Sea (Bothnian Bay) declined from 5400 per year in the mid-1960s to approximately 100 or less in 1975. The average age of the population in the northern Baltic increased during the past two decades from 10 to 12 years in males and 13 to 16 years in females. This change evidently was caused by exceptionally low productivity in females. The high incidence of pathological uterine blockages, beginning in the late 1960s as a result of heavy pollution and contamination of fish by DDT and PCBs, may have caused this change in the population composition. Ringed seals are now protected from hunting in Swedish and Soviet Baltic waters. Finland did not issue any licenses for hunting of ringed seals in 1988 or 1989. The Baltic ringed seal population numbered approximately 8500 in the late 1980s, down from several hundred thousand at the beginning of the century. The incidence of reproductive disorders apparently began decreasing in the 1980s; by the end of the decade there were hopeful signs that the Baltic ringed seal population was recovering.

Saimaa and Ladoga ringed seals face problems peculiar to their confinement in freshwater lakes. Saimaa seals have been protected from hunting since 1955 (there was a bounty on them before then), but substantial numbers, especially first-year animals, have been killed inadvertently in fishing gear. Plans to regulate water levels artificially in Lake Saimaa have caused concern about the effects of such regulation on birth lairs. If the drop in water levels during the birth and nursing period became too great, the shore ice could collapse against the bottom, destroying many of the lairs. In 1989 the population of the Saimaa seal was estimated as a mere 150 to 160 individuals, with only approximately 20 births each year. The seals are separated during the breeding season into four subpopulations by long, narrow sounds or densely inhabited human population centers. Almost all adult mortality is natural; therefore, the key to the population's survival is pup survival. In the early 1980s, nearly 25 percent of the pups drowned in fishing nets during the first 6 months of life. Recent conservation measures have involved restrictions on net fishing near most breeding areas and the development of captive breeding programs. Pollution also clouds the future of the Saimaa seal. Already the seals avoid the southern margins of the lake. Levels of mercury and nickel in the liver and hair of young Saimaa seals are high enough to cause concern that the rate of stillbirths is linked to these contaminants.

Lake Ladoga is heavily polluted by prewar cellulose and paper mills. However, since it is the principal source of Leningrad's city water supply, there is a strong incentive for monitoring and improving the lake's quality. The population of ringed seals in the lake has been reduced from about 15,000 early in the twentieth century to between 9000 and 11,000 at present.

FURTHER READING: Burns et al. (1985), Finley et al. (1983), Frost and Lowry (1981), Hammill and Smith (1991), Härkönen and Heide-Jørgensen (1990b), Helle (1980), Kelly (1988c), Kelly and Quakenbush (1990), Kingsley (1990), Lowry, L. F., et al. (1980b), McLaren (1958a), Ryg et al. (1990), Smith, T. G. (1973, 1976, 1987), Smith, T. G., et al. (1991), Yablokov and Olsson (1989).

Baikal Seal

Phoca sibirica
Gmelin, 1788

NOMENCLATURE: The specific name refers to this seal's relict distribution in Siberia. Its local name in Russian is *nerpa*.

DESCRIPTION: These seals are among the smallest pinnipeds, but there are problems with interpreting the available information on lengths and weights. Soviet authors report curvilinear, nose-to-tip-of-tail measurements, which are greater than the standard-length measurements to which westerners are accustomed. Female Baikal seals reportedly grow to 135 cm (curvilinear) and 63 kg, males to 142 cm and 66 kg. At birth they are 64 to 66 cm long and weigh 4 to 4.2 kg. Males are slightly heavier and longer than females throughout life. Males are, on average, 88 cm and 22 kg when weaned, and 99 cm and 31 kg when 7 to 8 months old. Females are 68 cm and 20 kg at weaning, and 97 cm and 30 kg when 7 to 8 months old.

The Baikal seal differs externally from the closely related ringed and Caspian seals mainly by the unspotted (or rarely spotted) appearance of its pelage. Also, its fore flippers and fore claws are larger and stronger than those of the other two species.

At birth, Baikal seals still have the whitish lanugo. After molting at 4 to 6 weeks of age, the pups are silvery gray. Adults are normally silver gray on the back and yellowish white on the belly.

DISTRIBUTION: Baikal seals are essentially confined to Lake Baikal (Ozero Baykal), a freshwater body 636 km long and an average of 48 km wide in the eastern U.S.S.R. just north of Mongolia. They occasionally wander up rivers flowing into the lake, and a sighting has been recorded 400 km down the lower Angara River, which flows out of the lake's southern end. An old report of seals in Lake Oron, some 400 km northeast of Lake Baikal, cannot be confirmed; they definitely are absent from Lake Oron today.

Lake Baikal is the oldest and deepest lake in the world, with a maximum depth of 1620 m (average, 740 m). Its total volume

A young Baikal seal hauled out on the Ushkany Islands, Lake Baikal, Russia. (August 1990: Brent S. Stewart.)

of water, some 23,000 cubic kilometers, is approximately that of the combined North American Great Lakes. Lake Baikal contains one-fifth of the world's freshwater reserves, exclusive of the polar ice caps. It represents more than 80 percent of the U.S.S.R.'s supply of fresh water. It also has a remarkable degree of endemism: at least three-quarters of its plant and animal species are endemic.

NATURAL HISTORY: The confinement of Baikal seals, as well as Caspian seals, to a freshwater lake system raises interesting zoogeographical and ecological questions. How did they become isolated? How do their behavior and natural history differ from those of their marine relatives?

Two main hypotheses concerning the origins of the Baikal seal have been discussed. One is that it (or its precursor) was pushed southward from the Arctic Ocean by Pleistocene glaciers, eventually moving up the Yenisey River and reaching Lake Baikal about 500,000 years ago. The other is that it evolved along with the ringed and Caspian seals from a common ancestor in the Paratethyan Basin of southeastern Europe, moving northward through glacially formed lakes and rivers and finally becoming established in Lake Baikal. Most Soviet experts accept the first of these hypotheses, that Baikal seals were isolated from ringed seal ancestors of the Arctic Ocean, and evidence from their internal and external parasites supports this view.

Like ringed seals, Baikal seals are adapted to the ice. They maintain breathing and haul-out holes in the ice that covers the lake to a depth of up to 1.5 m (average, 80 to 90 cm) in winter. Adult males have a primary hole surrounded by as many as 10 auxiliary holes; mothers and pups use a central hole and several auxiliary holes. Juveniles usually have only one hole. Although the seals are essentially solitary during winter, haul-out holes are sometimes shared. In spring, seals may concentrate along ice edges; in summer, groups of molting seals haul out on some offshore rocks and rocky banks, particularly at the Uskang Islands. Baikal seals spend 20 to 30 days hauled out on the ice before the molt begins in late May or early June. Molting can continue into July, but the time required varies individually. In fall, concentrations of seals occur in embayments as they haul out on newly forming ice.

Pups are born in lairs between mid-February and the end of March (the peak is mid-March). The lactation period is 2 to 2.5 months. The first molt takes place over a 2 week period at 1.5 to 2 months of age, while still in the birth lair. Pups begin to appear outside the lairs in early April. The spring breakup of ice occurs first in the southern part of Lake Baikal, and this causes the nursing period to be somewhat shorter and the pups smaller at weaning there than in the northern part. When a human approaches a pup, the mother does not attempt to defend or rescue it. She only calls to it from a haul-out hole, making a cowlike sound.

Mating occurs in the water near the birth lairs, at approximately the time mothers wean their pups. Implantation is delayed; the actual gestation period was believed until recently to be about 9 months. New data have suggested a gestation period of 11 months, with a 3.5 month delay in implantation. There is no evidence that males are territorial. The sex ratio at birth is approximately 1:1, but among older seals there are more females than males in each age group–an average of 42 males for every 58 females.

Females begin breeding as early as 3 years of age and no later than 6. They can give birth annually until they are 43 years old or older, but usually 10 to 20 percent of mature females are nonparous in any given year. When ice conditions are unfavorable, as they were in 1981, the number of nonparous females in the following year may be much greater. In 1982, for example, 63 percent of mature females did not pup. Males reach sexual maturity by 7 to 10 years.

Approximately 10 percent of the Baikal seal population at a given time is older than 20 years, although the age structure is highly sensitive to the intensity and selectivity of the hunt. Ages of seals younger than 6 to 8 years old are estimated by counting the number of ridges on the outside of their fore-flipper claws. Soviet scientists believe that the thickness of the last growth layer indicates the population's condition. If the layer is thin, then the ice broke up early, limiting the seals' foraging opportunities. A broad layer indicates late ice breakup and good population condition. V. D. Pastukhov of the Limnological Institute in Irkutsk, a lifelong student of the Baikal seal, has estimated that the maximum life span of the species, based on cementum layers in the canine teeth, is 56 years for females and 52 years for males. Baikal seals are physically mature when 20 to 25 years old, but their body mass may continue to increase slowly for another 7 years.

Since 1970, the average body and blubber mass has declined. Pastukhov believes that this trend has been caused by the recurrent early breakup of ice and the tendency of sealers to kill the largest seals. The surviving smaller seals apparently produce smaller pups.

In their freshwater inland environment, Baikal seals are spared from predation by killer whales and polar bears, but they are attacked occasionally by brown bears. There is little or no competition with other species for living space or food.

Baikal seals have been said to dive for up to 43 minutes in the wild, and newly captured, frightened seals have made dives of 40 to 68 minutes. These times are very long in comparison to the documented dive times of ringed seals. Stewart and Soviet colleagues from the Limnological Institute attached satellite-linked radio transmitters to four Baikal seals in summer 1990 to study their movements and diving patterns. Preliminary findings were that most dives lasted less than 10 minutes. Most daytime dives were to depths of 100 to 200 m, while at night the seals usually reached depths of only 10 to 50 m. The deepest dives recorded were to depths of 200 to 300 m, and the longest lasted for 10 to 20 minutes.

The Baikal seal's diet consists mainly of fish. At least 29 different species are consumed during the year, but golomyankas and sculpins are the most important prey at all seasons. During spring and autumn, golomyankas may constitute as much as 95 percent of the diet. Fortunately, Baikal seals do not eat significant quantities of omuls, the most valuable commercial fish in the lake, in spring and autumn. The seals may eat omuls more frequently in summer, but scientists generally believe that such predation does not have a significant deleterious effect on the omul population.

Baikal seals have been brought into captivity on a number of occasions. The first live Baikal seals to reach Great Britain were a pair sent by the Moscow Zoo to the London Zoo in 1959. Captive Baikal seals eat up to 1 kg of fish per feeding and 5 to 6 kg per day; that is, roughly 4 to 6 percent of their body weight each day.

HISTORY OF EXPLOITATION: Baikal seals have a long history of exploitation. They were hunted several thousand years ago by the region's prehistoric human inhabitants. In the early 1900s, some 2000 to 9000 seals were taken annually. One technique formerly used was to cover auxiliary breathing holes, then club the seal as it surfaced in its main hole. Today, most are shot or netted.

There are three major hunting seasons. In early spring (April), some 2000 to 3000 pups are taken on the shore-fast ice. Later (May

A young Baikal seal investigates a scale before being weighed, instrumented with a satellite transmitter, and released back into the lake. (August 1990: Brent S. Stewart.)

through June), another 1500 to 2000 seals, including molted pups, molting juveniles, and adults, are shot on the northern ice. In October and November, hunters net approximately 1000 seals in freshly frozen bays. Between 1977 and 1986, an average of 6000 seals were killed each year.

Products of the hunt include pelts, blubber, and meat. The pelts of pups are used for hats and coats; those of adults, for boots. Souvenirs are also made from the seal skins. The skinned carcasses are used primarily to feed animals on fur farms. During the nineteenth century, seal blubber was a delicacy along the shores of Lake Baikal. Seal oil is considered to have medicinal value.

In the past, when seals were believed to have been very abundant, managed seal harvests were favored by fishermen, who believed that Baikal seals were a burden on commercial fisheries.

Conservation Status: Censuses of haul-out holes and birth lairs in 1977 resulted in a population estimate of some 77,000 seals, not including pups. By 1986 the population had declined by more than 17 percent, to approximately 58,000 seals 1 year old or older. During the period 1977 through 1986, 50 to 60 percent of the approximately 6000 seals taken annually were young. Such intensive removal of younger seals had a significant impact on the age structure of the population. In 1986 only about 19 percent of the population consisted of seals 3 to 6 years old. Since this age group is vital to the population's replenishment, the trend caused alarm among Soviet scientists. A reduction in the official sealing quota was recommended. There is extensive poaching, particularly during the pup harvest in April. Also, during May and early June when seals are shot from boats, many wounded seals are not recovered. Mortality from hunting is therefore much higher than the managed commercial harvest might suggest. The undocumented kill may add 20 to 40 percent to the official catch in any given year.

Another concern is that molting seals are often disturbed by boaters and gunners when hauled out on coastal beaches. In mild years when the ice on the lake clears early, the seals cannot complete their molt on the last spring ice, so large numbers haul out on lakeside rocks instead. In 1981, for example, the molt took several months rather than the usual 2 weeks. This interfered with the seals' feeding regimen and made them more vulnerable to poaching and harassment. As a consequence, some 10,000 pups were lost in 1982.

The number of seals using the southern part of the lake reportedly has decreased during the last several decades as the human population living there has increased. Since the late 1950s, a large wood-processing center along the southern shore of the lake and an industrial complex on the Selenga River, the lake's largest feeder river, have caused serious pollution. The recent opening of the Siberian oil fields and the completion of the Baikal-Amur railroad have spurred additional industrial development of the Baikal region.

In autumn 1987, weakened seals were observed crawling out of the lake onto its icy shores, their hind extremities paralyzed. By October 1988, several thousand Baikal seals had died of a viral disease (a morbillivirus) closely related to canine distemper. This mass die-off was the first of its kind documented for the lake.

FURTHER READING: Grachev et al. (1989), Nijhoff (1979), Pastukhov (1989), Thomas et al. (1982).

Caspian Seal

Phoca caspica
Gmelin, 1788

NOMENCLATURE: The specific name refers to the Caspian Sea, where this seal lives.

DESCRIPTION: The Caspian seal is very similar externally to the closely related ringed seal. It reaches a maximum length of 1.5 m and a weight of 86 kg. Females are slightly smaller than males. The birth length is 64 to 79 cm, and neonates weigh about 5 kg.

Pups lose their long, white lanugo and acquire a coat of short, dark hair at about 3 weeks of age. Adults are deep gray on the back and grayish white on the belly. Most individuals are liberally spotted at least on the back. Males usually are spotted laterally and ventrally as well. There are no "rings" on this seal's pelage as there are on the ringed seal's.

DISTRIBUTION: As the name implies, these seals are isolated in the Caspian Sea, a large (1280 km long; 438,690 km²) lake in the south-

A white-coated Caspian seal pup on a sandy island. (Southern Caspian Sea, January-February 1984: Sasha Zorin.)

The Caspian seal is a relict species confined to the Caspian Sea on the border between Iran and the former Soviet Union. (Southern Caspian Sea, January 1984: Sasha Zorin.)

western U.S.S.R. between approximately 36°30′ N and 47° N. The southern end of the lake forms the international boundary with Iran. The northern third of the lake is shallow (less than 100 fathoms) and freezes over in winter. Seals move into the northeastern corner (less than 10 fathoms deep) in autumn. The ice that forms there is wind-driven and becomes ridged or hummocky. In this stable, rough ice the seals maintain breathing and haul-out holes. By early May, after molting, they spread southward to pass the summer in the cooler, deeper portions of the lake.

Early reports of these seals inhabiting the Aral Sea, about 400 km east of the Caspian Sea, have been discounted.

NATURAL HISTORY: Caspian seals pup mainly from late January through early February. Unlike ringed and Baikal seals, Caspian seals are not known to construct birth lairs. Pups are born on the exposed ice surface. Lactation lasts 4 to 5 weeks. Mating takes place from the end of February to mid-March. Adult females begin molting toward the end of the lactation period, followed by the adult males who begin molting at the end of March. Juveniles have completed the molt by mid-April, whereas adults continue molting into May and early June. Ages at sexual maturity are 5 years for females and 6 to 7 for males. Most females do not give birth until they are 6 to 7 years old.

Caspian seals eat a variety of fish. Sculpins, gobies, and small crustaceans are eaten in autumn and winter; sprats, herring, sand smelts, roach, and carp, in spring. In the Ural River estuary in October, seals consume large quantities of carp, roach, and pike-perch.

Chief predators, apart from humans, are said to be eagles and wolves. Wolves inhabit the islands in the northern part of the Caspian Sea. It was estimated in 1974 that 17 to 40 percent of the pups and perhaps 1 percent of the nursing females from this area were killed by wolves. Some pups die when they are crushed by shifting ice.

The oldest reported Caspian seal was approximately 50 years of age.

HISTORY OF EXPLOITATION: Caspian seals have been heavily exploited for at least 200 years. As many as 160,000 were killed per year before 1803. During 1867 to 1915, the average catch was 115,000, and catches of this magnitude continued through 1936. In winter 1929 more than 60,000 seals were taken by the sled hunters from a single village. The highest 1-year kill on record was in 1935, when more than 227,000 Caspian seals were taken. Of the 776,000 seals reportedly killed on the pupping rookeries between 1935 and 1939, 143,000 (more than 18 percent) were parturient females. The average annual catch before 1965 was 70,000. A regulated commercial harvest of 60,000 to 65,000 pups continues to be made each year.

The traditional method of sealing in the Caspian Sea was to club the seals on island rookeries in spring and fall. Mainly juveniles were taken in this manner. During the nineteenth century or earlier, a spring hunt was initiated, in which seals were taken on ice floes, reached on sleds and in small boats. By the twentieth century, the hunting on island rookeries had been largely replaced by the ice hunt.

CONSERVATION STATUS: The total current population is estimated to be more than half a million seals, and the population is considered essentially stable. Adult females are protected during the breeding season.

There are serious environmental problems in at least the northwestern part of the Caspian Sea. Dams in the Volga River divert much water for large-scale rice growing, reducing flow into the sea. Moreover, the river is polluted by pesticides and industrial effluents. Major decreases in catches of sturgeon and salmon have been blamed on a deteriorating aquatic environment. At least some of the impetus for continued sealing comes from the belief that the seals consume large quantities of commercially valuable fish.

FURTHER READING: Popov (1982). A monograph, "New Aspects of the Biology and Ecology of the Caspian Seal," edited by B. I. Krylov, was published in Moscow (in Russian) in 1990.

Harp Seal

Phoca groenlandica
Erxleben, 1777

Nomenclature: For a long time and until recently, the generic name of the harp seal was *Pagophilus,* meaning "lover of ice." The specific name refers to Greenland, which is approximately at the center of the distribution of the species. The vernacular name refers to the dark-colored, roughly horseshoe-shaped saddle, actually more reminiscent of a lyre than of a harp, on the back of adult males. Harp seals are also sometimes called Greenland or saddle-back seals. Norwegian names for the harp seal are *grønlandssel* and *salesael.* The German name is *sattelrobbe.* In most other European languages, the name is a literal equivalent of Greenland seal.

Description: The average size of adults is 1.6 to 1.7 m and 130 kg. Males are somewhat larger than females. The length at birth is about 85 cm. Blubber thickness of adults is 4.7 to 5.0 cm; maximum girth, about 128 cm.

The pup's hair is yellowish at birth, stained by amniotic fluids and blood. Within 3 days, it becomes the familiar fluffy white that gives pups the fitting name, whitecoats. During their first molt, which takes place soon after weaning (at about 3 weeks of age), the pups become known to the sealers as raggedy-jackets. By about 4 weeks of age, a short, silvery coat flecked with small dark spots develops, and the pups become known as beaters, in reference to their awkward manner of swimming. The 1-year-old beaters gain the characteristic spotted pelage of bedlamers (possibly from the French *bête de la mer,* or "beast of the sea") after their first molt. With successive annual molts thereafter, the faint outlines of a harp pattern emerge. At 4 to 7 years of age, they sport a "spotted harp" coat. The mature pelage, attained by age 7 (males) to 12 (females), consists of a dark harp pattern (actually two dark bands uniting across the shoulders and bending toward the abdomen and back near the tail) on a silvery gray undercoat. The background color is silver to steel blue when wet and pale gray when dry. The face and tail are black, the fore flippers and belly whitish. Those few males that become very dark just before maturity are called smutty harps or sooties. In contrast to males, females sometimes retain their spotted bedlamer coats until age 8, and their intermediate markings for much longer. The mature markings of females are generally more muted and less sharply demarcated than those of males.

DISTRIBUTION: There are three stocks of harp seals: the White Sea (East Ice) stock, numbering approximately 800,000 during the mid-1980s; the Jan Mayen (West Ice) stock, numbering approximately 300,000 with some 60,000 pups born annually in the late 1980s; and the northwest Atlantic (Newfoundland) stock, by far the largest today, numbering close to 2 million seals and producing more than 500,000 pups per year during the mid-1980s. There has been some uncertainty about whether the seals that pup in the Gulf of St. Lawrence comprise a separate stock from those that pup at the Front off southern Labrador and Newfoundland. Gulf harp seals pup about a week earlier (mean date: 2 March). Breeding females tend to return each spring to the area where they were born, but seals born in the Gulf and at the Front seem to mix freely as juveniles. Also, in years when ice conditions in the southern Gulf are unsuitable for pupping, some of the Gulf females probably pup instead at the Front.

The northwest Atlantic stock migrates between the High Arctic and Newfoundland or the Gulf of St. Lawrence. After leaving the Arctic in late September, the first migrants reach Labrador and northern Newfoundland by December. About a third of the mature seals enter the Gulf of St. Lawrence through the Strait of Belle Isle. This subpopulation generally breeds near the Magdalen Islands. The rest of the mature seals remain off the northeast coast of Newfoundland; they give birth and breed mainly on pack ice

Immature harp seals, such as the one on the left, do not have the well-defined "harp" pattern on the back characteristic of adults (right). (Gulf of St. Lawrence, March 1978: Wybrand Hoek.)

off southern Labrador. Most immature harp seals overwinter at the Front, although some remain at higher latitudes. Individuals rarely wander south along the Canadian and U.S. east coasts; an adult was captured at Cape Henry, Virginia, in March 1945.

Most of January and February is spent feeding, as the seals store energy to meet the demands of pupping and molting. Pupping begins in late February in the Gulf of St. Lawrence and on about 8 March at the Front. Weaned pups drift with the ice from mid- to late March, those born at the Front congregating to feed in White and Notre Dame bays, those from the Gulf moving toward Cabot Strait. Beaters in the Gulf swim north along the coasts of Newfoundland from late March to May. From May to November, most of the population is north of Hamilton Inlet, Labrador. Groups of 10 or more harp seals are very common along the west coast of Greenland and in the Lancaster and Jones sound regions in summer. They reach the southwest coast of Greenland by early summer. Most move north along the west coast, but some go south around Cape Farewell and head north along the east coast of Greenland as far as Angmagssalik. Younger seals tend to go to Greenland while older animals remain in Canadian waters. Also, older individuals tend to reach higher latitudes than younger ones, at least in Canada. Many harp seals enter Hudson Strait, and some reach Hudson Bay and Foxe Basin during the late summer and autumn, although they are not as common there as they are in the High Arctic. Those seals that migrate to near the extremities of this stock's range may have made a round-trip of some 8000 km during the year.

The White Sea stock migrates in summer into the Barents and Kara seas. Seals presumably from the White Sea stock have been observed in eastern Svalbard at 79° N in the summer. West Ice seals disperse to eastern Greenland, northern Iceland, and northern Norway. A large influx of harp seals along the coast of Norway in the winter of 1986–87 apparently involved animals from both the West Ice and the White Sea. This "invasion" of Norwegian coastal waters may have been caused by a major decline in capelin stocks in the Barents Sea.

Natural History: Harp seals are gregarious on their pupping and molting grounds. Concentrations of tens of thousands can be seen strewn across the pack ice. The pupping (often called whelping) patches cover 20 to 200 km^2, and the density of seals can be as high as 2000 per km^2. The sounds made by a whelping and breeding herd of harp seals have been likened to a noisy barnyard. The loud cacophony may help advertise the herd's presence since it can be distinguished from ambient noise at a considerable distance. Their gregarious nature is also evident when these seals are encountered in groups of tens or even hundreds while migrating, resting, or feeding in open water during the summer. Christian

Vibe (1950) described their occurrence in Greenland's Thule district as follows: "Such a flock will tumble through the water in a way which reminds of dolphins, since at short intervals they will break the surface rising half over it and fling themselves down again beating the water round them." At the pupping grounds near the Magdalen Islands, diving photographers have observed individual harp seals "hauled out" underwater on horizontally oriented shelves of ice protruding from pressure ridges.

The mean age at first reproduction and the fertility rate are density-dependent to some extent. The mean age at sexual maturity in females of the northwest Atlantic stock declined from 6.2 years in 1952 to 4.5 years in 1979, and the female fertility rate increased from 85 to 94 percent. Over that 27-year period the size of the stock was reduced by nearly 50 percent. Males reach maturity at 7 to 8 years of age.

Most mating takes place in mid- to late March. Males assemble in the water near the ice floes where pups are being nursed. Underwater calls, emitting air bubbles, and pawing gestures seem to be part of a courtship display. Harp seals appear to be most vocal during mid-March, circumstantial evidence that acoustic behavior plays a significant role in courtship. The occasional approaches made on the ice by males are generally rebuffed by females; copulation usually occurs in the water soon after the pups are weaned. Suspended embryonic development lasts for at least 2 to 3 months (until early June). The embryo may not attach to the wall of the uterus until August. Births occur mainly from late February through mid-March. Females may be able to delay giving birth for several days, either as a response to severe storm conditions or when unsuitable ice conditions in the usual pupping areas force them to seek alternate sites. (Other pinnipeds, for example, the gray seal, probably have this ability as well.)

The sex ratio at birth is 1:1. Pups weigh 10 to 12 kg at birth and about 35 kg when weaned. Male and female pups grow at similar rates. Lactation lasts for an average of 12 days. The mother's milk is about 43 percent fat. When 7 days old, pups weigh approximately 25 kg. They are 94 cm long and about 72 cm in girth; their blubber is about 2.3 cm thick. Pups gain some 2.3 kg per day, then lose about 0.6 kg per day after weaning while they fast and molt. Beaters enter the water when the ice breaks up or melts around them, and thereafter they feed themselves.

Mothers defend their pups against approaches by other seals by growling, posturing, and sometimes biting. During the brief nursing period, in which the mother fasts or eats very little, a remarkable transfer of energy is accomplished. Nursing females lose about 3.1 kg of body mass, mostly blubber, each day. The efficiency of mass transfer from mother to pup is approximately 77 percent.

Harp seals begin molting in early April, with adult males and juveniles starting a week or more earlier than adult females. The females have a short, intense feeding binge after weaning their pups. During the 4 week molt, the seals fast or at least eat very little, and their fat stores become depleted.

Harp seals feed heavily on decapod, amphipod, and euphausiid crustaceans and pelagic fish such as capelin and herring while migrating in the spring and fall and during the winter months prior to pupping. In the summer at high latitudes, arctic cod and polar cod are important prey. Bottom fish such as cod, redfish, American plaice, and Greenland halibut are also eaten. The diet of recently weaned animals is dominated by capelin and euphausiids, and harp seals of all ages feed intensively on capelin in early summer. The high-energy diet provided by arctic cod at high latitudes in summer probably accounts for the harp seal's long annual migration. Adult harp seals can dive to feed at depths of at least 250 m.

Harp seals are sometimes crushed by shifting ice and killed by polar bears on the pupping grounds. Walruses prey on females and pups in the White Sea. They can live to ages of at least 35 years.

HISTORY OF EXPLOITATION: Residents of coastal areas in northeastern North America, Greenland, and northwestern Europe have always exploited harp seals for subsistence to some extent. Northern Greenland's Polar Eskimos took only about 50 harp seals per summer during the late 1930s. The meat was eaten, and the skin was used to make boots, tent covers, and kayaks. Traditionally, most Polar Eskimo kayaks were made from five ringed seal skins and one harp seal skin. The latter was placed around the bow end, where its larger size and greater thickness and strength retarded wear and perforation. South of the Thule district, Greenlanders traditionally caught many more harp seals. Harp seals also contributed significantly to the subsistence of the Norse colonies in Greenland from the tenth to the fifteenth century.

The commercial hunt for harp seals was probably the most highly publicized form of marine mammal exploitation, with the possible exception of commercial whaling. An intensive commercial seal hunt was carried out at Jan Mayen in the eighteenth century by northern Europeans, mainly British, Dutch, Germans, Danes, and Norwegians. By the 1860s, only ships from Scotland and Norway remained active, and the declining hunt soon was left entirely to the Norwegians.

Harp seals also were hunted commercially in the western North Atlantic in the late eighteenth century. By 1832, the catch, including a small percentage of hooded seals, had reached 744,000, with 300 ships and 10,000 men involved in the hunt; in 1857, 370 vessels and 13,600 men participated. The catch exceeded half a million seals 11 times between 1825 and 1860. In spite of the improved technology for reaching the seals on the ice, including the introduction of steam-powered vessels in 1863, catches declined during the second half of the nineteenth century. The average catch was

A "whitecoat" on the pack ice of the Gulf of St. Lawrence. The valuable pelt of the newborn harp seal supported an intensive commercial hunt in the North Atlantic for more than a century. (March 1986: Fred Bruemmer.)

approximately 341,000 (harp and hooded, combined) from 1863 to 1894; 249,000 (harp only) from 1895 to 1911; 159,000 from 1912 to 1940. A reduction in hunting pressure during World War II was followed by a resumption of intensive sealing. Norwegian vessels, which had joined the northwest Atlantic hunt in 1938, became more heavily involved after the war. At one time or another, ships registered in Denmark, France, the United States, Canada, and the U.S.S.R. participated in this hunt. The commercial catch by Canada and Norway averaged 162,000 seals per year during 1971–80; during this same time, nearly 10,000 harp seals were taken each year by hunters in Greenland and the Canadian Arctic. Residents of the north shore of the Gulf of St. Lawrence and of the coasts of Newfoundland and Labrador have exploited harp seals for centuries, using stationary traps and nets as well as working from small boats.

The commercial vessel fishery in the White Sea was begun by the Norwegians in 1867, and by the turn of the century their annual catch had reached 40,000. They were joined by the Russians in the early twentieth century. The largest single-year catch was 460,000 seals in 1925. Catches remained greater than 200,000 seals in most years through 1950, the war years excepted. Unable to regain their sealing concession in the White Sea after the war, the Norwegians redirected much of their sealing effort to the western North Atlantic. The Soviet catch in the White Sea was first regulated by a quota (100,000 seals) in 1955; this quota was reduced to 60,000 by 1963. In 1965, the total catch was limited to 34,000, with 20,000 allocated to Soviet landsmen and 14,000 to Norwegian ships that hunted in the molting grounds of the southeastern Barents Sea.

Adult females on the pupping grounds were protected as of 1963, and the Soviet catch of seals other than pups was stopped in 1965. The total quota in 1979 was 50,000 seals.

The North Atlantic commercial seal hunt produces mainly oil, fur, and leather. The oil rendered from seal blubber is colorless, odorless, and tasteless, and it has been used for lighting, heating, cooking, and lubricating. Pelts are used for making coats, jackets, hats, boots, shoes, handbags, and belts. In addition, flippers and carcasses are sometimes retained and sold for human food in Newfoundland.

Conservation Status: The hunt by Native peoples of the Canadian Arctic and Greenland continues; as many as 10,000 harp seals are still taken each summer, most of them in west Greenland. As with most seal hunts that involve shooting, some harp seals are lost after being wounded or killed.

The harp seal hunt in eastern Canada was unregulated until 1961, when a closing date was introduced for the commercial hunt. Norway was stopped from sealing in the Gulf of St. Lawrence in 1965. Also in that year, adult females were accorded protection on the pupping grounds, and Canada set a quota of 50,000 seals for the commercial hunt in the Gulf. A long-overdue quota system finally was established in 1971 to limit kills at both the Gulf and the Front. The quota was reduced from 245,000 seals in 1971 to 150,000 in 1972. It was further reduced to 127,000 in 1976, but in that year the actual kill was more than 165,000. In 1977 the quota was raised to 160,000; in 1978, to 170,000, where it remained through 1981. The loss of markets for seal products beginning in 1982–83, together with Canada's decision in 1987 to end the large-vessel offshore hunt for pups, allowed the northwest Atlantic population to grow during the late 1980s. During the mid- to late 1980s, the Newfoundland media contained increasing calls for a cull. Large numbers of harp seals have been killed in gill nets set for groundfish along Newfoundland's west and south coasts. More than 10,000, mainly beaters, died in this way in 1988.

In November 1989, a new canned meat product appeared in Canadian markets under the name Landsmen Harvesters, produced by the Canadian Sealers Association. The seal meat comes from harp seals harvested by some 9000 Newfoundlanders who hold sealing permits. Federally imposed quotas permit 186,000 seals to be killed each year.

During the late 1980s the number of breeding females in the White Sea stock declined from approximately 140,000 to about 85,000. In winter 1986–87, large numbers of harp seals appeared along the Norwegian coast, and 60,000 to 100,000 were killed in fishing gear from Finnmark south to Skagerrak. The same phenomenon, but on a a smaller scale, occurred in 1987–88 and at Finnmark during 1978 to 1984.

The West Ice stock was declining until the mid-1960s but has increased since then. Aerial surveys in the late 1980s resulted in an estimate of about 60,000 pups born per year in the Greenland Sea.

Several approaches have been used to estimate trends in harp seal population size and productivity, from aerial photographic censusing to mathematical modeling. Annual pup production is the most frequently used index of population status. From 1950 to 1970, the western Atlantic stock apparently was reduced by 50 percent, that is, from nearly 3 million to 1.5 million (animals 1 year or older). In 1983, an estimated 530,000 harp seals were born in the western North Atlantic, suggesting a total population in this sector of around 2.3 million seals. The conservation measures applied to all three stocks have had a positive effect, although the world population of harp seals probably is still less than what it was when commercial sealing began.

Further Reading: Bowen (1985), Kovacs and Lavigne (1985), Lavigne (1978), Lavigne and Kovacs (1988), Sergeant (1976, 1991), Stewart, R.E.A. (1986), Stewart, R.E.A., and Lavigne (1984), Stewart, R.E.A., et al. (1989), Worthy and Lavigne (1983a, 1983b).

Ribbon Seal

Phoca fasciata
Zimmerman, 1783

NOMENCLATURE: The specific name is from the Latin *fascia*, for a "band," referring to the white, ribbonlike bands on the adult's pelage.

DESCRIPTION: These seals grow to 1.5 to 1.75 m long and weigh about 90 kg; the largest individual measured, a pregnant female, was 1.8 m long and weighed 148 kg. Pups are close to 90 cm long and weigh about 10 kg at birth. They remain in a thick, white, woolly lanugo for the first 4 to 5 weeks. By the end of this period, the natal pelage has been replaced by the short, straight juvenile hair that is dark bluish gray dorsally and silver gray ventrally. A banded pattern, similar in males and females, emerges during the second year. At 3, both have the full adult color pattern: white or yellowish white bands encircling the neck, hips, and fore flippers, with the rest of the coat being reddish brown to black in males and lighter in females. The banding is paler and less distinct in females than in males.

An adult male ribbon seal in an alert pose in the heavy pack ice of the northeastern Bering Sea. (June 1978: Kathryn J. Frost.)

The fore flippers and neck are rather long and the trunk slender compared with those of other phocids. Internally, adult males have a well-developed, inflatable air sac that branches off the posterior end of the trachea on the right side of the body and extends over the ribs. This air sac, which may function in sound production, is poorly developed and often absent in females.

The skull is short with a broad cranium. The 34 teeth are relatively widely spaced.

DISTRIBUTION: Ribbon seals are found principally in the Bering, Chukchi, and Okhotsk seas. Individuals occasionally wander south of the Aleutians into the North Pacific Ocean and north of the Bering Strait into the Chukchi Sea. A few reach the western Beaufort Sea and the East Siberian Sea. They are most common in the Okhotsk and Bering seas, where they live among pack ice floes in winter and spring. They rarely haul out on shore-fast ice and almost never haul out on land. They display a strong preference for moderately thick "clean" ice. Ribbon seals are common near the coast of northern Hokkaido when the pack ice is exceptionally heavy or extends farther south than usual. In spring 1984, ribbon seal pups appeared along the coasts of southern Hokkaido and northeastern Honshu, apparently due to the unusually strong development of the cold, southward-flowing Kuril Current that winter. A 1.3 m adult male was captured on a beach near Morro Bay, California, in 1962.

In late spring and summer, when the Bering and Okhotsk seas have become mostly free of ice, some seals migrate north with the receding ice edge. Most are presumed to be pelagic, remaining south of Bering Strait through the summer and autumn, although recent sightings indicate that some ribbon seals do summer in the Chukchi Sea. Few ribbon seals are seen during these seasons, and their distribution then is poorly known. Ribbon seals are among the least often seen large mammals in the Northern Hemisphere.

Few ribbon seals are seen in the Okhotsk Sea in summer, so much of the Okhotsk population may move into the Bering Sea at that season. The extent of mixing among seals from the two major centers of ribbon seal distribution, the Okhotsk and Bering seas, is unknown.

NATURAL HISTORY: Females give birth on pack ice far offshore (except when the ice edge is near St. Matthew and St. Lawrence islands), mainly in early and mid-April. Some pups are born as late as early May. The sex ratio at birth is 1:1. Lactating females evidently leave their pups unattended on ice floes while they feed. Pups are suckled for 3 to 4 weeks and weigh 27 to 30 kg when weaned. By early June, a weaned pup will have lost some of its fat reserves and weigh perhaps 22 kg.

Nothing is known of the breeding system. Females are impregnated after giving birth in April and May, probably near the time when their pups are weaned. Attachment of the blastocyst is delayed 2 to 2.5 months, until after the molt is finished in June or July. Subadults molt in April to mid-May, whereas adults of both sexes begin molting during the first half of May. Males are sexually mature when 3 to 5 years old, females when 2 to 4, although many do not give birth until they are 4 to 5 years old. A high percentage (close to 95 percent) of mature females conceive in successive years.

These seals may live for up to 30 years, but most are believed to live for only about 20. One specimen was estimated to have been 26 years old. Mortality during the first 4 years of life is nearly 58 percent. One seal lived at Kamogawa Sea World in Japan for 8 years, growing from 24 kg to 56 kg.

During the spring molt, these seals evidently feed less than in other seasons. Their blubber thickness may decline by 50 percent. Almost all information on diet comes from seals killed in the spring. In the Okhotsk Sea, ribbon seals mostly eat walleye pollock, cephalopods, and to a lesser extent shrimps (especially *Pandalus goniurus*) and Pacific cod. In the Bering Sea, the spring diet is mostly pollock, eelpout, and capelin, but sand lance, herring, saffron cod, shrimps, mysids, crabs, and cephalopods are also eaten. There appear to be regional differences in diet in the Bering Sea: walleye pollock is the major prey in the southcentral part, arctic cod in the northern part.

Two types of underwater vocalization have been recorded: long, intense, downward-frequency sweeps and short, broadband puffing noises. These sounds may be produced by seals during mating or during underwater defense of their territories.

John Burns (1981b) likened the ribbon seal's mode of locomotion across the ice to that of the crabeater seal. It can move over short distances "as fast as a man can sprint." The fore flippers are extended alternately, while the hind flippers are held together above the ice.

Natural predators are believed to include killer whales, walruses, and sharks, although there are no direct observations. Ribbon seals generally occur south of the normal limit of polar bears. Their unwary behavior while hauled out bespeaks a lack of predatory pressure. Some ribbon seals drown after becoming entangled in commercial gill nets in the Bering Sea and North Pacific.

History of Exploitation: Natives of Alaska and Siberia have killed small numbers of ribbon seals since prehistoric times. Harvests still occur in some areas, mostly near St. Lawrence and Little Diomede islands and along the Chukchi coast. Alaskan Natives kill less than 100 ribbon seals each year for subsistence.

Soviet sealers began harvesting ribbon seals in the Bering Sea

Adult male ribbon seals, banded with starkly contrasting areas of black and white, are among the most handsome of all seals. (Northern Bering Sea, June 1977: Lloyd F. Lowry).

in 1961 for oil, skins, fertilizer, and food for mink and fox on fur farms. From 1961 through 1967, around 13,000 seals were harvested annually. The catch declined to 6290 in 1968, and a quota of 3000 was imposed in 1969. The population of ribbon seals in the Bering Sea may have been reduced from 80,000– 90,000 during the early 1960s to 60,000 by 1969.

Conservation Status: From surveys made in the mid-1970s, the world population of ribbon seals was estimated at close to 240,000, with perhaps 90,000 to 100,000 of this number living in the Bering Sea. Soviet investigators estimated that there were 117,000 ribbon seals in the Bering Sea in May 1987. Because of its offshore distribution, the ribbon seal is not likely to be threatened by coastal hunters. However, commercial hunting from ships, industrial development offshore, and overharvesting of prey species are among the potential threats.

Further Reading: Burns (1981b), Frost and Lowry (1980), Kelly (1988b), Naito and Konno (1979), Naito and Oshima (1976), Sakurai et al. (1989), Watkins and Ray (1977).

Bearded Seal

Erignathus barbatus
(Erxleben, 1777)

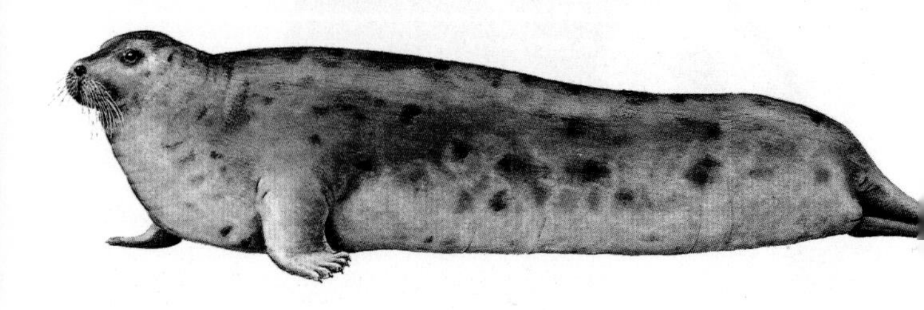

NOMENCLATURE: The generic name is from the Greek intensive prefix *eri* in combination with *gnathos,* for "jaw," referring to the relatively deep mandibles and large cheek cavity (compared with those of other phocids). Both the scientific (from the Latin *barba,* for "beard") and common names of the species refer to the array of long, white vibrissae that dominate the animal's face. Another widely used English name is squareflipper, referring to the shape of the fore limbs, on which the third digit is the longest (in other phocids the first digit is the longest). The Russian name is *morski zaits,* meaning "sea hare," apparently referring to the leap forward made by an alarmed bearded seal as it dives off the ice into a hole or crack of open water. The name *laktak* is generally used in the Soviet Far East. The Yu'pik of southwestern Alaska, St. Lawrence Island, and the southern Chukchi Peninsula call this seal *mukluk;* the Inupiat of the northern Chukchi Peninsula and Alaska and the Inuit of northern Canada and Greenland call it *oogruk.* The Norwegian name for the bearded seal is *storkobbe.* The Danes call it *remmesael.*

Two subspecies have been proposed: *E. barbatus barbatus* of the Laptev and Barents seas and North Atlantic Ocean, and *E. barbatus nauticus* of the Bering and Okhotsk seas and Arctic Ocean. However, the geographical limits of these putative subspecies are vague, and the morphological distinctions are meager.

DESCRIPTION: Size at birth is about 130 cm and 34 kg. Most adults are 2.1 to 2.4 m long and weigh 200 to 250 kg, with females slightly longer than males. From late fall through early spring, bearded seals are at their fattest, and at this time they can weigh up to 425 kg. The girth can be as much as 83 percent of the body length, the blubber layer more than 7 cm thick.

The white lanugo is shed before birth, so pups are born with a soft coat of silver bluish to brownish hair. The face is lighter colored, and there are several light bands across the crown and back.

A newly weaned bearded seal pup in the pack ice of the northern Chukchi Sea. The markings on top of the head are characteristic of pups. (May or June 1978: Budd Christman.)

Adults are more or less evenly pigmented, although they are darker on the back. The ground color varies from shades of gray to tawny or dark brown. The face and fore flippers are often rust or reddish brown. Young individuals can have blotches of irregular size, intensity, and spacing on the body. The hair of juveniles and adults is short and straight.

The bearded seal can be distinguished from other northern phocids by several features in addition to the prominent moustache and squared flippers. It is the largest member of this group but has a proportionally small head. One early author called this the "long-bodied" seal; indeed, it has a long-bodied appearance in part because of the small size of the head. It has four rather than two retractable teats (a feature shared with monk seals). The long whiskers are not beaded (again as in monk seals) and are straight when wet, but curled, especially at the tips, when dry. The bearded seal has strong claws on the fore flippers and slender, pointed ones on the hind flippers.

DISTRIBUTION: Although present in much of the Arctic and Subarctic, the bearded seal has a patchier distribution and occurs in much lower overall numbers than the sympatric ringed seal. Its movements are tied to changing ice conditions, which, to an extent, determine access to shallow feeding banks. Bearded seals generally associate with unstable (moving) pack ice, sometimes riding ice floes over considerable distances. They also overwinter in the fast ice bordering polynyas in the High Arctic, maintaining breathing holes in relatively thin ice. Access to shallow water for feeding is, if not a requirement, at least a preferred habitat characteristic. Some populations are thought to be essentially sedentary, while others migrate with the pack ice.

The Bering-Chukchi Platform, a vast shelf underlying roughly the northern half of the Bering Sea and the entire Chukchi Sea, provides excellent habitat for bearded seals. Most of the Bering-Chukchi population probably migrates south through Bering Strait and overwinters in the Bering Sea, moving north through the strait in spring (mid-April through June). In winter, the largest concentrations of bearded seals are near St. Lawrence Island, in the ice 60 to 100 km north of the ice's frontal zone, west of St. Matthew Island, and in the southern Gulf of Anadyr. In summer, the seals are scattered across the broken margin of the multiyear ice that covers much of the continental shelf in the Chukchi and western Beaufort seas. Juveniles are sometimes found in open water, and they are known to enter bays and ascend rivers.

Bearded seals rarely haul out on land, except in the Okhotsk, White, and Laptev seas where the ice melts in summer or moves outside the limits of the shallow waters where bearded seals forage. In such areas they prefer gravel beaches.

In Canada, bearded seals occur in high densities in western Hudson Bay, from Churchill to Southampton Island, and in northern Foxe Basin. They are relatively abundant in Ungava Bay and Roes Welcome Sound. Although excluded by fast ice from much of the High Arctic Archipelago during the winter, bearded seals are scattered throughout many of the inlets and fiords from July to October. At least in some years, bearded seals overwinter in the North Water of Baffin Bay, over deep (200 to 500 m or deeper) water. In the Beaufort Sea, high densities are found from Sachs Harbour to Norway Island (off the west coast of Banks Island). The underwater sounds of adult males have been recorded in an area of annual ice more than 400 km from the nearest open leads.

Bearded seals occasionally wander outside their normal range, which extends from as far north as 88° N south to Hokkaido in the Pacific and to northeastern Newfoundland in the Atlantic. The

Bearded seals haul out on ice floes, like this one 20 mi. south of King Island, near Bering Strait. (June 1978: Louis Consiglieri.)

effective range in Europe is no farther south than northern Norway, although individuals have appeared as far south as Portugal. Bearded seals commonly appear on the north coast of Iceland in winter.

NATURAL HISTORY: Bearded seals are essentially solitary. Even in areas where they occur in high densities, individuals generally haul out on separate ice floes. When several are on a single large floe, they are well spaced. While hauled out, the seals always remain oriented toward the water and at the edge of the ice floe or shoreline. Before the ice breaks up in the spring, they generally avoid stable land-fast ice; they also appear to avoid areas that are used heavily by walruses. They seem to prefer less stable ice, especially when breakup occurs early in the spring.

Like ringed seals, bearded seals use the strong claws of their fore flippers to maintain breathing holes in ice 20 to 30 cm thick. The breathing holes of bearded seals sometimes can be distinguished by the presence of bottom soil frozen into the ice dome. Their holes also are generally larger than those of ringed seals and have a cupola of clear ice formed over them. The cupola is re-formed between the seal's visits to the hole. An individual bearded seal usually has two or three breathing holes within a radius of a few hundred meters, and it surfaces at them alternately.

Unlike ringed seals, bearded seals appear to give birth only on the surface of the ice; they are not known to construct or use subnivean birth lairs. Most births take place from mid-March to early May. The peak in the Bering Strait region is the last third of April; in the southern Bering Sea and Sea of Okhotsk, early to mid-April. Pups enter the water within a few days but are not weaned until they are 12 to 18 days old.

Males reach sexual maturity at 6 to 7 years of age; females at 3 or more years. Virtually all females are mature by 8 years. Mating takes place during or at the end of lactation. Males are in breeding condition from mid-March through mid-June, but most effective copulation occurs during the first 3 weeks of May. Implantation is delayed for about 2 months, resulting in a gestation period of approximately 11 months. There is some uncertainty about the frequency of pupping. Alaskan studies have indicated that up to 85 percent of adult females can be pregnant, suggesting that the interbirth interval is usually 1 year. Some other studies have indicated that female bearded seals give birth only in alternate years, but this may be a local feature caused by relatively low benthic productivity in some areas.

Molting in bearded seals has not been studied closely, but it appears to take place mainly from April through August, with a peak in May and June. This is also the period when the seals spend the most time hauled out.

Although the social behavior and organization of bearded seals are not well understood, they appear to be territorial during the fast-ice season. John J. Burns (1981a), who has studied bearded seals in Alaska for many years, has suggested that the deep, parallel scratch marks often found on the rear third of the body may be made by the claws of other bearded seals rather than by polar bears, as is usually assumed.

As largely benthic and hyperbenthic feeders, bearded seals show a marked preference for shelf waters (less than 130 m deep). The maximum diving depth is thought to be about 200 m. Crustaceans (crabs and shrimps), mollusks (clams and whelks), fish (especially arctic and saffron cod, flounders, and sculpins) and octopuses are the staples of the bearded seal's diet. The types of prey that dominate the diet vary among areas and according to the seal's age. In some areas, such as the Bering and Chukchi seas, crabs, shrimps, and clams are the most important prey. In other areas, such as northern Baffin Bay and the Canadian High Arctic, bottom fish such as sculpins and pelagic schooling fish such as arctic cod may be more important than invertebrates. Off Alaska, young animals eat proportionately more shrimp than do older animals. Judging by the types of food they eat, bearded seals may compete for food resources with ringed seals, harp seals, narwhals, and white whales. Walruses are probably their most serious competitors for mollusks. An apparent decline in the consumption of clams by bearded seals in the Bering Strait region has been taken as evidence that the growing walrus population has depleted the clam resources locally or regionally. Bearded seals are often common in the same general areas as walruses, but not in the immediate vicinity of large groups of walruses.

Bearded seals can live for at least 31 years, but probably few live for more than 25 years.

Male bearded seals are among the most vociferous marine animals. Their eerie but melodious songs often dominate tape recordings made under the spring sea ice. The song is stereotyped and repetitive. It can last longer than a minute. As they sing, bearded seals dive slowly in a loose spiral, releasing bubbles and finally surfacing in the center of their area of activity. Native hunters in kayaks formerly stalked bearded seals partly by listening for them. The song is audible in air at close range, but it can be heard fully and clearly by placing a paddle in the water and pressing an ear against the butt end. The function of the song is assumed to be related to courtship, possibly as advertisement or territorial marking.

Bearded seals are important prey of polar bears, although probably secondary in importance to ringed seals. Bears certainly kill larger numbers of ringed seals, but a single bearded seal would provide the bear with much more food energy than would a single ringed seal. A bear has been seen to crush a bearded seal's skull with a single blow of its paw, then grasp the seal's neck in its jaws and drag it away from the edge of the ice floe. Bearded seals are also killed and eaten by killer whales, and pups are eaten occasionally by walruses.

History of Exploitation: For thousands of years the bearded seal has been an important subsistence resource for some human communities in the Far North. During the 1940s, the Polar Eskimos of Greenland's Thule district caught only some 50 bearded seals per year, but the species was nonetheless considered vital to their subsistence. In addition to meat and blubber, bearded seals provide high-quality leather for lines, whips, boat and tent coverings, trousers, and water buckets. Bearded seal skins are considered essential for the soles of kamiks (boots) in parts of the Canadian Arctic, and some skins are traded among villages as a result. In Alaska, open boats used for whaling (umiaks) are traditionally made of bearded seal skins stretched over a wooden frame. The small intestine of this seal, when frozen or boiled, is a great delicacy. In the past, a part of the intestines was used to make rain gear and translucent windows. Bearded seal blood was used to make waterproofing compounds and dyes. The reported total catch of bearded seals in Greenland has exceeded 900 in some recent years. The annual catch in the Canadian Arctic during the late 1970s and 1980s has been crudely estimated at 2500 to 3000.

It has been estimated that during the mid-1800s the Eskimos of Chukotka, U.S.S.R., caught a total of 350 to 400 bearded seals per year. After the introduction and proliferation of firearms, the secured catch along the coast of Chukotka (by Eskimos and Chukchis, combined) averaged 1822 per year from 1915 to 1937 and 608 from 1940 to 1980. The estimated annual catch in Alaska was 1784 from 1966 to 1977. The documented harvest at five villages in the

Bering Strait region from August 1985 to June 1986 was 791 bearded seals. In general, bearded seals have come under increasing hunting pressure in Alaska in recent years. Most bearded seals are taken in late spring and early summer from small, outboard-powered boats working in the leads among scattered ice floes close to shore.

In addition to subsistence hunting by coastal residents over much of the range of the species, some commercial hunting of bearded seals is done, particularly by Soviet sealing ships. During 1947 to 1956, from 2000 to 6000 bearded seals were taken each year in the Sea of Okhotsk. The annual catch increased to 9000 to 13,000 between 1957 and 1964. Bearded seals began to be harvested from Soviet ships in the Bering Sea in 1961, and the combined Okhotsk and Bering seas catch was 8000 to 10,000 per year from 1964 to 1967. After a suspension of ship-based sealing during 1970 to 1975, the reported Soviet commercial catch in the Bering and Chukchi seas during 1976 to 1985 ranged between 1130 and 2053 per year. In addition, Soviet commercial sealers in the White and Barents seas reported taking 52 to 378 bearded seals per year during the same period. The reported Soviet harvest, mainly from the Bering Sea, was 1881 in 1988 and 1418 in 1989. The seals are used as food for people, dogs, and fur-farm animals.

Sinking loss is a serious problem. Bearded seals generally have thinner blubber and a higher specific gravity than ringed seals. When they are hunted in open water or at the floe edge in spring, more than half the animals shot may be lost.

CONSERVATION STATUS: There are at least several hundred thousand (probably more than half a million) bearded seals in the world. However, their solitary habits and association with moving ice make accurate counting extremely difficult. Alaskan researchers have estimated that there are more than 300,000 bearded seals off the coast of Alaska. A major decline in the abundance of bearded seals in the eastern Beaufort Sea and Amundsen Gulf from 1974 to 1975 was inferred from aerial surveys. This decline is thought to have been caused by heavier than normal ice conditions.

We are unaware of any conservation measures taken explicitly for bearded seals, apart from quotas set by the U.S.S.R. for the Sea of Okhotsk and Bering Sea. The bearded seal population in the Sea of Okhotsk was probably overexploited during the 1950s and 1960s, but its status may have improved with the introduction of quotas. Considering their wide distribution and wary behavior, there is no immediate cause for concern for the survival of the species. However, regional populations subjected to intensive harvesting may be depleted. Also, the small breeding population of bearded seals that existed until recently in waters off southeastern Labrador and northeastern Newfoundland may have been largely or entirely extirpated by the commercial sealers who took them as a by-catch in the large-scale harp and hooded seal fishery.

Further Reading: Benjaminsen (1973), Burns (1981a), Burns et al. (1985), Finley and Evans (1983), Kelly (1988a), Lowry, L. F., et al. (1980a), McLaren (1958b), Ray, C., et al. (1969), Ray, C. E., et al. (1982), Vibe (1950).

Hooded Seal

Cystophora cristata
(Erxleben, 1777)

NOMENCLATURE: The generic name is from the Greek *kustis,* "a bladder," and *phoros,* "carrying," in reference to the inflatable nose and nasal sac that are the most distinctive features of adult males. The specific name comes from the Latin *crista,* "a crest," again in reference to the proboscis. The hooded seal is often called the bladdernose or crested seal. Its Norwegian name is *klappmyss,* its German name *klappmütze,* its French name *phoque à capuchon.* The Danes call it *klapmyds.*

DESCRIPTION: These are large seals, with males reaching a maximum size of about 3 m and 400 kg; females, 2.4 m and 300 kg. The average adult size is 2.5 m and 300 kg for males and 2.2 m and 230 kg for females. Pups are 90 to 110 cm long and weigh 20 to 30 kg at birth. They weigh approximately 50 kg at weaning, having taken on a remarkable amount of fat in an astonishingly brief time.

The most distinctive external feature of the hooded seal is the adult male's prominent nasal appendage. This "hood," which is really an enlargement of the nasal cavity, begins just above and behind the eyes and extends in front of the mouth. When relaxed or uninflated, the sac appears flaccid and wrinkled. Once inflated, it resembles a taut, leathery, bilobed football. The animal can move air back and forth between the front and rear parts of the hood.

In addition to the hood, adult males have the ability to extrude a red, membranous "balloon," mainly from the left nostril. This is accomplished by closing one nostril and blowing air into the hood. The membrane dividing the nostrils eventually bulges from the pressure and presses through the open nostril, forming the balloon. The two bizarre structures–hood and balloon–are inflated in response to disturbances and as part of the courtship display. The hood also can be inflated underwater and may function in underwater agonistic behavior of males competing for access to estrous females.

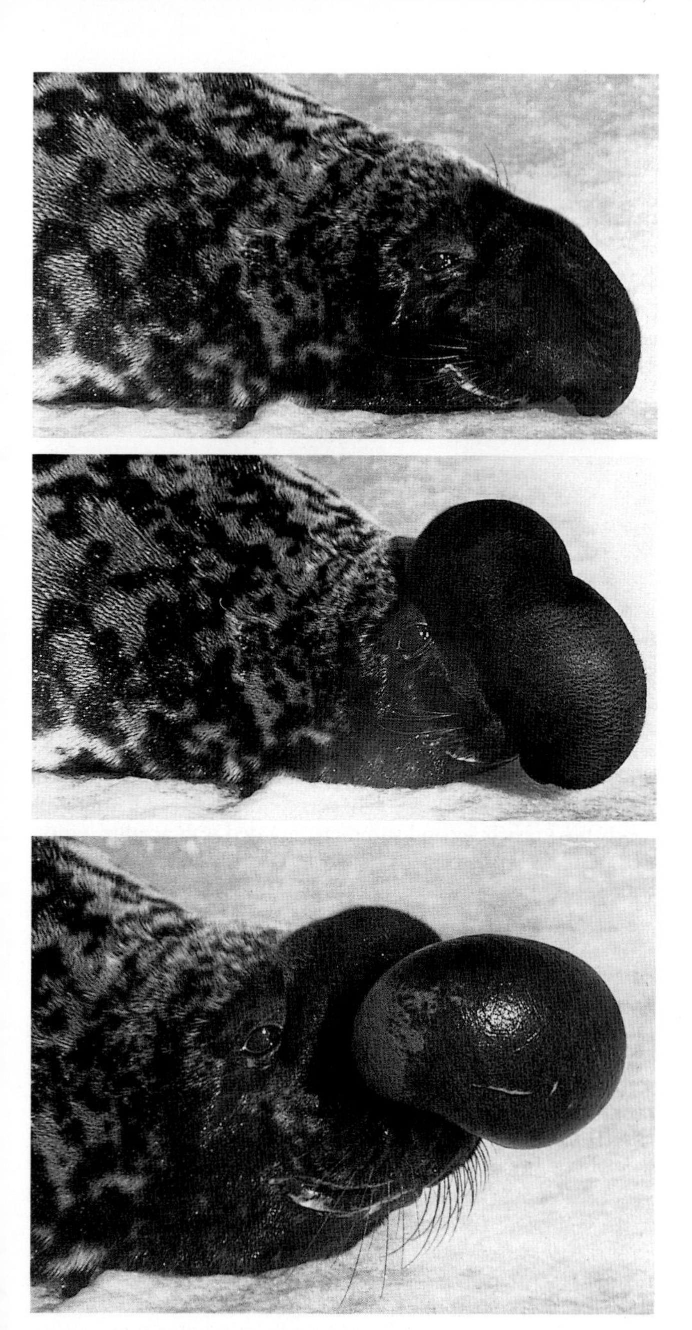

A male hooded seal with its proboscis or "hood" flaccid (top), inflated (center), and inflated with its nasal septum extruded (bottom). (Gulf of St. Lawrence, Canada, March 1986: Fred Bruemmer.)

A newly weaned hooded seal pup or "blue-back." The pup's lustrous blue gray pelage made it a target of commercial sealers. (Gulf of St. Lawrence, Canada, March 1986: Fred Bruemmer.)

Hooded seals shed their lanugo in the uterus and are born with a lustrous hair coat, blue gray dorsally and cream white ventrally. The face and flippers are dark, and there is a fairly sharp division between the dark and light parts of the pelage. After molting (probably in their second summer), hooded seals acquire a mottled pelage, best described as silvery or bluish gray with irregular dark spots and splotches. The face and flippers remain dark.

DISTRIBUTION: Hooded seals live over the deep waters of the North Atlantic, especially near the outer edge of the pack ice. They are associated with drifting pack ice during much of the year. For pupping and mating, they congregate on and around thick, drifting ice floes. Winter distribution is poorly known, although at least some hooded seals are present off Labrador and northeastern Newfoundland, on the Grand Bank, and off southern Greenland.

There are three main areas where whelping patches form in late winter and early spring: the West Ice near Jan Mayen Island southeast of Spitsbergen; the Front off southeastern Labrador and in the Gulf of St. Lawrence; and the middle of Davis Strait at about 62° to 64° N, 56° to 60° W. These pupping groups have been treated as separate stocks for management, although seals from all three regions probably molt in the summer in Denmark Strait. The Denmark Strait molting ground actually consists of two areas off southeastern Greenland used by the molting seals. A southern area centered at 63° N was used before World War II. Since then, seals have been found mainly in the northern area at 66° to 68° N. The only other known molting grounds are on the pack ice of the Greenland Sea north of Jan Mayen Island and east of Greenland at about 72° to 78° N. Seals from the Front probably molt in Denmark Strait while those from the West Ice probably molt off northeastern Greenland.

Biochemical studies have detected no major differences among the hooded seals breeding at Newfoundland and Jan Mayen and those molting in Denmark Strait. Studies of skulls from Newfoundland, Denmark Strait, and Jan Mayen revealed no differences and suggested that there is substantial genetic mixing among the animals from these three areas. Tagging results have demonstrated that Newfoundland breeders molt in Denmark Strait, but the evidence that the other two breeding groups molt there is only circumstantial. The synchronous pupping and breeding seasons of hooded seals in all areas is one circumstance favoring the hypothesis of one rather than several stocks.

In summer, hooded seals are common along much of the Greenland coast, on the west as far north as Cape York. They are uncommon but occur regularly in the Thule district of northwestern Greenland and in Lancaster and Jones sounds. Hooded seals are abundant in late summer and early fall in the pack ice between Greenland and Svalbard. They are also present throughout the Norwegian Sea and in much of the Barents Sea.

As great wanderers, hooded seals often appear in unexpected places, far from the normal breeding and foraging range of the species. Juveniles in particular are seen occasionally on the European coast as far south as Portugal and on the U.S. coast as far south as Florida. Individuals sometimes move up the St. Lawrence River as far as the Montreal area. Pups have been born on the Norwegian Sea coast of Norway and in Maine. Until recently, the eastern limit of the hooded seal's documented occurrence was the Yenisey River in Siberia; the western limit, Herschel Island in the eastern Beaufort Sea. Several young males were observed in the western Beaufort Sea in the 1970s, and on 23 July 1990, a fairly healthy, 160 kg female, probably 3 years old, came ashore in San Diego, California, presumably having traveled more than 13,000 km through the Northwest Passage, Bering Sea, and northern North Pacific.

NATURAL HISTORY: Hooded seals are regarded as solitary animals, except in the breeding and molting seasons. They generally seem less gregarious than harp seals throughout the year.

The birth season is in March and early April, with the peak in the last 2 weeks of March. In any given area, the entire breeding season, encompassing birth, nursing, weaning, and subsequent impregnation, spans only 2 to 2.5 weeks. Adults evidently do not feed during this time. Females in the Gulf of St. Lawrence haul out in loose aggregations on large, thick ice floes and usually give

An adult male hooded seal attempts to chase away an interloper. Bulls are aggressive in protecting their exclusive access to an estrous female. (Gulf of St. Lawrence, Canada, March 1986: Fred Bruemmer.)

birth near the center of the floe well away from open water. The seals in Davis Strait and the Front pup near open water on ice that has been broken into smaller chunks by oceanic swells. An average distance of 50 m is maintained between the pupping females in the Gulf of St. Lawrence, but spacing between mother and pup pairs varies greatly, depending, at least to some degree, on ice conditions (for example, large, flat pans as opposed to heavily rafted and broken ice). Mothers can be aggressive in defending their pups. The pupping habitat is generally unstable and transient; this may help to explain the brevity of the period of pup dependence.

The hooded seal has the shortest lactation period known for any mammal. On average, pups are weaned after only 4 days of nursing. They gain weight 2.5 to 6 times faster than other phocid pups: an average of 7.1 kg per 24 hour period from the day after birth to weaning. Hooded seal milk is 60 percent fat; it has the lowest percentage of protein content of any mammalian milk. The sex ratio at birth is 1:1.

Mating takes place soon after the pups are weaned. The tight reproductive schedule calls for estrus to begin on day 5 after parturition. Adult males appear to be serially monogamous; in other words, the mating system is polygynous. A large, powerful individual often attempts to chase away other would-be suitors. As many as 10 males have been observed near a single female. Serious fighting among males is indicated by the frequency of bloody wounds. Some males remain with a single female during her lactation period, then presumably mate with her when she enters the water after weaning the pup. Other males exhibit considerable mobility, staying for short periods with different females. Since the mating period for the population spans 2 weeks, and females nurse their pups for only 4 days, even the males that stay with a given female have the potential for mating several different females sequentially. Although the triads seen on the pupping grounds have often been called "family groups," this term implies a fidelity that probably does not exist. Copulation usually takes place in the water.

Implantation is delayed for up to 16 weeks. After giving birth and mating, the adults leave the pupping and breeding grounds and begin moving toward the molting grounds. Some young seals apparently do not join the adults on the molting patches in their first year. Molting takes place in summer, mainly during June and July.

Approximately 50 percent of females are sexually mature when 3 years old and bear their first pup at 4. Most of the rest give birth at 5 years of age. The pregnancy rate is high, with more than 90 percent of the adult females becoming pregnant each year. Sexual maturity in males is attained at 4 to 6 years of age, but they may not contribute to reproduction until they are several years older.

Relatively little is known about the food habits of hooded seals since adults generally do not eat during the pupping or molting seasons when most specimens have been collected. They are known to eat redfish, Greenland halibut, capelin, arctic cod, crustaceans, octopus, and squid. Hooded seals are believed to feed in deep water, perhaps at depths greater than 200 m. Resting metabolic rates of captive hooded seals were found to be nearly the same as those of similar-sized terrestrial mammals.

At least three types of low-frequency, pulsed sounds, described as "grung," "snort," and "buzz," are made by male hooded seals underwater. Females, and occasionally pups, produce low-frequency calls in air, with the mouth open. In-air sounds of low intensity are produced by adult males as they inflate and deflate the proboscis.

Little is known about predation, but polar bears can do considerable damage on the pupping grounds by killing large numbers of young. Presumably, killer whales take hooded seals occasionally, but there is little direct evidence for this.

Maximum age is about 35 years. Generally, hooded seals appear not to reach ages older than the late 20s.

Hooded seals have been brought into captivity infrequently. They have been used in a number of live animal experiments. In one of these, the heart rate of pups was found to decline from about 100 to 10 beats per minute during forced submersion.

History of Exploitation: There is a long history of commercial exploitation of hooded seals on their pupping and molting grounds. Norway has been the principal nation involved in the hunt, although Great Britain, Newfoundland (later Canada), and the U.S.S.R. have also participated. Much of the exploitation of hooded seals developed in tandem with that of harp seals. However, hooded seals at the Jan Mayen pupping grounds were more difficult to reach in sailing vessels because they were on thicker ice. With the exception of the large harvests of hooded seals on the molting grounds in Denmark Strait beginning as early as 1874, large catches usually have been made in conjunction with even larger ones of harp seals. The Norwegian hunt in Denmark Strait also involved catching Greenland sharks for liver oil.

Female hooded seals have a reputation among sealers as aggressive and dangerous. However, researchers have found that they can examine and weigh pups with little fear of being attacked by the mother. (Gulf of St. Lawrence, Canada, March 1986: Fred Bruemmer.)

Commercial harvesting of hooded seals has been encouraged since the mid-twentieth century principally by demand for the pup's "blue-back" pelt, which has a higher market value than that of any other phocid. Prior to World War II, pups and adults were taken mainly for oil and leather. The largest hooded seal catch at the Front may have been in 1901 when close to 62,000 were killed.

An aspect that has complicated management of the pup harvest is that mothers aggressively defend their pups. Sealers often have insisted that it is necessary to kill the adult in order to take the pup safely and efficiently. Thus, large numbers of mature females have been killed. However, researchers are able to handle and weigh pups, then return them to their mothers, on a routine basis. Regulations limiting the percentage of adults allowed in the catch were progressively tightened, so that in the last years of the commercial hunt in Canada, relatively few adult females were killed.

The average annual catch by Norway during 1934 to 1938 was 53,140. There was essentially no commercial hunting of hooded seals during World War II. The highest catch in the northeast Atlantic after the war was in 1951, when 131,028 seals were taken, just over a third of them in Denmark Strait and the rest at the West Ice. Soviet sealers began sealing at the West Ice in 1955. From 1946 to 1976, the total catch at the West Ice ranged from 11,565 (1946) to 83,421 (1951), with an average during this period of just under 39,000 per year. The killing of hooded seals in Denmark Strait was greatly reduced in 1961; in most years since then the catch has been less than 1000. Published catch figures for Denmark Strait do not include seals that sank after being shot.

Greenlanders kill several thousand hooded seals each year, mainly along the southern half of the island between Qaqortaq (Julianehåb) in the west and Angmagssalik in the east. Their landed catches at the end of the nineteenth century were as high as 10,000 to 15,000 per year. In recent years, their reported catches have averaged 4000 to 6000. Most of the Greenlandic catch consists of animals 2 to 6 years old. Relatively small numbers of hooded seals (probably less than 100 in most years) are taken by Native hunters in northern Canada. Seals shot in summer in open water or on small ice floes often sink, and their deaths go unrecorded.

CONSERVATION STATUS: Norway and the U.S.S.R. signed a bilateral agreement to manage the harvest of hooded seals at Jan Mayen Island and Denmark Strait in 1958. Under the terms of this agreement, both hunts had opening and closing dates, and a catch limit of 15,000 seals per year was set for the Norwegian hunt in Denmark Strait. This hunt was terminated in 1961. The West Ice (Jan Mayen) stock declined considerably between the 1950s and 1970s. Quotas on hooded seal catches in the West Ice were introduced in 1971. The quota was 30,000 seals from 1971 to 1975. In 1972, the opening date was changed from 20 to 23 March. In 1976, Norwegian ships were allowed to take 34,000 hooded seals of all ages at the West Ice; the Soviet quota was 5500. Beginning in 1977, the overall quota was broken down into pups, adult females, and adult males. In 1980, the Norwegian quota was 16,700 pups. Since 1976, steps have been taken to reduce the killing of adult females. With the ban on blueback imports by the European Economic Community (EEC) beginning in 1983, Norway greatly reduced its take of hooded seals at the West Ice. More than 60,000 pups were born at the West Ice in the late 1980s, suggesting a total population of seals 1 year old and older of perhaps a quarter million.

Aerial surveys of the Front off Newfoundland in the mid-1980s resulted in an estimate of approximately 60,000 pups, suggesting a total population of 300,000 seals. More than 2000 pups were born in the Gulf of St. Lawrence in 1990. The Davis Strait stock was estimated in the mid-1980s as about 18,600 pups and 93,000 seals 1 year and older. Estimates of population size are very approximate due in large part to the difficulty and expense of surveying for these seals.

Canada and Norway agreed on an opening date of 12 March in 1968, then delayed to 20 March in 1974, for the hooded seal hunt at the Front. The closing date varied from as late as 5 May in 1961 to as early as 24 April in 1968 and 1971. A catch quota was set at 15,000 seals per year beginning in 1974, and this was reduced to 12,000 in 1982 and to 6000 in 1983. The allowable percentage of the catch to be comprised of adult females was set at 10 percent in 1977 and reduced to 5 percent in 1979. The EEC import ban, which was extended indefinitely in 1989, effectively stopped most of the commercial exploitation of hooded seals by Norway and Canada.

The Davis Strait stock was only discovered in 1972 and has not been exploited (except for research catches) on its pupping grounds in the twentieth century. However, seals from this stock are hunted by Greenlanders in summer, and they probably were hunted on the Denmark Strait molting grounds in years before 1961. If there is significant mixing of the Davis Strait and Front herds, then the former may have served as a buffer against the latter's depletion by commercial sealing.

In addition to the continued direct harvesting by Greenlanders for food for dogs and humans, hooded seals die in substantial numbers in fishing gear off the coast of Norway.

FURTHER READING: Boness et al. (1988), Bowen et al. (1985, 1987), Campbell (1987), Kapel (1975), Kovacs (1990), Kovacs and Lavigne (1986), Lavigne and Kovacs (1988), Sergeant (1974, 1976), Wiig (1985), Wiig and Lie (1984).

Gray Seal

Halichoerus grypus
(Fabricius, 1791)

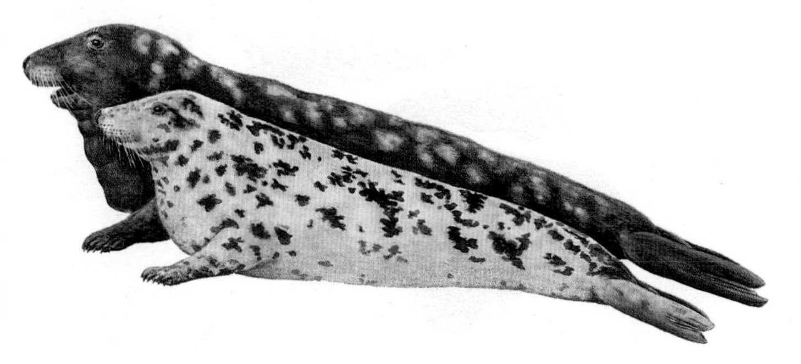

NOMENCLATURE: The gray seal's scientific name comes from the Greek *halios,* "of the sea," *khoiros,* "a pig," and *grupos,* "hook-nosed," in reference to its high Roman nose. Thus, it is the hook-nosed sea pig. In Atlantic Canada the adult male often is called the horsehead, again in reference to its distinctively shaped head. Regional names in Canada include hopper in Newfoundland, hodge in Labrador, *tête de cheval* or *phoque gris* in Québec, and cowmore among English-speaking residents of the Magdalen Islands. The last of these apparently derives from the French name of Deadman Island, Île de Corps Mort, the site of a gray seal pupping rookery. The gray seal is sometimes known as the Atlantic seal in Great Britain. The alternate spelling, grey, is used by most non-American English authors. Norwegians and Danes call this seal *havert* or *grasael.* Germans call it *kegelrobbe,* Icelanders *útselir.*

DESCRIPTION: Gray seals in Canadian waters have an average body mass about 20 percent larger than those in British waters. Males in the western Atlantic grow to lengths of 2.3 m or more and weights of up to 350 kg; females, to 2 m and 200 kg. The difference in size between the sexes is actually more pronounced than these extreme dimensions indicate. An individual adult male may weigh up to three times as much as a small adult female. The length at birth is 90 to 110 cm; the weight, 11 to 20 kg. On average, male neonates are somewhat larger than female neonates, and they grow faster and are larger at weaning.

In addition to the pronounced sexual dimorphism in size, adult males and females differ in the shape of the head and neck. The male's arched snout is more elongate and broader. Also, the male's neck and shoulders are much more massive; the skin is often thrown into folds and wrinkles, marked by wounds and scars from fighting.

Since the species are partially sympatric and generally similar

Many gray seals in the Gulf of St. Lawrence give birth in winter on the newly forming pack ice. (Eastern Canada, January 1984: Wybrand Hoek.)

in appearance, it can be difficult at times to distinguish gray seals (especially young ones) from harbor seals. In gray seals, the eyes are set farther back from the nostrils, and when the face is viewed head-on, the nostrils appear well separated, giving the impression of the letter *W*. In harbor seals, the face is more doglike, the eyes are positioned farther forward, and the nostrils form a V or heart shape.

Gray seals are usually born with a silky, cream white lanugo, stained yellowish for the first few days of life. This fetal coat sometimes has a smoky tinge from the pigmented tips of the hairs. After the first molt, at 2 to 4 weeks of age, the spotted adult pigmentation is established but muted. There is considerable individual variation in the color of gray seals, with tones ranging from black, brown, and dark gray or olive to silver, tan, and even white. In most cases, the back is darker than the belly. Males usually have a darker overall appearance than females. In general, the background (continuous) pigmentation of males is dark, interrupted by lighter spots and irregular patches; that of females, light, interrupted by darker markings. Some males appear almost entirely black, while some females are cream white with little spotting. It is not unusual for gray seals to have a rust or red to orange tinge on the snout, neck, flanks, or flippers. British researchers have developed an automated system for matching photographs of female gray seals since the color pattern on the sides of the head and neck, consisting mainly of dark blotches against a light background, is distinctive for each individual.

Gray seals have five long, slender claws on each of the fore flippers.

Distribution: Gray seals are confined to the northern North Atlantic, where they occupy subarctic and temperate waters. At least three populations are recognized: western, eastern, and Baltic. Seals of the western stock grow larger and live longer than those of the eastern stock. Throughout their range gray seals are sympatric with harbor seals.

The western population is centered in the Canadian Maritimes, including the Gulf of St. Lawrence and the Atlantic coasts of Nova Scotia, Newfoundland, and Labrador. Gray seals are rare north of Hamilton Inlet, Labrador. Small numbers occur regularly in the lower Bay of Fundy and the Gulf of Maine south to Cape Cod. The principal pupping rookeries on land are at Sable Island, the Basque Islands, Camp Island east of Nova Scotia, Amet Island in Northumberland Strait, and Deadman Island in the Gulf of St. Lawrence southwest of the Magdalen Islands. Large numbers of gray seals also pup on the newly forming pack ice in the eastern portion of Northumberland Strait and between eastern Prince Edward Island and western Cape Breton Island. Anticosti Island in the Gulf of St. Lawrence and Sable Island are major spring molting sites. A few females may still pup on Muskeget and Tuckernuck islands in Nantucket Sound, Massachusetts, and on Monomoy Island south of Cape Cod.

The present-day range of the Baltic population is considerably reduced from what it was historically. For example, the island of Anholt in Denmark's Kattegat was once an important pupping site. There are only some tens of gray seals in the Kattegat today, and these may be immigrants from the expanding eastern Atlantic population. The famous Harstena seal skerries in Sweden had thousands of gray seals as recently as the 1930s, but these are now essentially deserted. Although individuals and small groups still occur throughout much of the Baltic, particularly in summer, the only sizable groups (rookeries with more than 150 or so seals) are in the northeastern Gulf of Finland, in the archipelago separating the Gulf of Bothnia from the Baltic proper, and off southeastern Sweden. Recent counts from Estonia show that there is a larger gray seal population in Soviet Baltic waters than previously believed.

The eastern Atlantic population is distributed from Iceland to northern Norway, with occasional individuals reaching as far east as the mouth of the White Sea. There are breeding rookeries off the Sør-Trøndelag coast and north to Finnmark in Norway. The largest rookery is at Froan (230 to 270 pups per year in the early 1990s). Scattered individuals reach as far south as Rogaland and, less often, the Swedish border. Many of the gray seals in southern and southwestern Norway are from the breeding rookeries in Great Britain. Gray seals are abundant around the Faroe Islands. The largest rookeries are on the British Isles, particularly the Outer and Inner Hebrides, North Rona, and the Orkney, Shetland, and Farne islands. There are many smaller rookeries scattered around the coasts of southwestern England, Wales, and Ireland. Gray seals are seen occasionally in Brittany (France) and in the Wattenmeer of Schleswig-Holstein (Germany). There are at least three records of gray seals from Portugal, as far south as Lisbon.

NATURAL HISTORY: Gray seals are gregarious, forming large rookeries during the pupping, breeding, and molting seasons. They do not undertake well-defined, long-distance migrations, but newly weaned pups in particular are known to disperse widely and rapidly away from the breeding rookeries. One pup tagged on Sable Island on 5 February was recovered at Barnegat Light, New Jersey, on 2 March, having traveled a straight-line distance of 1280 km in 25 days (an average of at least 50 km per day). Another pup marked in Scotland's Firth of Forth was recovered 9 days later at Karmøy, Norway, 580 km distant (an average of about 65 km per day). During the nonbreeding season, gray seals tend to be seen most often around rocky, wave-affected shores. One individual in the North Sea whose movements were tracked continuously for 9 days by a combination VHF radio transmitter and underwater sonic device, traveled at about 4.5 km per hour. Its course took it from the Farne Islands to Dundee, with stopovers at the Isle of May and the Firth of Forth.

The breeding season differs among the populations. For the species as a whole, it is from late September through early March. However, each stock, and to some extent each colony, has its own season. The seals in the British Isles breed earliest, with most births in September, October, or November. The peak pupping season in Iceland and Norway is also October, and some pupping apparently occurs on the Finnmark coast sometime after 20 November. Most pupping in the Baltic Sea is on the drifting ice in February or March, although historical evidence indicates that the extirpated population in the Kattegat pupped mainly in January. In Canada, the first pups are born in late December, the last in early February, with the peak in mid-January. In addition to the variable timing of their pupping and mating seasons, gray seals exhibit opportunism in the types of substrates used for hauling out. Although most give birth on rocky mainland shores and small offshore islands, others use sandbars, ice, or even sea caves. Remote, uninhabited islands tend to have the largest gray seal colonies. Some gray seals climb hills to find suitable haul-out sites. On North Rona off Scotland, a few individuals ascend a steep, 80 m high hill to pup on a grassy ridge overlooking the main rookery.

Gray seals are polygynous. Males fight for access to females who come into estrus toward the end of lactation. However, males do not defend discrete territories, nor do they sequester females in the same way that some otariids do. Rather, they attempt to monopolize access to groups of females that have gathered together for pup-

A group of gray seals of various ages on a sandy beach. (Sylt, Germany, 16 July 1988: Hans Reinhard.)

ping and nursing. Displays usually include open-mouthed threats accompanied by "hooting," a term applied to the exchange of long, quavering calls. Although threat displays are often adequate, fights do occur, and participants can be left with bloody neck wounds. Copulation takes place in the water or on land or ice and lasts up to 45 minutes (average is 20 minutes). Attachment of the resting embryo and resumption of fetal growth is delayed for 3 or 4 months.

Pups consume about 3.2 kg of milk per day. The milk is low in fat (about 40 percent) early in lactation. The fat content increases sharply during lactation to as much as 60 percent, while the water content declines. About 60 percent of their ingested energy is converted to fat and lean tissue. They gain at least 1.2 to 1.5 kg per day until weaned (up to 2 kg per day in Canada). The average age at weaning is 17 days. Pups begin molting at 9 to 18 days and have finished by 11 to 27 days of age. Weaned pups may fast for 1 to 4 weeks before they begin feeding at sea. During this fast, the pups lose weight rapidly, although their metabolic rate declines by 45 percent. Blubber, and to a lesser extent protein, is catabolized to permit the pups to endure this long fast. Adult females mate 15 days postpartum, on average.

After being abandoned by the adults at close to 3 weeks of age, weaned pups may remain on shore for several more weeks, gathering in small bands of four to five individuals and often sheltering in depressions in the sand or behind logs or other beach litter. Once their molt is complete, these juveniles leave their birthplace and become wanderers. Other pups may begin their wandering immediately after weaning. Some head inland instead of seaward;

they have been known to be struck by vehicles on roads in the Canadian Maritimes.

Pup mortality ranges from 14 to 35 percent. It is especially high on narrow beaches where crowding is intense. In such circumstances, separation from mothers is common. At high tide and during storms, pups can be washed away and drown. Otherwise, the main causes of pup mortality are starvation and infections, especially pneumonia.

Females attain sexual maturity by 3 to 5 years of age, or possibly as early as 2 years on rare occasions. Males become sexually mature by 4 to 8 years but are not likely to be successful breeders until 10 years of age or older, when they are capable of holding their own with other dominant bulls. Successful bulls weigh 200 to 350 kg. Their success may be due as much to their ability to stay ashore for a long period without eating as to their fitness for combat.

Gray seals in Canada molt from early May to early June. Baltic gray seals molt on land in April, May, and June; those in the British Isles, between January and March (females) or from March to May. During the molt, the seals may haul out for considerable periods, but they also enter the water regularly.

Gray seals eat a variety of organisms, but mainly juvenile schooling or aggregated fish, squid, and octopus. Seabirds are eaten occasionally. In the Gulf of St. Lawrence, the average size of fish eaten is about 18 cm. Herring, capelin, flounder, cod, haddock, and other commercially valuable fish, including salmon, are certainly consumed, but so are less valuable species such as sand lance, saury, lumpfish, eelpout, and silver hake. Fish weighing less than 1 kg are swallowed whole, but larger prey are often held in the fore flippers and torn into chunks for easier swallowing. Newly weaned seals often eat shrimp and crabs. Although fishermen complain about gray seals opening and emptying lobster traps, the seals apparently are most interested in obtaining the rancid bait. There is little evidence that they eat lobsters. Gray seals are known to forage on the bottom at depths of at least 70 m. A trained gray seal dived to 225 m.

Considerable attention has been given to the energetics of gray seals because their impact on fisheries is seen as a major management problem. As much as 6 liters of fish remains have been found in the stomach of a single gray seal. One adult male killed in Great Britain had 10 kg of salmon flesh and bones in its stomach. On average, gray seals are believed to consume 2 to 3 percent of their body weight in food per day, although any such figure obviously depends on the caloric value of the prey organisms. Adults fast during the breeding season, females for at least 3 weeks and males for as long as 6 weeks. Heat, and particularly water, conservation during such long fasts is facilitated by counter-current heat exchange mechanisms in the nasal passages.

Gray seals are vulnerable to predation by sharks and probably killer whales. Young pups separated from their mothers are subject to attacks by gulls that peck at their eyes and may even kill them.

In the wild, female gray seals can live to be more than 40 years old. Most, however, do not live longer than 35 years. Males generally die younger, with a maximum documented age of about 30 years.

Gray seals are commonly displayed in zoos and aquaria, and they have bred in captivity. Important studies of diving physiology and sleep have been conducted using captive gray seals.

HISTORY OF EXPLOITATION: The gray seal has been exploited intensively over most of its range. Swedish and Finnish hunters killed some 30,000 gray seals, mostly pups, during 1932 to 1939 in the northern Baltic Sea (Gulf of Bothnia). As many as 10,000 pups were killed in the Baltic in 1913. The hunt for gray seals was encouraged by bounties introduced in most Baltic countries at around the turn of the century.

In Norway, gray seal remains are common in archaeological sites, indicating that the species has been hunted there for thousands of years. Beginning in 1980, the Norwegian government authorized culls of gray seals in specific localities. The total quota during 1980 to 1986 was 1270, but of this number only 797 seals were documented as killed.

There is a long history of commercial and subsistence hunting of gray seals in Great Britain. Also, beginning in 1962 a series of government-sponsored culls, involving mainly mothers and pups, began at some of the outer islands. The culling program was suspended in 1979 in the face of strong public protest.

Although gray seals were regularly hunted by settlers in Nova Scotia in the 1670s, and they were hunted in the Magdalen Islands during the 1800s, the presence of the species in Canada was essentially unrecognized until 1949, when scientists noted that some of the jaws being submitted for the harbor seal bounty were from gray seals. Beginning in 1949, the gray seal was excluded from the bounty program. However, by 1967, the gray seal population was considered large enough to require culling. From then until 1983, an average of about 1000 gray seals were killed each year at the breeding rookeries other than Sable Island. Most (close to 80 percent) of the seals culled were pups. A bounty was introduced explicitly for gray seals in 1976. From then until 1983, some 720 gray seals, mainly adults and subadults, were bountied each year. Considering that approximately half the seals shot for the bounty are not secured, the combined total of gray seals removed by the cull and bounty programs was probably more than 2000 per year during 1976 to 1983. In 1984 and 1985, when there was no cull, the combined research and bounty kills in Canada were 580 and 446, respectively.

The remains of gray seals have been found in Native American middens in Maine and southern New England dating back several thousand years. Also, a written description exists of subsistence hunting for this species in Maine during the early seventeenth century. Bounties on seals (species not specified) encouraged the killing of gray (and harbor) seals in Maine (1891 to 1905 and 1937 to 1947) and Massachusetts (1888 to 1908 and 1919 to 1962). About 40 gray seals were bountied at Muskeget Island, Massachusetts, during a 5-year period of the late 1940s and early 1950s, and at least 25 more were killed during 1958 to 1964.

More than 7200 pups and an unknown number of adults were killed in the commercial hunt at Iceland between 1962 and 1981. From 1982, when a bounty system was introduced, through 1985, the average annual kill of pups and adults, combined, was nearly 1800. The pups were taken by clubbing them to death on the coastal rookeries.

In many areas, gray seals, especially juveniles, die when they become entangled and drown in fishing nets.

CONSERVATION STATUS: Nigel Bonner estimated the world population of gray seals in 1980 as approximately 120,000. More recently, the Natural Environment Research Council estimated it at close to 200,000, nearly half of which breed in Great Britain.

Pup production in eastern Canadian waters was estimated as more than 12,000 per year during the mid-1980s. The large rookery on Sable Island has been growing at a rate of more than 12 percent per year since the early 1960s. In 1990, more than 11,000 pups were born on the island. Estimates of the total gray seal population in the western North Atlantic are in the range of 85,000 to 115,000.

The large British stock of gray seals that breeds in autumn on a few uninhabited islands and along the Scottish mainland coast is monitored closely by the Sea Mammal Research Unit of the Natural Environment Research Council. The total population was estimated as 86,000 seals in 1989. Gray seals were protected in Great Britain by the Grey Seal Protection acts of 1914 and 1932. Although the gray seal population in British waters had grown steadily for the last 30 years, pup production declined by about 20 percent in 1988, the year of the phocine distemper (PDV) epizootic in Europe. By autumn of that year, most breeding gray seals in Great Britain had antibodies to PDV or a closely related virus, although only several hundred gray seals were found dead.

A critical factor in the gray seal's recovery in Great Britain has been people's abandonment of some of the outer islands. For example, gray seals are thought not to have bred at North Rona before 1844, when the human population was evacuated. After that

The grassy, windswept beaches of Sable Island are the site of a thriving gray seal colony. Births there have been increasing by more than 12 percent a year since the 1960s. (Sable Island, Nova Scotia, Canada, January 1967: Fred Bruemmer.)

time, in spite of continued hunting, the seal population on the island grew rapidly. By 1939, an estimated 1500 pups were being born there each year.

Gray seal numbers have increased along the Norwegian coasts since 1973, when a royal decree gave them additional protection. There are at least 3100 gray seals in Norwegian waters, possibly many more. The species is protected year-round in southwestern Norway, from 1 May to 31 October in central Norway, and from 1 May to 30 November north to the Soviet border. However, during the closed seasons, Norwegian fishermen are allowed to shoot gray seals to protect fishing gear and aquaculture impoundments. At Froan, the number of births increased rapidly through 1985 but may have declined somewhat since, evidently because of culls on the breeding grounds.

Gray seals were given explicit protection in the Nantucket area of Massachusetts beginning in 1965. Some pups have been born at Muskeget Island and a few other sites since then, but appreciable numbers of gray seals are not known to have been born in U.S. waters since 1973. Small numbers disperse regularly southward from Sable Island, so gray seals are seen occasionally in parts of New England and on Long Island, New York. About 100 gray seals were hauled out in March 1986 at Wasque Shoals, a tidally exposed sand shoal southeast of Martha's Vineyard.

The total population of gray seals in the Baltic Sea has declined from perhaps 100,000 early in the twentieth century to approximately 2500 to 3000 at present. The decline of Baltic gray seals was caused by a long period of overexploitation. Pathological changes in reproductive organs, thought to have been caused by organochlorine pollution, have slowed recovery of this population. Gray seals have been protected from hunting in Denmark since 1967, in Sweden since 1974, in the U.S.S.R. since 1970, and in Finland since 1982. However, substantial illegal killing occurred in the Gulf of Riga during at least the 1970s. Sweden, Denmark, and the U.S.S.R. have seal sanctuaries where human activity is restricted. Most gray seal colonies in Sweden are in seal or bird sanctuaries. To promote the gray seal's recovery, a few animals are kept in an enclosure northeast of Stockholm. Pups born there are released soon after weaning. Since 1979, attempts have been made along the French coast to rehabilitate and release young, sick seals that come ashore, in the hope that breeding colonies will eventually be reestablished there. On the Farne Islands off the North Sea coast of England, gray seals damaged vegetation and exacerbated the problem of soil erosion begun by colonies of burrowing puffins. The culling of mother seals and their pups was deemed necessary to prevent further environmental damage.

The most immediate threat to the gray seal is probably its unpopularity with fishermen. It offends them in at least three ways: (1) by preying on commercially valuable fish such as salmon, (2) by removing or mutilating fish in nets and traps (and often damaging gear in the process), and (3) by serving as the primary host of the codworm (recently renamed "sealworm" by the Canadian government). These parasitic nematodes live as larvae in cod and other fish, then mature in the stomachs of seals that have eaten infected fish. The presence of codworm larvae in fillets reduces their market value. As a result, the larvae must be removed by hand, at considerable expense. The outrage of fishermen has resulted in bounty or culling schemes in those countries with large gray seal populations. It seems unlikely that the pressure to control their populations will diminish, although the methods used may change.

FURTHER READING: Anderson, S. (1988), Anderson, S. S., et al. (1979), Bonner (1972, 1981, 1989a, 1989b), Corbet and Harris (1991), Davies (1957), Fedak and Anderson (1982), Harwood and Greenwood (1985), Harwood and Prime (1978), Hook and Johnels (1972), Mansfield (1988), Mansfield and Beck (1977), Miller and Boness (1979), Prime (1985), Wiig (1986), Wiig et al. (1990), Yablokov and Olsson (1989).

Crabeater Seal

Lobodon carcinophagus
(Hombron and Jacquinot, 1842)

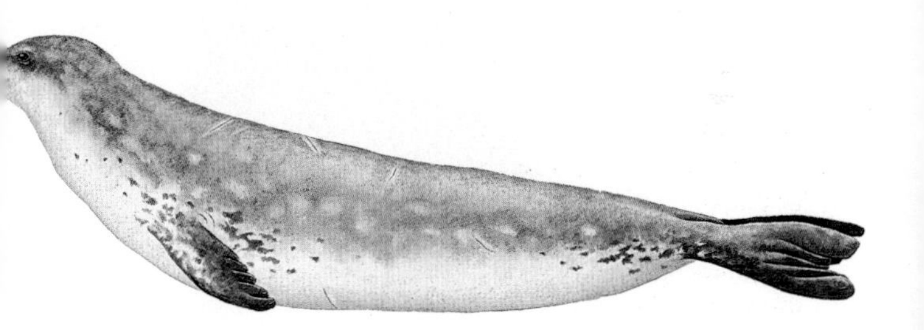

NOMENCLATURE: The name *Lobodon* refers to this seal's extraordinarily complex dentition. The Greek *lobos* means "lobe"; *odōn,* "tooth." The name *carcinophagus* comes from the Greek *karkinos,* "a crab," and *phagein,* "to eat." The Frenchmen who coined the name of the species were on the right track since this seal eats mostly crustaceans. However, it is shrimp (krill) rather than crabs that dominate the crabeater seal's diet.

DESCRIPTION: These seals are at least 114 cm long and weigh an average of 36 kg at birth; they are 1.5 to 1.6 m long and weigh 80 to 110 kg at weaning. They reach 198 to 208 cm at 1 to 5 years of age, 221 cm at 5 to 8 years, and 226 cm at 12 years or older. Maximum length is close to 2.6 m, and weights of up to 225 kg have been reported. There is little difference in size between the sexes, but females may be slightly larger than males, on average.

The body is long and slender. Crabeater seals have a noticeably elongated snout with a distinctly upturned aspect in profile. When agitated, crabeater seals typically foam at the nose (as do Weddell and southern elephant seals).

Crabeater seals molt in January and February, and they continue to swim and feed during this period. The new pelage is

A young crabeater seal sinuously works its way across the ice. (Weddell Sea, March 1986: Brent S. Stewart.)

basically dark gray dorsally, shading to silver ventrally. At times, the border between the cape and the ventro-lateral coloration is quite distinct. There are many irregular dark and light brown patches or rings on the sides and back. As the coat ages, the color lightens, eventually becoming almost entirely blonde or creamy white. The flippers and snout often remain darker. When they are sleeping on the ice, crabeater seals that have recently turned over tend to be half dark and half light, the dark half being wet from the melting ice under the dozing animal. In contrast to the coarse adult hair, the newborn pelage is soft and woolly in texture, grayish brown in color. The first molt begins about 2 weeks after birth.

Most crabeater seals are prominently scarred. Many of the straight, parallel scars along the sides and back are from the teeth of leopard seals; far fewer are from the teeth of killer whales. The scars on the face and around the fore and hind flippers are from wounds inflicted during the breeding season.

The dentition of these seals is unusual. There are two upper pairs and one lower pair of small incisors in addition to the small upper and even smaller lower canines. The postcanines are multi-

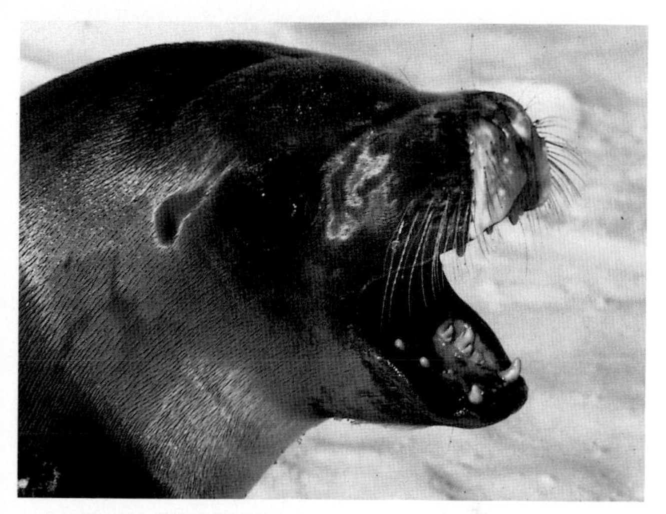

The cheek teeth of crabeater seals are ornate in comparison with those of other species. The lobed cusps are adaptations for straining krill, their main food. (Ross Island, Antarctica, November 1973: Frank S. Todd.)

cuspid, the cusps forming separate, well-defined lobes. Upper and lower postcanines interlock to form a kind of sieve, allowing the seal to filter krill from seawater. A bony flange behind the last lower cheek teeth keeps food from escaping from the corners of the mouth.

DISTRIBUTION: The distribution of crabeater seals is circumpolar in the pack ice around Antarctica. Coastal waters of the Antarctic Peninsula and the southern Ross Sea are among the areas where crabeater seals abound.

Long-distance wandering by individuals has produced some novel records. A carcass was found on the Ferrar Glacier, 1100 m above sea level. A live male pup was discovered in December on the Crevasse Valley Glacier, 113 km from open water and 920 m above sea level. Crabeater seals also appear occasionally far north of their normal range, along the coasts of New Zealand, South Australia, Tasmania, Heard Island, southern Africa, and South America (to near the Tropic of Capricorn).

In general, crabeater populations are believed to move south in summer as the ice breaks up and north again as the ice edge advances in winter. However, in southeastern Africa, vagrants appear most often between late December and early March – the austral summer.

NATURAL HISTORY: Most observations are of solitary individuals, pairs, and groups of three. As many as several hundred seals some-

Although they might prefer larger and more buoyant floes, in late summer crabeater seals may remain tenaciously on small, rotting floes. Most crabeater seals have scars on the back and sides from attacks by leopard seals. (Inside the caldera at Deception Island, Antarctica, August 1982: Frank S. Todd.)

times are seen hauled out on the same ice floe, but it is unclear whether such groups represent a social alliance or are merely adventitious. Hundreds of crabeater seals may be found in the water, feeding together. Although they are definitely primarily pack ice animals, crabeater seals are occasionally seen hauled out on shore.

Pups are born mainly in late September and October, and crabeater seals are found in triads in the pack ice during October and November. A typical group consists of a male, a female, and a pup. Other males may be in the water near the floe occupied by the family group, and they compete for access to the female. The male aggressively repels other male crabeaters and will even attempt to chase away leopard seals and approaching humans. Since the pup is likely not his own, such apparently protective behavior most likely represents the male's determination to mate with the female when she comes into estrus.

Lactation lasts 14 to 21 days. During this time, the mother loses about 5.6 kg per day and the pup gains about 4.2 kg per day. Newly weaned pups weigh 80 to 110 kg.

Once the pup is weaned, the male and female remain together as a mated pair, and copulation probably occurs within 2 weeks. The male and female presumably separate afterward and begin foraging to replenish their depleted energy stores.

At least 80 percent of adult females give birth each year. Sexual maturity is reached in both sexes at an age of 2.5 to 6 years. The mean age at sexual maturity of females declined in some areas from about 4.5 years in 1945 to about 2.5 in the 1960s, evidently in response to the greater availability of krill after the numbers of large krill-eating baleen whales had been greatly reduced.

Krill is the main, and often the only, food of crabeater seals. In some areas they also eat fish such as antarctic silver fish seasonally. A captive crabeater seal was observed to suck prey items into its mouth with one gulp from distances of up to 50 cm. The food was retained inside the mouth by pressing the tongue against the palate, while the seal raised its lips to release seawater. Feeding may be most intensive at night, when the invertebrate prey are near the surface. In late summer, the seals remain on ice floes during the day and enter the water as the sun approaches the horizon. Much of the night is spent diving, presumably to feed. In the Weddell Sea, dives as deep as 430 m and lasting as long as 11 minutes have been documented by Stewart and colleague John Bengtson of the U.S. National Marine Mammal Laboratory. Most feeding dives, however, were to shallower depths (about 25 m) and lasted less than 5 minutes.

As major consumers of krill, crabeater seals may compete with baleen whales, penguins, other krill-eating seals, and humans. (They consume an estimated 63 million tons of krill in the Antarctic each year.) With the severe reduction of Southern Hemisphere whale populations by commercial whaling, it appears that the crabeater seal population, itself virtually unexploited, has expanded and grown during the last half-century.

Unlike the loquacious Weddell and Ross seals, crabeaters are usually silent underwater. When startled or frightened on the ice, they hiss and blow through the nose.

Crabeater seals are frequent prey of both leopard seals and killer whales. The maximum life span is 40 years; the average is about 20 years. An incident of mass mortality was documented in August 1955, when some 3000 crabeater seals were found trapped in sea ice off the west coast of Graham Land, 5 to 25 km from open water. Within 2 months, most of them were dead. No obvious cause of death could be determined, but a viral disease was suspected.

Only a few crabeater seals have been held in captivity.

History of Exploitation: In 1986–87, a Soviet commercial sealing expedition in the Antarctic took more than 4000 crabeater seals. Small numbers have been killed in some years by scientists from the United States, West Germany, Great Britain, and other countries for the purpose of investigating their diet, demography, physiology, and anatomy. Substantial numbers (low thousands) also have been taken to feed sled dogs, and some or many of the 13,223 seals killed by the Norwegian sealing expedition in 1893–94 may have been crabeaters.

Conservation Status: Crabeater seals are the world's most abundant pinnipeds. Their total population may be well over 10 million; some estimates have been as high as 70 million. Census data obtained in 1983 were compared with similar data from 1968–69 in the western Weddell Sea and in 1973–74 in the Pacific Ocean sector of the Antarctic. Significant declines in density had occurred in both areas. However, the causes of these declines are unknown.

Like all antarctic seals, the crabeater is protected absolutely on land or ice shelves by the Antarctic Treaty. Seals can be taken only for scientific, educational, or display purposes, under permit, and for essential food for dogs or humans. At sea or on floating ice, they are protected under the Convention for the Conservation of Antarctic Seals. This convention provides for the setting of conservative kill quotas, area closures to sealing, and the granting of scientific permits to kill or capture seals.

FURTHER READING: Bengtson and Laws (1985), Bengtson and Siniff (1981), Bengtson and Stewart (1992), Klages and Cockcroft (1990), Laws (1984), Lowry, L. F., et al. (1988), Shaughnessy and Kerry (1989), Siniff et al. (1979).

Ross Seal

Ommatophoca rossii
Gray, 1844

NOMENCLATURE: The genitive form of the Greek *omma,* "eye," was used to denote this genus because of the very large orbits of the Ross seal's skull. James Clark Ross was commander of the HMS *Erebus,* which entered the Ross Sea during the British Antarctic Expedition of 1839 to 1843. Ross secured two specimens of the seal that bears his name; these were described and named by J. E. Gray of the British Museum. Other names occasionally applied to the Ross seal are big-eyed seal and singing seal.

DESCRIPTION: The maximum length of these seals is probably between 2.3 and 2.5 m. Body weights occasionally exceed 200 kg. Females may attain a somewhat larger size than males. Pups are approximately 1 m long and weigh about 16 kg at birth.

The Ross seal's body differs in a number of obvious ways from those of other seals. The head is proportionally small but broad; the snout is short. There are 15 to 17 well-developed mystacial (facial) vibrissae (whiskers), 10 to 42 mm long, on each side of the snout. These are probably important sensory organs. The eyes, although prominent, are surprisingly normal looking, considering the extraordinary size of the orbits. Ross seals have a small mouth, but when approached they often lift the head high off the ice and open the mouth wide to vocalize. Thus, in many photographs, the mouth becomes a noticeable feature. Both the fore and hind flippers are proportionally long, the latter up to 20 to 22 percent as long as the body.

Their hair is the shortest of any phocid. The color pattern is basically streaked rather than spotted, with longitudinally oriented stripes particularly on the sides of the neck and on the throat. There is much variability among individuals, but generally the back is dark and the belly silvery. Spotting often occurs in the zone of transition between the dorsal darkness and the ventral lightness. Some of the streaks on the throat have been described as approaching

The small but sharp teeth of Ross seals are well suited for snagging and holding squid, their principal prey. (Weddell Sea, January 1987: Brent S. Stewart.)

chestnut or chocolate brown. The streaking pattern on the face can suggest a mask.

The incisors and canines are small and recurved; the cheek teeth are much reduced, often barely piercing the gum.

DISTRIBUTION: Ross seals have a circumpolar distribution in the Antarctic, with centers of abundance in the King Haakon VII Sea and possibly the Ross Sea (especially near Cape Adare). They tend to occupy areas of heavy, consolidated pack ice, although they can be seen on large or small smooth ice floes.

Individuals rarely wander outside the Antarctic, although there are records of Ross seals being seen in South Australia and Heard Island.

NATURAL HISTORY: Ross seals are essentially solitary, at least during summer when most observations have been made. Very little is known about their social and other behavior. They probably give birth and mate in spring (November) in the pack ice, with most pups being weaned in December. Ross seals molt in January and February, and recent data suggest that the attachment of the fertilized egg to the uterine wall is delayed until after the molt. The diet is known to include vertically migrating squid and midwater fishes, which may be hunted mainly at night when they approach the surface. Ross seals also eat krill. Stewart and colleague J. L. Bengtson documented the diving behavior of one Ross seal in the Weddell Sea. Most dives were deeper than 100 m and lasted around 6 minutes. The deepest was to 212 m; the longest was for 10 minutes.

Their in-air sounds are of two kinds: an explosive, unvoiced exhalation emitted with the mouth somewhat open, and a voiced sequence of pulsed "chugging" sounds and tonal "siren" calls usually emitted with the mouth and nostrils closed. Unlike those of most pinnipeds, the underwater calls of Ross seals are similar to their in-air sounds. The siren call in particular is sufficiently conspicuous and distinctive to be useful for locating Ross seals.

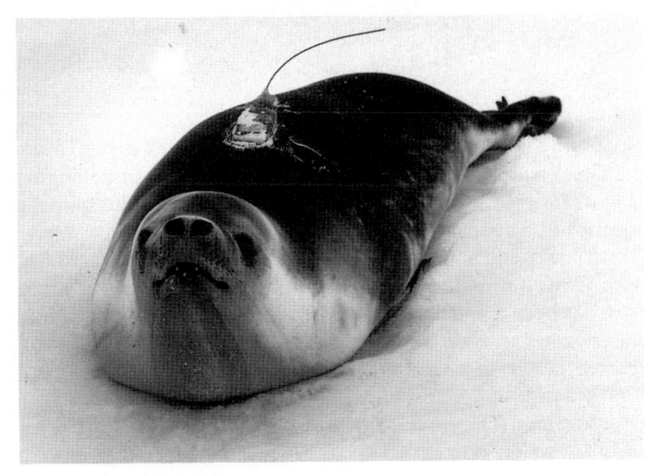

The Ross seal's inaccessibility has made it the least known antarctic phocid. New technology, such as the microprocessor-based dive recorder on the back of this adult female, is helping researchers learn about behavior, movements, and habitat use. (Weddell Sea, January 1987: Brent S. Stewart.)

HISTORY OF EXPLOITATION: Only small numbers of Ross seals have been killed for research. In 1986–87, 30 were killed for commercial purposes under a special permit. The species has never been exploited to a significant degree.

CONSERVATION STATUS: Guesses at the total number of Ross seals in the Antarctic have ranged from a few tens of thousands to several hundreds of thousands. However, the truth is that no satisfactory census has been made. Ross seals are thought to be the least abundant of the antarctic phocids, but this impression could be an artifact of their relative inaccessibility.

FURTHER READING: Laws (1984), Ray, G. C. (1981), Skinner and Westlin-van Aarde (1989).

Leopard Seal

Hydrurga leptonyx
(de Blainville, 1820)

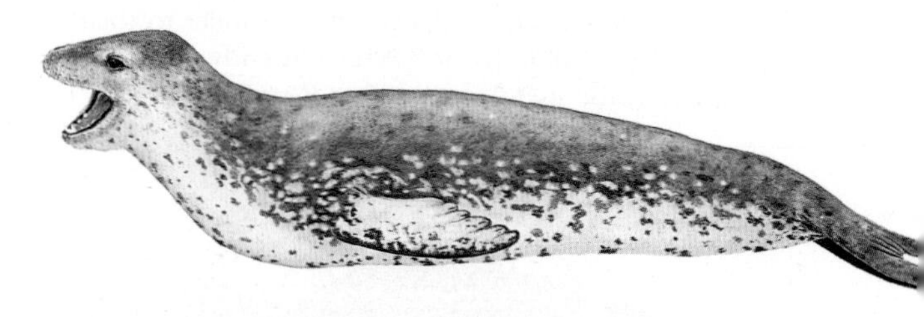

NOMENCLATURE: The name *Hydrurga* means "water worker," from the Greek *hudōr* for "water" and the Greek *ergō* (a misspelling of "I work") or the Latin *urgeo,* "I drive," a reference to the animal's aquatic life. The species name *leptonyx* comes from the Greek *leptos,* "small" or "slender," and *onux,* "claw." The name sea leopard has been used in some early references to the species, but the same term has been applied to the Weddell seal and even to the harbor seal in some areas of the Northern Hemisphere because of their spotted pelage patterns.

DESCRIPTION: The newborn leopard seal is more than 1 m long and may weigh close to 30 kg. Females grow larger than males, reaching maximum lengths of nearly 3.6 m and 3.15 m, respectively. Most adults are about 3 m long. Very large individuals can weigh at least 450 kg.

This long, slender seal has a huge head, with massive jaws and a tremendous gape. The general appearance is often described as reptilian. When the jaws are agape, it appears as though the whole animal has opened up. There are few vibrissae. The fore flippers are proportionally long, almost reminiscent of wings.

The leopard seal's color pattern is bipartite, with the dorsal part much darker than the ventral. The demarcation zone is obvious but rather diffuse. Along the midline of the back, the ground color is dark gray. It fades through various shades of gray to almost blue on the sides and silver on the belly and throat. On the head, the color changes at approximately the level of the eyes. There are many dark spots and blotches that are especially conspicuous on the lateral and ventral parts of the body. The nature and degree of spotting vary considerably. Pups have a soft, thick pelage with a color pattern resembling that of adults. In fact, pups look essentially like miniature adults, including the huge head.

The leopard seal's dentition appears formidable, mainly because of the great length of the canines and the massiveness of the postcanines. The postcanine teeth are similar to those of the crabeater seal except that these have only three rather than four

The origin of the leopard seal's common name is apparent in this close-up view, but the most striking characteristic of this species is its reptilian head. (Paradise Bay, Antarctica, 24 February 1990: Stephen Leatherwood.)

or five cusps. The occlusion between the incisors and canines is precise; that between the postcanines is less so.

DISTRIBUTION: Leopard seals have a wide distribution in high latitudes of the Southern Hemisphere. They are present from the edge of the antarctic pack ice to the Antarctic continent, as well as on and near many subantarctic islands. Adults appear to concentrate near the Antarctic continent during the reproductive season from November to late December. Heard Island in the southern Indian Ocean has a large year-round population. The abundance of leopard seals there is greatest in midwinter (July) and lowest in summer (November and December), suggesting that the seals move southward with the pack ice. Leopard seals occur at the islands of South Georgia and Kerguelen almost year-round, but they are generally only seasonally present at Macquarie Island (June to December, peak in August; regular visitors in winter and spring) and the Falklands (spring and early summer). They also occur annually at the Auckland Islands and Campbell Island.

Outside their normal breeding and foraging range, leopard seals often wander into cold temperate latitudes, appearing unexpectedly on the coasts of Africa and South America (almost as far north as Buenos Aires on the east coast) as well as Tristan da Cunha. A few even wander into tropical waters; the most northern record is from the Cook Islands (20°45′ S). Although not abundant, they

An emaciated leopard seal on the outer reef at Tubuai, Austral Islands, French Polynesia. Strays such as this one appear occasionally on beaches far north of the leopard seal's usual range. (August 1981: Don Travers.)

are regular visitors, mainly in winter or spring, to the mainlands of New Zealand and southern Australia as well as their offshore islands (including Tasmania).

NATURAL HISTORY: Leopard seals are basically solitary; it is rare to encounter more than one on an ice floe. However, Nigel Bonner reports having seen 8 on one floe and 11 more on a bergy bit near Signy Island during early January 1985. Judging by their scats, the seals had been eating krill and penguins.

Although data are limited, it appears that leopard seals are less productive than many other phocids. The annual pupping rate (percentage of adult females giving birth) has been estimated as within the range of 47 to 61 percent. Pups are born mainly in the pack ice in November and December, slightly later than the births of crabeater seals, which are important prey of leopard seals in many regions. Lactation may last a month. Males are rarely seen near the pupping and nursing sites, and their whereabouts during this period are unknown. However, mating is believed to take place in the water from mid-November through December. Some researchers have suggested that there is no delay in implantation and that mating occurs in late December or January. Others have suggested that there is a delay, with mating taking place in November.

Males become sexually mature at 3 to 6 years of age; females at 2 to 7 years. Adults molt between January and June.

It has been said that leopard seals will attack and eat just about anything that moves. Krill, presumably strained from seawater in much the same manner as is done by crabeater seals, are consumed in large quantities, particularly during the non-summer months. Cephalopods are eaten year-round, but especially in January. Antarctic silver fish are important fish prey.

Leopard seals can be the bane of a penguin's existence. Here, one Adelie penguin (top) is about to be consumed, while another (bottom) tries to escape from a leopard seal that has broken through thin ice in the "no man's land," between the lip of the ice shelf and open water. (Cape Crozier, Ross Island, Antarctica, November 1976: Frank S. Todd.)

Some leopard seals also prey on penguins, especially Adelies, when the young are fledging and beginning to enter the sea in February and March. The Adelie rookeries at Cape Crozier, on Ross Island, are among the best sites to observe this. As they enter the water for the first time, the young chicks experience the terror of being greeted by hungry leopard seals (most of them males) that patrol just off the beach. Adult penguins also are chased and caught underwater or grabbed as they fall back into the water after attempting to leap onto the ice. Leopard seals sometimes lunge onto an ice floe after a penguin, or explode through soft ice near a rookery early in the season, attempting to take the penguins walking above. They grasp a bird and literally smash it to pieces on the water surface; the resultant smacks can be heard at a distance of a kilometer or more. It may take as little as 4 to 7 minutes, or as long as 15 minutes, for a leopard seal to devour an Adelie penguin. One seal was seen to catch and devour six penguins in 70 minutes. Another had 79 kg of penguin remains in its stomach.

Large numbers of birds usually gather around a penguin kill–giant petrels, Wilson's storm petrels, skuas, Cape petrels, and the like–hoping to snatch a morsel.

Seals (mainly crabeater seals but also southern elephant seals, antarctic fur seals, and probably Ross and Weddell seals in areas where they are sympatric), seabirds other than penguins, and fish round out the leopard seal's varied diet. Crabeater seals, mostly young pups, are important prey from November through January. After January, the older surviving pups are better able to escape, and the frequency of successful predation declines.

The stomach of one male leopard seal taken near Sydney, Australia, contained an adult platypus! Another leopard seal regurgitated a sea snake. Leopard seals have been known to snap at the legs of unwary researchers standing at the ice edge and to stalk others working on ice floes. Some of this behavior might be better interpreted as curiosity rather than aggressiveness. Nevertheless, their ability to lunge far out of the water should be borne in mind by anyone standing near an opening in the antarctic sea ice.

Leopard seals are themselves preyed upon to some extent by killer whales.

Leopard seals are fairly vocal. Many of their underwater calls are at relatively low frequencies. They make a long, steady drone, described by one researcher as haunting and sonorous, in the range of 300 to 3500 Hz. The lowest-frequency drones can be heard in air and even felt through the ice. Other sounds, apparently made in air, have been described as gargling and grunting noises; high-pitched, birdlike chirps; a musical sighing, crooning, or whistling; and a throaty alarm note made by vibrating the tongue as air is expelled from the mouth.

Quite a few leopard seals have been maintained in captivity, especially at Taronga Park Zoo in Melbourne, New South Wales, but they have not bred.

History of Exploitation: Leopard seals have never been exploited on a major scale. However, some have been killed for research, and in 1986–87, two Soviet commercial sealing vessels took 649 of them as part of a commercial venture.

Conservation Status: Most estimates of the total population of leopard seals in the Antarctic are well above 100,000; 200,000+ may be a reasonable approximation. Their extensive, partly pelagic distribution and solitary habits may make them, as a species, less vulnerable to the effects of intensive exploitation or regional habitat degradation than are some other pinnipeds.

Further Reading: Kooyman (1981a), Laws (1984), Lowry, L. F., et al. (1988), Siniff and Stone (1985), Thomas et al. (1983).

Weddell Seal

Leptonychotes weddellii
(Lesson, 1826)

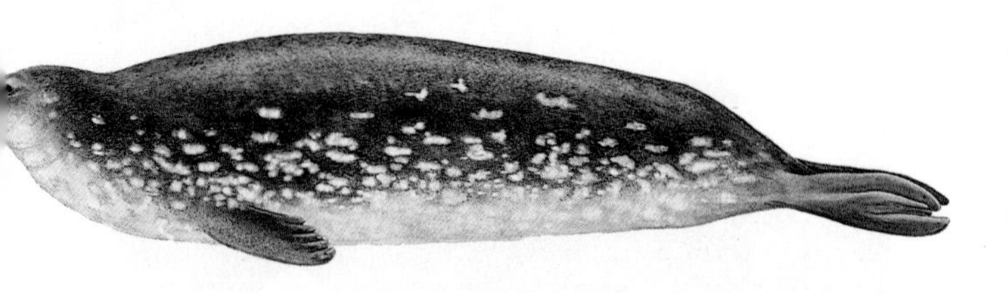

Nomenclature: The name *Leptonychotes* refers to the small size of the claws on this seal's hind flippers. The Greek *leptos* means "small" or "slender"; *onux,* "claw"; *otēs* is a suffix indicating possession. The name *weddellii* honors Captain James Weddell, whose book on an antarctic sealing voyage in the 1820s contains a description and an illustration of *L. weddellii.*

Description: Weddell seals are large (males reach 2.8 to 2.9 m, females, 3.0 to 3.3 m) and weigh as much as 400 to 600 kg. At birth, they are about 1.5 m long and weigh about 29 kg. Very young pups can be recognized by the presence of the umbilicus, which generally falls off at about 10 days of age. The pups gain an average of 10 to 15 kg per week, reaching about 113 kg at weaning. The body length at weaning is close to 140 cm.

Their rotund body dwarfs the small head, with its short muzzle. The large brown eyes and upturned ("smiling") mouthline give the face a catlike impression. The vibrissae are strongly curled, sometimes almost 360°. The head appears smallest early in the austral spring, just before pupping, when the seals are extremely fat. By the end of lactation, mothers have become much thinner, and their head appears proportionally somewhat larger.

The adult pelage is short (the hairs are about 1 cm long). Both the color and the pattern of the Weddell seal's pelage vary greatly. In general, the background pigmentation is dark bluish black dorsally grading to gray ventrally. There are many silver or white streaks and splashes, increasing in number and size from back to belly. The dark fur often fades to a rust brown color by the time another molt begins. Pups are born in a soft, woolly gray or light brown lanugo, with a dark stripe along the back. Their hair is much longer than that of adults. The lanugo is shed during the first month, after serving an important thermoregulatory function.

The teeth are adapted for keeping holes open in sea ice. The powerful canines and incisors in the upper jaw project forward at a greater angle than in other seals.

A young female Weddell seal on Paulet Island, off the northeastern Antarctic Peninsula. (22 February 1990: Hans Reinhard.)

DISTRIBUTION: Weddell seals have a circumpolar distribution in the Antarctic. The distribution is continuous, except for populations associated with islands such as Signy Island in the South Orkney group. They are closely associated with land-fast and pack ice. A relict colony in Larsen Harbour, South Georgia, produces some 20 to 30 pups, and another 15 to 20 pups are born elsewhere in the southeastern part of the island each year.

NATURAL HISTORY: Adult Weddell seals are not consistently gregarious, although a number of them may use the same pool of open water for breathing, and loosely associated groups of pregnant females often haul out on the same stretch of sea ice prior to pupping. For example, several hundred individuals form a kind of "rookery" at Hutton Cliffs on Ross Island each spring.

Adult seals do not appear to be thigmotactic. While hauled out, they usually remain a discrete distance apart. However, at times as many as 30 or 40 can be seen hauled out on snowfields of the Antarctic Peninsula, some lying close together. This may be due to a scarcity of suitable substrate rather than a need to conserve or share heat.

The habitat of Weddell seals is generally on and along the land-fast ice close to the Antarctic continent. They occasionally are seen hauled out on pack ice, but rarely associate with crabeater seals. Pupping takes place mainly in early spring, when the females haul out along cracks in the ice. Births at McMurdo Sound peak during the last week of October, whereas they peak in early September farther north at Signy Island. Although most pupping occurs on the ice, a few pups are born on land, particularly on the north-

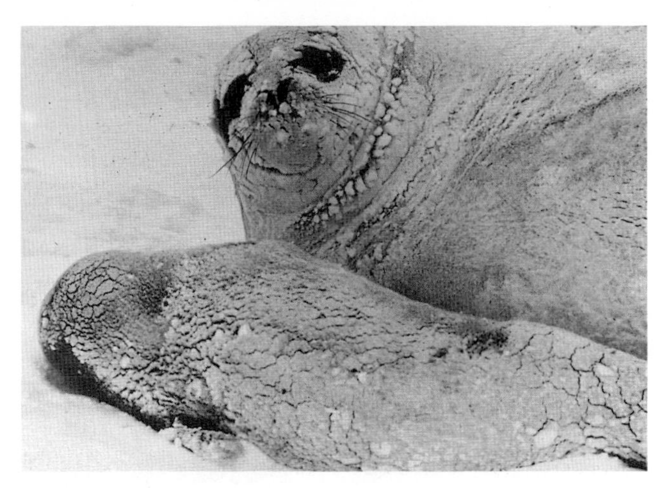

A female Weddell seal and her pup are coated with ice after an antarctic blizzard. (Hutton Cliffs, Antarctica, November 1973: Frank S. Todd.)

ern islands of Antarctica where less fast ice is available. Pups are nursed for 50 to 55 days, on average. The mother remains on the ice with the pup for approximately the first 12 days, then begins spending at least a third of her time in the water, making characteristic feeding dives. These are probably not particularly rewarding, however, and she presumably feeds little while lactating and loses considerable body mass as a result. The pup, on the other hand, grows rapidly, doubling its weight in about 11 days and becoming four to five times heavier at weaning than at birth. The milk, approximately 60 percent fat, is consumed at a rate of 7.5 to 8.75 liters per day, for an average daily weight gain by the pup of about 2.2 kg. The efficiency of weight transfer from mother to pup averages 48 percent; larger females have been found to be less efficient than lighter females in transferring weight to pups. Some pups may enter the water during their second week of life. The first molt begins at approximately 18 days of age and lasts some 17 days.

Pupping begins earlier at lower latitudes, with a documented gradient in the peak season from mid- to late October in the Bay of Whales at 78° S, to late August and early September near the South Orkneys (60°43′ S) and South Georgia (55° S). Lactation lasts for 6 to 7 weeks.

Males are territorial beneath the ice cracks, as they seek to control a 20 m diameter area. At McMurdo Sound, rutting males are most territorial (that is, most fighting occurs) in mid-November, just before they begin mounting females. Copulation takes place in the water under the ice. A male may mate with any receptive female that ventures into its territory.

Weddell seal female and pup at the base of Hutton Cliffs, Ross Island, McMurdo Sound, Antarctica, an area where much of the research on this well-studied species has been conducted. (November 1973: Frank S. Todd.)

By the end of the pupping and breeding season, adult males and females have lost considerable weight. They begin an intensive period of foraging as they disperse away from the breeding sites. Weddell seals molt during the summer. The winter is spent mostly in the water, feeding. During this time of year it is usually warmer in the water than on the ice.

Aspects of the life history of Weddell seals have been studied closely at McMurdo Sound, the Vestfold Hills, and Signy Island. At McMurdo Sound, 84 to 97 percent of the pups survive to weaning. The mean annual survival rate of adult females is 85 percent; of adult males, 76 percent. At Signy Island, adult females and males both have a mean annual survival rate of 80 percent. The average age at first reproduction is 6 to 8 years at the three sites. Approximately 80 percent of adult females give birth each year. The females at Signy Island show little fidelity to their birth sites. In contrast, there is little exchange of tagged seals on opposite sides of McMurdo Sound, only about 90 km apart, suggesting a high degree of fidelity to specific breeding colonies.

The diet of Weddell seals consists mainly of small- to moderate-sized nototheneiid fish, but they also eat various cephalopods and crustaceans. They are versatile feeders, taking both slow-moving benthic fishes and more active demersal and midwater species. The Antarctic silver fish, a herring-like pelagic schooling species, is an important prey in many areas at particular seasons. Adult silver fish measuring 14 to 19 cm are most frequently eaten in the southern Weddell Sea. Decapod crustaceans are significant prey of the seals near Vestfold Hills but not in other areas such as Mawson and McMurdo Sound. Cephalopods are also important prey at Mawson, and octopus are eaten at Vestfold Hills. Frank Todd, a longtime penguin watcher, told us of two well-documented records of Weddell seals taking chinstrap penguins in the water, beating them to pieces at the surface in the manner of a leopard seal, and eating them.

Weddell seals dive routinely to depths of 300 to 400 m and at least occasionally to 600 m. Most dives, even to these considerable depths, do not last longer than approximately 15 minutes, but a 73 minute dive has been documented. Their navigational capabilities seem extraordinary. They manage to return to a specific breathing hole after traveling at least 2 km under ice 2 m thick.

The vocal behavior of Weddell seals has attracted much scientific interest. Polygynous males defend their underwater territories, in part, by using vocal threats or displays, and they also use sounds to attract females. These seals produce a variety of underwater sounds, from low-frequency (1 to 6 kHz) buzzes, trills, or whistles, to higher frequency (25 to 70 kHz) chirps. The underwater vocalizations of the seals in McMurdo Sound differ from those of the seals along the Palmer Peninsula. The calls of Weddell seals can be heard easily from above the surface, even through ice several meters thick. Sometimes the vibrations created by calling seals can be felt through the ice as well. In a defensive mode, seals on the ice often snap their jaws together, causing a distinctive snapping sound.

Weddell seals experience very little predation by killer whales and leopard seals. Nevertheless, they are not particularly long-lived. The maximum documented age is 22 years. It has been suggested that the rigors of maintaining breathing holes in the winter fast ice are costly to the seals and that significant mortality is due to tooth abrasion and consequent dental pathologies.

History of Exploitation: Small numbers of Weddell seals have been killed for research, and in the past they were killed to feed sled dogs at research stations. The Norwegian sealing expedition of 1893–94 (see Crabeater Seal account) may have killed a large number of Weddell seals. A Soviet commercial sealing venture accounted for 107 Weddell seals in 1986–87.

Conservation Status: Weddell seals are widely distributed and abundant, but no good estimate of population size is available. There probably are at least a quarter million in the entire Antarctic, with the largest concentrations centered in the Weddell Sea.

Large harvests of seals to feed U.S. and New Zealand sled dogs during the mid-1950s apparently depleted the population in McMurdo Sound. During the first 2 years of harvesting by the U.S. station, nearly 25 percent of the 2000 seals that lived in McMurdo Sound were killed for dog food. Nearly all were adults. Kills of

75 to 150 per year continued from the late 1950s through 1982. Immigration of juveniles fueled population growth, but there was another unexplained decline during the 1970s, with the population reaching its lowest point in 1976 to 1978. It seems to have been essentially stable since then; the resident population was about 1500 seals in 1983. Some 500 pups were born in the Mount Erebus colony in 1984.

Their importance to science is considerable. Weddell seals show no fear of humans and occur in relatively accessible areas. When approached closely, they respond by rolling onto one side with the flipper extended, and they are generally unaggressive. They can be caught and transported safely and inexpensively. Also, in areas where the ice is thick enough, temporary research facilities can be situated near the breathing holes.

FURTHER READING: Croxall and Hiby (1983), Kooyman (1981b), Laws (1984), Plötz (1986), Siniff et al. (1977), Testa and Siniff (1987), Testa et al. (1985, 1989, 1990), Thomas and Stirling (1983).

Northern Elephant Seal

Mirounga angustirostris
Gill, 1866

NOMENCLATURE: *Mirounga* is derived from *miouroung,* the Australian Aboriginal name for the southern elephant seal. The specific name is derived from *angustus* (Latin), meaning "narrow," and *rostrum* (Latin), meaning "snout," referring to the fact that the snout of this species is narrower than that of the southern elephant seal.

The elephant seals were commonly referred to as sea elephants by early naturalists, whalers, and sealers. In Mexico, the northern elephant seal is called *elefante marino.*

DESCRIPTION: Adult males can grow to 4 m in length and 2000 kg in weight; adult females to 3 m and 600 kg. Newborns average 125 cm and 35 kg; newly weaned pups are about 145 cm long and weigh up to 200 kg.

Adult male northern elephant seals are easily recognized by the proboscis and the broad, calloused, scarred chest. Male growth accelerates at puberty, at 3 to 4 years of age. Thereafter, males can be distinguished from females of similar size by differences in the shape of the nose. Also, the skin on a male's neck and chest becomes thickened and creased, a process reinforced by scarring due to wounds sustained in battles with other males.

The hair is relatively sparse in both sexes, and its color varies with age and sex. Newborns retain a black pelage until they are weaned, either prematurely when permanently separated from their mothers or naturally at 3 to 4 weeks postpartum. Once nursing is terminated, pups begin shedding this black natal pelage, which is replaced by a shiny light gray or silver one. The color of this hair gradually fades while the pup remains ashore fasting for another 6 to 8 weeks. By the time the pup departs the rookery at 2 to 3 months of age, its pelage is brown dorsally and pale yellow ventrally. Adult females vary in color from nearly blonde to dark brown; males are generally darker brown. Both sexes are countershaded, being slightly lighter ventrally than dorsally.

The canine teeth are sexually dimorphic in size and shape. The upper canines of males may grow to 15 cm long, with as much as 6 cm exposed above the gums. They are functional in fighting other males during the breeding season. Northern elephant seals have 30 teeth at most. This is similar to southern elephant seals and hooded seals but different from all other pinnipeds. The postcanines are small and weakly developed; they are evidently not functional in feeding.

Distribution: Northern elephant seals are restricted to the northeast Pacific Ocean. Their historical distribution and abundance are poorly known. Indians hunted them on the southern California Channel Islands in prehistoric times, perhaps as early as 11,000 years ago; this exploitation may have limited their abundance on some islands. During a brief period in the early and mid-1800s, Russian, European, and U.S. sealers, whalers, and sea otter hunters eliminated the elephant seals on Año Nuevo, Southeast Farallon, and the Channel Islands and most of those on Mexican islands; indeed, the species was believed to be extinct by the 1880s. A few were found on Guadalupe Island, in Mexican waters off western Baja California, in 1892. Most of those were killed for scientific collections, so it is a wonder that the species survived into the twentieth century. In 1911, the Mexican government finally prohibited further killing. The population has increased exponentially for nearly a century, and colonies have been established since the 1950s in U.S. waters on Santa Barbara, San Nicolas, San Miguel, Santa Rosa, Año Nuevo, and Southeast Farallon islands and on the mainland at Año Nuevo Point and Point Reyes. Colonies in Mexican waters on Guadalupe, San Benito, and Cedros islands evidently have grown very little since the 1970s. A few elephant seals give birth on Natividad, San Martin, and the Coronado islands in western Baja California waters and on San Clemente Island in U.S. waters.

During the nonbreeding season, elephant seals range along the coasts of Oregon, Washington, and British Columbia; adult males reach as far north as the Gulf of Alaska and the eastern Aleutian Islands. Elephant seals are now seen regularly in the coastal waters of British Columbia and Washington, especially in Puget Sound, each year. Anomalously, two young elephant seals tagged at Año Nuevo Island were observed at Midway Island in the northwestern Hawaiian Islands in 1978, and one young male was discovered on a small island near Japan in 1989. Stewart and colleague R. L. DeLong of the U.S. National Marine Mammal Laboratory recently have used small computers glued to the seals' hair to docu-

The northern elephant seal has made a spectacular recovery from near extinction. (Tyler Bight, San Miguel Island, California, 29 April 1981: Brent S. Stewart.)

ment the seasonal movements of adult male and female northern elephant seals. Adult males and females remain apart when they are at sea, and both sexes make two foraging migrations each year, returning to San Miguel Island briefly to molt and to breed. Males migrate to the Gulf of Alaska and along the Aleutian Islands, whereas females remain farther south off Oregon and Washington. Some may travel as far west as 173° W. The round-trip migrations take females about 65 days and males approximately 125 days in spring. The second (post-molt) migration takes males about 125 days and females about 240 days.

Elephant seals were hunted traditionally by Indians along the Washington coast and in and near the Strait of Juan de Fuca in prehistoric times. Elephant seal remains have been found in Indian kitchen middens at Cape Alava, Washington, and along the Oregon coast. This evidence indicates that elephant seals made

By the time they are weaned, at some 3 weeks of age, northern elephant seal pups have ballooned to 100 kg or more. The pup in the foreground, a "superweaner," is exceptionally large because it has been nursed by more than one female. (San Miguel Island, California, February 1983: Brent S. Stewart.)

northward seasonal migrations prior to their extirpation at most southern breeding colonies in the 1800s.

Studies at several California rookeries have demonstrated seasonal changes in the sex and age composition of seals ashore. In general, there are three seasonal peaks in abundance: one in late January at the height of the breeding season; one in late April and early May when adult females and juveniles are ashore molting; and one in October when resting females, pups of that year, yearlings, and some juveniles haul out briefly. Relatively few seals are ashore in June, July, and August when subadult and adult males haul out to molt.

Weaned pups leave San Nicolas and San Miguel islands in late winter and spring; most move northward, arriving at Año Nuevo Island and then the Farallon Islands in autumn and winter. A few remain near their birth sites or move south during the first year of life. Yearlings and 2-year-olds generally return south in the spring, often to their islands of birth where they haul out to molt. The distribution of young elephant seals at sea is poorly known, but recent data on adult diving patterns suggest that they feed in deep waters seaward of the continental slope. Elephant seals dive repeatedly with few interruptions, so it is not surprising that they are seen rarely at sea.

NATURAL HISTORY: Elephant seals are polygynous but not territorial. They are gregarious during the breeding season to the extent that females congregate on sand and gravel beaches, which

A female northern elephant seal nuzzles and sniffs her newborn pup. Although now velvety black, at approximately 3 weeks the pup will shed its natal coat for a coarse silvery pelage. (Point Bennett, San Miguel Island, California, January 1986: Brent S. Stewart.)

are limited in supply, to give birth in winter. During the nonbreeding season, elephant seals are less aggressive toward one another and may form dense groups on some beaches.

A female gives birth to one pup approximately 6 days after hauling out between December and March. She then nurses the pup for about 27 days before weaning it abruptly. Twins have never been observed. Some rare premature births occur in October and November. The sex ratio at birth and at weaning seldom varies from 1:1. Most females nurse their own pups exclusively until weaning, although the number of orphaned and adopted pups may be influenced by the timing and severity of winter storms and, perhaps, by crowding on rookeries.

Mortality rates of pups prior to weaning vary considerably. About 3 to 4 percent die at San Nicolas Island, 5 to 10 percent at Guadalupe and San Miguel islands, 13 to 48 percent at Año Nuevo Island, and 7 to 76 percent at Southeast Farallon Island. Such mortality may be habitat-dependent, but there is no evidence that it is influenced by the density of the breeding seals, at least under the prevailing densities on most rookeries. Most neonate deaths result from starvation of pups separated from their mothers during the first week of life. Only rarely do pups die from being trampled by adult males.

Females can give birth as early as their second year of life, although they rarely do. Most give birth for the first time when 4 years old. Between 8 and 20 percent of females may not give birth in successive years. Survival of females to ages 1, 2, and 3 years has

Northern elephant seal bulls threaten each other vocally and visually to gain access to estrous females. (San Nicolas Island, California, January 1983: Brent S. Stewart.)

been estimated at 35 percent, 30 percent, and 20 percent, respectively. The oldest known female (one tagged at birth) was 18.

Males may be sexually mature when 6 to 7 years old and may breed when 7 or 8, although many do not until they are 9 or 10. Males may live for 15 to 16 years.

Aggression dominates the atmosphere during the breeding season from December through March. Males compete with one another for access to estrous females. They establish dominance hierarchies through the use of stereotyped visual and vocal threats and, occasionally, physical combat. Most threats are visual, and aggressive encounters rarely escalate to physical battles. Such battles, when they occur, are spectacular to witness. The primary vocal threat is a loud, pulsed, clapping sound, called a clap-threat, made as a male raises himself up on his fore flippers with his elongated nose dangling in his mouth.

Females threaten one another vocally to protect their pups. They also threaten, bite, and toss pups that stray too far from their own mothers.

Estrus is not synchronized among females; consequently, copulation may take place any time between late December and late February. However, most occurs in early through mid-February. Copulatory activity reaches a peak on about 14 February, Valentines Day. A few copulations occur in the water near shore, either with virgin females, arriving mature females that did not give birth that season, or departing females that have weaned their pups. These departing females have, with few exceptions, already been mated by one or more bulls.

Most females depart the rookeries during the first 2 weeks of February and remain at sea feeding for about 75 days before returning to land to molt. Those females that have not successfully weaned pups spend less time at sea than do successful mothers. Females gain about 75 kg during this period, partially replacing the weight lost (up to 45 percent of initial breeding-season body mass) while nursing their pups. At sea, they dive continually, typically to depths of 370 to 480 m for 13 to 22 minutes.

Adult males depart the rookeries in February, having lost up to half of their body mass during the breeding season, and few remain ashore by mid-March. They remain at sea, diving continually, for 110 to 145 days before returning to land to molt in late June and July. They weigh 1000 to 1200 kg in early March and gain approximately 25 percent of their body mass by the time they haul out in summer. They are capable of diving longer, up to 80 minutes, than females. The record for the deepest dive by an air-breathing vertebrate belongs to a bull elephant seal whose dive to 1581 m was documented by a small microprocessor-based time-depth recorder in 1989. Elephant seals rarely spend more than 4 minutes at the surface between dives of 20 to 35 minutes or more during their many months at sea.

The molt is catastrophic: the upper layer of skin is shed together with the hair in large patches, a trait shared only with the southern elephant seal and monk seals. Juveniles haul out on land in April for about 2 weeks to molt; they are followed by adult females in late April and May, adolescent males in June, and adult males in July and August.

The most comprehensive data on diet were obtained by G. Antonelis of the U.S. National Marine Mammal Laboratory and colleagues after they chemically immobilized newly arrived juveniles and adult females on San Miguel Island in spring and sampled their stomach contents using a technique called lavage. This involved pumping seawater into a seal's stomach and then pumping the mixed slurry out through a sieve to capture cephalopod mouth parts and fish otoliths; these were used to determine identities of prey species. Northern elephant seals feed mostly on mesopelagic cephalopods (at least 13 species of squid and 2 of octopus), fish, and, occasionally, small sharks.

Northern elephant seals have no significant predators, although great white sharks and killer whales occasionally kill and eat a few in some areas.

A small number of bacteria and viruses cause skin lesions, and goose and stalked barnacles occasionally attach themselves to the skin of northern elephant seals, particularly young ones. Cookie-cutter sharks sometimes remove small plugs of skin and blubber; parasitic copepods, perhaps transferred from cookie-cutter sharks, occasionally are imbedded in the tissue surrounding these crater wounds.

HISTORY OF EXPLOITATION: See "Distribution," above. The commercial exploitation of northern elephant seals began in about 1818 and ended some 50 years later, by which time the species was commercially extinct. In the early 1880s, scientists, believing that the northern elephant seal would soon be biologically extinct, collected seven of eight individuals thought to be the last survivors of their species.

CONSERVATION STATUS: Northern elephant seals are protected by Mexican law in the waters off Baja California; Guadalupe Island, the site of the largest rookery, is a wildlife sanctuary. Killing or harassment of elephant seals in U.S. waters is prohibited by the MMPA, although permits are issued to scientists for research and to commercial fishermen to kill small numbers incidentally during fishing operations. The population was estimated to number about 115,000 in 1991. More than 21,000 pups were born at California rookeries in 1991, about 85 percent of them on San Nicolas and San Miguel islands.

FURTHER READING: Antonelis et al. (1987), Bartholomew (1952), Cooper and Stewart (1983), DeLong and Stewart (1991), Huber (1987), Le Boeuf (1972), Le Boeuf et al. (1989), Shipley et al. (1981), Stewart, B. S., and Huber (in press).

Southern Elephant Seal

Mirounga leonina
Linnaeus, 1758

NOMENCLATURE: The specific name is derived from the Latin *leoninus,* for "lion," possibly referring to the male's roarlike threat vocalization. Alternatively, the name may have been misapplied to this species because of confusion in early descriptions with the South American sea lion, so called because of its lionlike mane.

DESCRIPTION: Male southern elephant seals are the largest seals in the world. They grow to as long as 5 m and can weigh as much as 5000 kg. Females are considerably smaller, weighing as much as 800 kg (average postpartum weight is about 538 kg) and growing to 3 m in length. At birth, pups of both sexes are approximately 130 cm long and weigh 40 to 50 kg. They gain about 3.6 kg each day; so when weaned 20 to 23 days after birth, they weigh 96 to 125 kg, depending on the size of their mothers. Lactating females lose about 8.6 kg each day.

The pup's short, wavy black natal pelage is shed soon after weaning and is replaced by shorter, silvery gray hair that is slightly darker ventrally. During the first year, this pelage fades to a dull yellow. Adults are yellowish brown to dark brown most of the year, slightly lighter ventrally than dorsally, but during each annual molt they grow a new gray pelage.

Males develop an elongated nose that first becomes noticeable as they enter puberty; it is similar to but not as long or as well developed as that of northern elephant seal bulls.

In the large canine teeth, dentin and cementum layers are well defined, permitting accurate determination of individual age. The incisors are very small; they and the even smaller, peglike postcanines appear to be essentially nonfunctional.

A weaned southern elephant seal pup, its coat still clean before its first trip to sea. (Elephant Island, South Shetland Islands, December 1988: Brent S. Stewart.)

The eyes are large, and the retina contains pigments similar to those of deep-sea fishes. These pigments correspond in their sensitivities with light emitted by the bioluminescent mesopelagic and epibenthic cephalopods and by some of the fishes on which southern elephant seals prey.

DISTRIBUTION: Southern elephant seals range throughout the southern ocean around the Antarctic continent, but generally north of the edge of the pack ice. They occur on most of the subantarctic islands both north and south of the Antarctic Convergence; a breeding colony at Valdés Peninsula on the southern Argentine coast has been increasing rapidly in recent years. There are breeding colonies between 40° S and 62° S, and some individuals wander as far north as the Equator and as far south as the Antarctic mainland at 78° S.

Elephant seals may have occurred historically in the Indian Ocean as far north as the Seychelles, and they evidently once were common on the Juan Fernández Islands off the coast of Chile in the South Pacific. Southern elephant seals apparently bred along the northwestern coast of Tasmania 2000 years ago and were hunted there by Aborigines. They bred on islands, especially King Island, in Bass Strait off southeastern Australia until they were exterminated by commercial sealers in the early 1800s.

Breeding colonies now exist on South Georgia and the Falkland, South Shetland, South Orkney, and South Sandwich islands and on Kerguelen, Gough, Marion, Crozet, Heard, Macquarie, Campbell, and the Antipodes islands. Along the western side of the Antarctic Peninsula, they haul out nearly year-round near

Southern elephant seals are expert contortionists, capable of arching to touch nose to tail. (Grytviken, South Georgia, February 1988: Frank S. Todd.)

Palmer Station on Anvers Island. Most of the seals using these sites, particularly during the late summer molt, are adult or subadult males, although a few pups have been born there each year since 1983. The major populations are at South Georgia, and at Heard, Kerguelen, and Macquarie islands.

In recent years, there have been some sightings along the coasts of central Chile, Australia, Tasmania, and the South Island, New Zealand, where a few pups have been born. In 1986, a small group of elephant seals was found to be spending winters at Hawker Island, just off the Antarctic mainland station at Davis, from which they had some access to open water. A few seals, mostly males, also have been recorded near Mauritius and on the South African and Angolan coasts, particularly during the summer molting season; in 1982, a southern elephant seal gave birth on the Natal coast. Several hundred seals, mostly immatures, haul out to molt on the Antarctic mainland at Vestfold Hills and on the Windmill Islands in summer, evidently having dispersed primarily from Kerguelen and Heard islands. On 10 January 1989, a 5-year-old female southern elephant seal was shot by a fisherman at Sawqarah (18°07′ N, 56°32′ E) on the coast of Oman, the first record of the species in the Northern Hemisphere.

The distribution at sea is unknown, but some seasonal movement among islands has been documented, suggesting that seals from different rookeries feed in common areas. In most areas, movements of breeding seals among colonies are exceptional.

NATURAL HISTORY: Breeding males arrive at the rookeries in August, and pregnant females arrive in September and October. Males

Two male southern elephant seals display aggressively on a beach at South Georgia. (October or November 1980: D.W. Doidge.)

do not maintain territories but do establish dominance hierarchies near groups of females and compete for access to estrous females. The males threaten each other visually and vocally with what has been described as a roar; this sound is much different from the pulsed clap-threat of northern elephant seal males. The dominance hierarchies are structured primarily by age, secondarily by size, and, to some extent, by previous experience. Only the largest 2 or 3 percent of the males are likely to breed in a given year. A dominant male may mate with 100 females in a season. High mortality and intense competition prevent most males from breeding at all during their lives, and only a few individuals manage to breed for two seasons or more.

Males are sexually mature at 3 to 4 years at South Georgia and 4 to 6 years at Macquarie Island, but few breed before they are 10 years old. At Macquarie Island, no males younger than 10 years old and few 10-year-olds breed; 30 percent of 11-year-olds breed, whereas 55 percent of 12-year-olds do. Only about 17 percent of males survive their first 8 years of life; 5 or 6 percent live until 11, and only 1.2 percent live until 13 years old. Nearly 90 percent of males die before reaching social maturity.

In general, females attain sexual maturity when 2 years of age at Kerguelen Island, 4 at Macquarie Island, and 3 at South Georgia.

Approximately 70 percent of females at Macquarie Island have given birth by the time they are 6, compared with about 75 percent of the 4-year-old females at South Georgia. Once mature, a female may give birth annually for about 12 years.

Mortality during the second year of life for seals born at Kerguelen Island has been estimated at 64 percent, although it actually may be lower if dispersal (emigration) is taken into account.

Some researchers have suggested that greater growth rates of pups, larger sizes of bulls and cows at similar ages, and earlier breeding at South Georgia compared with Macquarie Island are due to the greater availability of krill and, consequently, fish and squid near South Georgia.

Each female gives birth to a single pup some 6 to 8 days after coming ashore in September or October and then remains ashore nursing her pup for another 23 days, on average. Because of a 3-month delay before attachment of the embryo, fetal development lasts about 8 months. The sex ratio at birth is 1:1. Approximately 80 percent of the births occur from 6 to 26 October. The postpartum mass of females at South Georgia averages 538 kg. During lactation, females lose about 8.5 kg per day for 19 to 27 days; their pups gain 40 to 45 kg to reach 135 kg when weaned. Pup mortality before weaning is 5 to 10 percent. Females enter estrus approximately 19 days postpartum, mate within a few days, and then depart several days later; on departure, they abruptly wean their pups. The pups remain ashore for another 50 days, fasting, and lose about 0.95 kg each day. Adults fast while ashore during the breeding season, females usually for about 1 month and males for up to 3 months. Once pups have lost about 70 percent of their weaning mass, they go to sea to feed for about 105 days before returning to shore in autumn for 10 to 20 days. After the breeding season, adult females are at sea feeding for some 70 days and gain about 1.1 kg each day before returning ashore to molt.

Maximum longevity is about 23 years for females and 20 years for males.

Adult females have been recorded to dive to depths of 1255 m; most dives during the day are to 500 to 800 m, those at night to 300 to 400 m. Adult males dive to 400 to 500 m, perhaps feeding on epibenthic prey. Dives last, on average, 20 to 27 minutes, but a dive of one female lasted 120 minutes. The time spent at the surface between dives is much briefer, only 2 to 3 minutes. Preliminary studies in 1988–89 indicated that adult males and females from Macquarie Island feed in different areas, males near the Antarctic mainland and mostly in epibenthic habitats and females in the mesopelagic zone farther north near the Antarctic Convergence. More recent data from 1990 indicate that some females from Macquarie and from South Georgia travel to areas near the Antarctic Peninsula to feed. Both males and females mainly eat cephalopods and fishes, particularly nototheniids, which prey on krill.

Elephant seals transform naturally occurring bogs into wallows. Some animals die when they are unable to escape from the quicksand-like mud. With its coating of mud, urine, and feces, there are few things dead that smell as bad as a molting elephant seal alive. (Grytviken, South Georgia, January 1980: Frank S. Todd.)

Juveniles haul out to molt in late November and December, whereas adult females return to molt in January and February and males mainly in March and April. Seals fast while ashore molting; adult females lose about 4 kg each day. Molting seals trample vegetation and create depressions in the ground called wallows. Although the soil and peat layers are eroded considerably, the seals' urine and feces add substantially to the nutrients available to plants.

Leopard seals occasionally attack and kill pups, and killer whales may prey on pups and older seals, but neither is believed to affect the population significantly.

At the Kerguelen Islands, most elephant seals breed along about 80 km of the southern and eastern coastlines of the Courbet Peninsula. The population at Kerguelen was estimated to number about 100,000 in 1960 and 210,000 in 1970. Since 1970, however, numbers have declined at annual rates of about 1 percent (adult females) to 2 percent (adult males). Some seals may move seasonally between Kerguelen and Heard islands and the Antarctic mainland coast near the Vestfold Hills.

In the South Shetland Islands, elephant seals breed and haul out at Elephant and King George islands, but there are no data on the size of these colonies.

At Signy Island, in the South Orkneys, some 80 pups were born each year during the 1950s; in recent years, however, there have been only 5 to 10 births each year. Furthermore, between 1950 and 1971 there was a 50 percent decline in the number of molting seals hauled out at Signy in summer.

The breeding population at Marion Island declined between 1974 and 1982 at about 11 percent (bulls) to 8 percent (cows) annually. This trend continued through 1985, although the annual rate of decline has been somewhat less (9 percent for bulls and 7 percent for females and pups) in recent years. The colony produced some 700 pups in 1985, yielding an estimate of colony size of 2450, excluding pups. Seals, particularly immatures, frequently move between Marion and Prince Edward islands (separated by about 23 km). A few tagged seals were found to disperse from Marion Island to Crozet Island, some 1000 km away, where they hauled out to molt in summer and autumn.

Numbers at Crozet Island evidently have declined at rates similar to those at Marion Island, but there are no recent estimates.

Some 13,000 pups were born at Heard Island in 1985, supporting an estimate of total abundance of approximately 46,000; this is a 50 to 60 percent decline in numbers since the 1950s. Movements of tagged seals have been documented among Heard Island, Kerguelen Island, and Antarctic mainland sites at Davis and Casey.

The colony at South Georgia was estimated at 357,000, including about 102,000 pups, in 1985; abundance seems to have remained stable since 1951.

In 1985, the size of the Macquarie Island colony was estimated at 82,000 seals, including about 21,000 adult females, a 40 percent decline since the 1950s. In 1990, the estimate for this colony had declined to 76,000. Some elephant seals have been documented to move between Macquarie Island, where they bred, and Campbell Island, where they molted in August.

HISTORY OF EXPLOITATION: Sealers at South Georgia began harvesting southern elephant seals for oil after antarctic fur seals there were reduced in the early 1800s. But the colony of elephant seals was reduced so quickly that by 1900, sealing was no longer commercially profitable. The colony recovered somewhat in the late 1800s and early 1900s; sealing resumed in 1909, but under regulations established to manage the harvest. Sealing continued on a profitable, sustained-yield basis through 1964, when shore-based whaling at South Georgia ended and sealing alone could not justify maintaining an oil-producing business there. Some 2000 to 3000 adult males were killed each year from 1910 to 1927, and some 5000 to 6000 each year from 1927 to 1965, accounting for a total harvest of approximately 260,000 bull elephant seals.

CONSERVATION STATUS: For unknown reasons, there has been a long-term, annual decline of 5 to 11 percent of the elephant seals at most colonies in the southern oceans. Stocks in the Indian Ocean may have declined as much as 60 percent during the past 40 years. However, at South Georgia and Kerguelen Island, colony sizes have been relatively stable; one colony at Valdés Peninsula on the Argentine coast has been increasing. Some researchers believe that the declines are at least partly due to commercial exploitation of prey stocks, causing survival rates of elephant seals, particularly males, to decline. The proportionately higher mortality of males subsequently may have caused a decline in pregnancy rates among females, thus leading to further declines. An alternate hypothesis is that recent declines at some rookeries are due to the populations' returning to pre-sealing levels after having recovered to abnormally high levels–a phenomenon sometimes called "overshoot." In 1990, the total population of southern elephant seals was estimated at 607,000, compared to 768,000 in 1985.

FURTHER READING: Aarde (1980), Bester (1980), Carrick and Ingham (1960, 1962), Condy (1979, 1981), Gales et al. (1989), Heimark and Heimark (1986), Hindell (1991), Jones (1981), Laws (1960, 1984), McCann (1980a, 1981, 1982, 1983), Pascal (1985), Skinner and van Aarde (1983).

Mediterranean Monk Seal

Monachus monachus
(Hermann, 1779)

NOMENCLATURE: Although he recognized it as a phocid and placed this seal in the genus *Phoca,* J. Hermann proposed the specific name *monachus* (from the Greek *monakhos,* "a monk, solitary") because he thought the smooth, round head resembled a human head covered with a hood, and the shoulders and short flippers extended like two elbows beyond a scapular covering a frock. J. E. King (1983) has suggested, alternatively, that the name refers to "the cowl-like effect of the rolls of fat of the neck, seen particularly when the head is drawn back." The name *Monachus* was applied to the genus by John Fleming in 1822. Italians call this seal *foca monaca,* Spaniards call it *foca fraile,* and French speakers call it *phoque moine.*

DESCRIPTION: There is considerable variation in the color of adults, although the pattern is generally dark brown on the back shading to lighter on the belly. There are all-black individuals and all-silvery white individuals, often with irregular blotching. A prominent white belly patch occurs on some, and light patches elsewhere on the body are not uncommon. The thick, woolly birth pelage of pups is black or dark brown, interrupted by a yellowish white patch on the belly. The first molt occurs at 4 to 6 weeks of age. The new pelage is silvery gray dorsally and light ventrally.

Size at birth is 80 to 90 cm and 15 to 26 kg. Adults can be close to 3 m long and weigh up to 400 kg, but most are 2.3 to 2.8 m and 240 to 300 kg. There is no documented major difference in size between the sexes.

DISTRIBUTION: The overall historic range extended from about 20°N along the Atlantic coast and offshore islands of northwestern Africa, into and throughout the Mediterranean Sea, and in at least the southwestern, southern, and southeastern Black Sea. Today, seals are present in only a small part of this region.

In the Atlantic, small groups regularly inhabit two parts of the Cape Blanc Peninsula in Mauritania. One area is at the tip of the cape, the other along Las Cuevecillas coast (Côte des Phoques). Four caves on the cliffbound coast near Las Cuevecillas were being used by about 70 monk seals in 1988. Most of the dozen or so seals at the tip of the cape were large, old males. It is likely that additional groups of monk seals are present to the north of Cape Blanc, between Cape Barbas and Guerguerat. However, because of the recent warfare in this part of the former Spanish Sahara, it has been difficult for scientists to visit these sites and confirm that seals are there.

Cape Blanc is the southern limit of their effective distribution, although individual seals are known to wander as far south as Dakar, Senegal, and possibly even to The Gambia. The Desertas Islands of Portuguese Madeira had a large population of monk seals in the nineteenth century, but their numbers have declined to about a dozen in recent years. Monk seals formerly inhabited the Canary Islands but are now absent there, apart from occasional solitary visitors.

Monk seals are thinly distributed throughout much of the eastern Mediterranean archipelago. The Aegean and northern Ionian seas have relatively high densities of seals, and efforts have been made to identify caves used by monk seals in both Greek and Turkish waters with a view to protecting them from disturbance. Seals occur throughout the Greek archipelago, mainly as isolated individuals or in groups of two to five. A few small groups survive along the Turkish and Bulgarian coasts in the southwestern Black Sea and Sea of Marmara.

Although monk seals were once present in much of the Adriatic Sea and especially along the Dalmatian coast, there are now only a few individuals living around islands off the coast of Yugoslavia. A few also survive along the Mediterranean coasts of Algeria, particularly on the Oran coast, and Morocco; on La Galite archipelago off northern Tunisia; and along the Cyrenaican coast of Libya.

The monk seal has virtually disappeared since World War II from the French, Italian, and Spanish Mediterranean mainlands and from most of the islands owned by these countries, as well as from Cyprus, Malta, Egypt, and Israel. Sightings in these areas today are sporadic and invariably involve very few individuals (in fact, usually just one or two a year). Several seals survived into the 1980s at Sardinia. A few seals may still be present in remote parts of Albania and Lebanon.

Natural History: Most observations of Mediterranean monk seals have been within 5 to 6 km of shore, and the species is con-

The area around Las Desertas, Madeira, is a marine mammal sanctuary watched over by three wardens. Access by land or sea is restricted for about half the sanctuary. In 1990, 12 monk seals, including 3 pups, were observed there. (Madeira: da Costa/Monachus.)

sidered predominantly coastal. However, the dispersal of seals to offshore islands demonstrates a willingness and ability to traverse deep stretches of ocean. Monk seals have been observed as far as 37 km from shore. Home ranges of individual seals in the Aegean Sea have been estimated to extend along 20 to 40 km of coastline. Movements along 600 km of shoreline have been documented. One large male whose movements were monitored in the Ionian Sea alternately used five different caves, as much as 20 km apart, for hauling out.

A single reliable observation of copulation has been reported. At the Desertas Islands on 1 August, a male approached five seals and mated with one of them underwater. In spite of this observation, the principal mating season is considered to be October to November. The pupping season extends over much of the year, from late spring (May) through mid-winter (December and January), with a peak in September and October. The gestation period is estimated to be 10 to 11 months. Females become sexually mature at a length of about 2.1 m. The estimated age at sexual maturity is 5 to 6 years for both sexes, although some females may become mature by 4 years of age.

Little interaction between individuals occurs while they are hauled out. Mediterranean monk seals, like other phocids, probably socialize more frequently in the water than out. Judging by their observed behavior on land, these seals would be considered among the least social phocids. However, Turkish scientists who observed monk seals in a sea cave reported episodes of vocal and tactile interactions between a mother and her pup as well as amicable social associations between the mother and an old female, between the mother and a juvenile, and between the old female and the pup. Mothers are reluctant to leave their pups alone during the first month after birth, although young monk seals can swim and dive proficiently by 2 weeks of age. Lactation presumably lasts for 16 to 17 weeks.

Probably because monk seals have been either exterminated on or otherwise excluded from virtually all the open sandy beaches potentially available to them in the Mediterranean, most hauling out and pupping occurs today in caves or grottos, some of which have underwater entrances. Seals use the sandy or pebbly beaches inside caves for hauling out and caring for their young, and adults have been seen floating and resting in the water inside caves. Most of the caves used by monk seals, at least in the Ionian Sea, are associated with cretaceous limestone, a soft rock readily eroded by wave action. One implication of this is that storms and earthquakes can dramatically alter conditions within the caves, making them more or less suitable for the seals.

Only on the northwest coast of Africa do the seals still haul out regularly on open beaches. Monk seals hauled out there can be approached and measured while they sleep. In this area, there is said to be a relatively friendly relationship between the local African fishermen and the seals. The fishermen sometimes throw fish to the seals, and they swim among the fishermen's lobster nets without being harassed deliberately. Adult monk seals in the Mediterranean are wary of people and boats, but juveniles are known to play with divers and swimmers.

The diet of Mediterranean monk seals consists mainly of fish and octopus. They apparently do not forage in depths greater than about 70 m, although they are believed capable of diving to 100 m. Most dives do not last longer than 10 minutes. Two captive monk seals in Tunisia did not eat consistently until they were offered eels. Several observations of both captive and wild monk seals suggest that large fish are eviscerated at the surface before being swallowed. This is accomplished by the seals holding the fish with their teeth and shaking them vigorously.

Although monk seals have been brought into captivity in many of the countries bordering their range, survival of captives has generally been poor. An exception is a female that died in 1978 after 24 years in a Lisbon aquarium. The life span of free-ranging monk seals is believed to be 20 to 30 years.

History of Exploitation: This seal's exploitation began in ancient times. Hides were used to make clothing. Magical properties were ascribed to seal products. For example, it was believed that a tent covered with sealskin would be impervious to lightning, and that a sealskin dragged around a field and then hung above a door would prevent damage by hailstones.

In Greece, the skins were formerly used to make sandals, and even in recent years the monks from Athos were said to be using the skins to make belts. Some Greek fishermen believe seal oil promotes healing when applied to open wounds.

If Mediterranean monk seals ever were subjected to a large-scale commercial hunt, we are unaware of it.

CONSERVATION STATUS: There is serious concern that human predation and habitat modification have disrupted the gene flow within the monk seal population. In particular, the seals in the Black Sea may well be isolated from those in the Mediterranean, with no recent evidence of movement through the Bosporus or Dardanelles. The seals in the Atlantic are unlikely to mix with Mediterranean animals since the species disappeared several decades ago from the Atlantic coast of Morocco. The aggregate population in all areas of present occurrence is probably less than 1000 animals, possibly no more than 500.

The prospects for survival of the Mediterranean monk seal seem poor at present. The Mediterranean Sea and its coasts are intensively used for fishing, shipping, waste disposal, and recreation. Fishermen in many areas regard the seals as pests, whether because of perceived competition for fish resources or because the seals steal catches and damage gear. Until recently, it was common for seals to be shot on sight. For example, in the neighborhood of Assos, Greece, one of the monk seal's few remaining strongholds in the Mediterranean, at least 10 seals were killed during approximately 1978 to 1987. With stricter enforcement of protective laws and heightened awareness of the seals' plight, the incidence of such shooting may have decreased. However, some seals die after becoming caught accidentally in fishing nets and lines. Deep, fine-meshed gill nets are especially dangerous for monk seals, but they also die in purse seines, trammel nets, and trawls. The rapid proliferation of gill nets throughout the Mediterranean Sea is a major threat to monk seals as well as to cetaceans, turtles, diving birds, and other marine organisms. In addition to direct killing by humans, monk seals appear intolerant of some kinds of human disturbance. The ever-expanding development of tourist facilities in the Mediterranean threatens to degrade and usurp what little suitable habitat remains for the seals.

Within the Mediterranean, only Greece has a sizable population of monk seals, estimated in the late 1980s as 150 to 300 individuals. Smaller groups still survive in Turkey and Yugoslavia. However, in spite of legal protection and the establishment of parks and reserves in many countries where the monk seal once thrived (for example, the Northern Sporades marine park in Greece), the pervasive trend of decline continued during the 1970s and 1980s. The population along the Oran coast of western Algeria declined during this period to some 40 animals distributed in about a dozen

Mediterranean monk seal hauled out. (Banc d'Arguin, Mauritania: World Wide Fund for Nature/J. Trotignon.)

localities. It will take a tremendous commitment of willpower and resources to ensure that a viable, wild population of monk seals survives in the Mediterranean into the twenty-first century.

The situation on the Saharan Atlantic coast, a region plagued with political and military instability over much of the past two decades, is also precarious. The total population in the late 1980s was estimated at no more than 150 seals, at least half of these centered at Cape Blanc. Some seals were killed by soldiers, but it is possible that the unrest worked to the seals' advantage by discouraging coastal development, especially fishing. With the end of hostilities among the warring factions in Western Sahara, monk seal mortality from accidental net entanglement may increase.

The caves in which most groups of seals now give birth and nurse are not always optimal sites for pup rearing. Flooding from storm surges, accumulation of debris (including oil) funneled to the head of the cave, and disturbance by adventuresome divers, could all reduce a cave's suitability for monk seals. Caves harboring monk seals have been known to collapse (as the famous one at Las Cuevecillas did in 1978), making them abruptly unavailable as haul-out and pupping sites.

Finally, it seems clear that the remnant populations of monk seals are highly vulnerable to outbreaks of disease. If an epizootic like the one that killed more than 18,000 harbor seals in northern Europe were to occur in the Mediterranean, it could rapidly undo any gains made by conservation measures.

FURTHER READING: Francour et al. (1990), Harwood (1987), Marchessaux (1989), Reijnders et al. (1988), Ronald and Duguy (1979, 1984), Sergeant et al. (1978).

Caribbean Monk Seal

Monachus tropicalis
(Gray, 1850)

NOMENCLATURE: The name *tropicalis* merely refers to the tropical distribution of this species, a feature that fails to distinguish it from the other monk seals. It also has been called the West Indian monk seal.

DESCRIPTION: The Caribbean monk seal closely resembled the other two monk seals. The adult pelage was brown on the back, with a gray tinge, lighter on the sides, and pale yellow or yellowish white on the undersides and muzzle. The vibrissae (whiskers) were mostly white. Neonates had long, soft, glossy black hair, with dark facial bristles.

Males grew to body lengths of 2.1 to 2.4 m, and females may have been slightly smaller. These seals were probably about a meter long at birth.

DISTRIBUTION: The earliest known written report of Caribbean monk seals came from Christopher Columbus's second voyage to the New World in 1494. His crew killed eight "sea-wolves" on a small islet off the south coast of Haiti. Surely these were monk seals.

In addition to the waters bordering Hispaniola, the seal's domain extended throughout much of the northern and western Caribbean Sea and the Gulf of Mexico. From the Bahamas, its range continued west to Mexico's Yucatán Peninsula, thence south along the Central American coast, and east to the northern Lesser Antilles. By Columbus's time, the seal's principal haul-out sites may have been confined to the sandy beaches of remote and uninhabited offshore islets and atolls, the animals having already been excluded from the mainland and large islands by Native hunting.

NATURAL HISTORY: Little was learned about the Caribbean monk seal's behavior and biology before it became extinct. The most useful firsthand observations were made by members of a collecting expedition that encountered one of the last colonies off the Yucatán

Peninsula in 1886. The expedition's brief stay among the seals was chiefly occupied with killing 49 of them for scientific specimens. Thus, much of the resulting commentary concerned the seals' reactions to the hunters:

> Upon first approaching them they appeared to have no dread of the human presence, lazily looking at us, perhaps uneasily shifting their position, and then dozing off in restless sleep. Upon advancing to within three or four feet they would somewhat rouse themselves, bark in a hoarse, gurgling, death-rattle tone, and easily hitch themselves along a few paces (Ward, 1887).

Obviously, these seals never learned to defend themselves against predators on land.

The above-mentioned expedition took place in early December, which proved to be at least a part of the pupping season.

HISTORY OF EXPLOITATION: Early visitors to the Caribbean were quick to take advantage of the monk seal's approachability and unaggressive behavior. A large but poorly documented seal fishery was carried on in the Caribbean by Europeans during the seventeenth and eighteenth centuries. At the Alacrán Islands north of Yucatán, there were in 1675 "such Plenty of Fowls and Seals (especially of the latter), that the Spaniards do often come hither to make Oyl of their Fat; upon which account it has been visited by English-men from Jamaica" (Allen, 1887). Fishermen in the Bahamas were known to take as many as 100 seals in a night during the early eighteenth century, selling the oil for burning in lamps. It is safe to assume that casual killing by sailors, buccaneers, whalers, turtle hunters, and fishermen over the course of several centuries contributed to the Caribbean monk seal's decline.

CONSERVATION STATUS: In 1887, J. A. Allen of the American Museum of Natural History wrote: "the few [Caribbean monk seals] that survive to the present day are merely the scattered remnants of once populous colonies." After this time, the few reports of the seals that reached print invariably described how these remnants were themselves being destroyed. The most recent reliable report refers to a small colony known to have been present in 1952 on Serranilla Bank, about halfway between Jamaica and Honduras.

The Caribbean monk seal is almost certainly extinct. No less than five surveys of former monk seal habitats have been conducted by trained naturalists since 1950; no definite evidence of living seals has been brought to light since 1952. As several sightings of seals in the Gulf of Mexico have proven on close examination to be of California sea lions escaped from captivity, the occasional reports of seals in the region require close scrutiny. The fact that hooded seals occasionally stray as far south as Florida, and that harbor seals and, less often, harp seals could do so as well, needs to be borne in mind. The view expressed by Karl Kenyon (1977) after his unsuccessful search for Caribbean monk seals in 1973 continues to ring true:

> My conclusion . . . is that the Caribbean monk seal has been extinct since the early 1950's. The fact that I saw no monk seals was not as important as the fact of ubiquitous human presence. . . . Even if a few old Caribbean monk seals had survived to the 1970's, all available evidence leads me to believe that there is no hope that the species can recover. Man has now dominated its environment.

FURTHER READING: Allen (1887), Kenyon (1977, 1981a), Le Boeuf et al. (1986), Rice (1973), Ward (1887).

Hawaiian Monk Seal

Monachus schauinslandi
Matschie, 1905

NOMENCLATURE: The specific name honors H. H. Schauinsland, a German scientist who in 1899 obtained from a guano collector on Laysan Island the skull used in the original description of this monk seal.

DESCRIPTION: Maximum adult size is about 2.4 m and 270 kg. Females grow larger than males. Pups are about 1 m long and weigh 16 to 18 kg at birth. After 35 to 40 days of nursing, the pup has at least quadrupled its weight but gained little in body length.

Hawaiian monk seals are born with a soft, woolly black pelage that molts by the time of weaning (about 6 weeks of age). The new juvenile coat, like the adult coat that follows it, is silvery gray on the back and sides and cream on the throat, chest, and belly. Over time, with exposure to sunlight and seawater, the color becomes dull brownish on the back and yellowish on the belly. Adult males generally skip the silvery gray stage; their newly acquired pelage is brown to black immediately after the annual molt. An interesting feature of monk seals is that the superficial layer of the epidermis comes off with the hairs in large, ragged patches during each annual molt (except the first postnatal molt). Elephant seals are the only other pinnipeds that molt this way. The monk seal's hair, 2 to 9 mm long, is shorter than that of cold-water phocines.

Monk seals, like bearded seals but no other pinnipeds, have two functional pairs of abdominal mammae. Pups are toothless at birth, and it takes several weeks for them to acquire the full complement of permanent teeth.

DISTRIBUTION: The modern distribution of this seal is limited to the small, mostly uninhabited northwestern Hawaiian Islands, often called the Leeward Chain. The principal rookery sites are on Nihoa and Necker islands, French Frigate Shoals, Pearl and Hermes Reef, Kure Atoll, and Laysan, Lisianski, and Midway islands. Pupping and pup rearing take place mainly on French Frigate

An alert Hawaiian monk seal, showing the typical coloration of juveniles and adults. (Laysan Island, July 1987: Stan Minasian, Earthviews.)

Shoals, Pearl and Hermes Reef, Kure Atoll, and Laysan and Lisianski islands. Midway Island was a regular pupping site 40 years ago but is now essentially abandoned. Individual monk seals appear occasionally at the main Hawaiian islands and at Johnston Atoll, but there is no evidence that these were ever part of the breeding range. A birth was recorded on Kauai, the westernmost of the main islands, in 1988. One tagged pup traveled from Laysan Island to Johnston Atoll, 1013 km distant, within a period of less than 5 months. Monk seals have been seen at Brooks Bank, nearly 140 km from the nearest exposed land. In recent years, a few have been observed along the north side of Oahu, and one pup was born on that island's north shore in 1991.

Natural History: Hawaiian monk seals do not migrate, although individuals may disperse over long distances. The hair of some has green algae growing on it when the seals arrive at the rookery, suggesting that they have been at sea for several weeks or longer. Hawaiian monk seals appear to be essentially solitary. Approaches by large, dominant males (or humans on foot) elicit submissive responses from other seals. These responses involve lying perpendicular to the aggressor and presenting the belly and throat, with the exposed flipper raised (Weddell seals react similarly, minus the flipper raising). Open-mouth displays and some snorting, "bubbling" (see below), and bawling may then ensue.

Monk seals are among the least productive phocids. Most females mature at 5 years of age or older, and only 60 to 70 percent of the adult females give birth in a given year.

Most births (about 95 percent) take place from March through June, although the entire pupping season lasts from late December to mid-August. A stable substrate is essential to successful pup

A weaned Hawaiian monk seal in the surf. (Kure Atoll, Hawaiian Leeward Islands, August 1988: Monte Costa.)

rearing. An area high on the beach, near vegetation, and clear of the high-tide mark is preferred. Females manage to give birth and suckle on sandy beaches with or without shade and even on the rocky shores of Necker Island, but those forced to haul out on shifting sandspits have considerable difficulty and often lose their pups. Mothers remain with their pups for 5 to 6 weeks and apparently fast for this entire time, losing as much as 90 kg of weight. They are extremely sensitive to disturbance, and they threaten or, if necessary, attack intruders. Aggressive interactions among females often lead to the pups switching mothers. However, the resultant fostering does not appear to reduce the survival chances of the pups or the reproductive performance of the mothers. Pups can swim weakly soon after birth, are reasonably adept when 4 days old, but remain ashore with their mothers until weaned at about 6 weeks of age. Pups gain about 57 kg as sucklings, then fast for about 2 months after being weaned. By the age of 1 year, they weigh about 45 kg, indicating that the first year of life at sea is difficult.

Adult males remain on or near the rookery islands during the pupping season, generally going to sea in the evenings, feeding at night, and returning to haul out at mid-morning. Sometimes they remain at sea continuously for several days and nights. While ashore, they often position themselves between a solitary female and the edge of the water, with their rear flippers actually in the water. They are constantly on the lookout for a female in estrus. Since males outnumber females by about 3 to 1 at some breeding sites, groups of aggressive males will sometimes "mob" a female, inflicting serious, even mortal, wounds in their eagerness to mate her. Studies of blood serum testosterone levels in males suggest that the mating season begins in May and ends around the time of molt, in September and October. Mating is presumed to occur in the water.

Most adults molt sometime from May to November, especially in June. Mothers do not molt until their pup is weaned.

These seals feed mainly on benthic and reef-dwelling fishes and invertebrates, including flatfish, scorpenids, eels, octopuses, and spiny lobsters. They often forage in depths of 10 to 40 m along the slopes of coral reefs. Dives of more than 120 m have been recorded, and these seals can stay submerged for at least 20 minutes.

Hawaiian monk seals, like their congeners, live in a tropical climate where the avoidance of overheating is at least as important as keeping warm, especially when they haul out. Their blubber is of a thickness similar to that of phocids living in polar regions. Monk seals do a number of things to keep from overheating while hauled out on sunbaked sandy beaches. They lie on and wallow in damp sand at the water's edge, assume postures that expose the pale-colored belly more than the dark back, and curtail vigorous movements while hauled out. They also reduce their breathing and heart rates to minimize metabolic heat production.

Shark predation is an important mortality factor. Many seals bear wounds or scars from shark attacks. Tiger sharks are thought to be the main predators, but gray reef sharks and reef white-tip sharks are also abundant around the rookeries. Preferred pupping sites tend to have very shallow water along the beach and coral ridges or large rocks near the shoreline. These characteristics have been interpreted as providing critical protection from large sharks.

On land, monk seals make "bubbling" sounds that have been likened to water being poured rapidly out of a jug or to repeated belching. Such sounds seem to emanate from deep in the seal's throat, and they often are made in the context of threat displays or disturbance responses. Females also "bellow," with the mouth wide open, when trying to drive away another seal or a person approaching a pup.

Hawaiian monk seals can live for up to 30 years, but most of the older individuals found dead are 20 to 25 years of age.

A few Hawaiian monk seals have been kept at the San Diego Zoo, Waikiki Aquarium, Sea World, and Sea Life Park. Several have been in captivity for 4 to 6 years.

HISTORY OF EXPLOITATION: Although the Leeward Chain was visited by prehistoric Polynesians, the islands apparently remained uninhabited by humans until their discovery by Europeans in the nineteenth century. Like the other two monk seals, the Hawaiian species was easy to find and kill. It was "genetically tame" and thus vulnerable to rapid depletion. In 1824 the sealing brig *Aiona* took what was thought to have been the last monk seal in the Pacific. However, some seals obviously survived, probably mainly on beaches difficult to approach from the sea. Whalers, guano diggers, and bird hunters undoubtedly killed monk seals but left little record of it.

Conservation Status: It is generally believed that there were never more than a few thousand Hawaiian monk seals during historic times. The first counts of the population were made in 1958, when 1206 seals were seen by Dale Rice. The aggregate population may have declined by 50 percent between 1958 and 1978. In the latter year, a beach count gave a total of only 502 seals. Major demographic changes, in addition to the decrease in absolute abundance, occurred during this time. Populations on the westernmost atolls (Kure, Midway, Pearl and Hermes Reef, Laysan and Lisianski islands) declined while the population on French Frigate Shoals increased greatly.

The most likely causes of the general decline are disturbance by humans, shark predation, and disease, but details of cause and effect are unknown. Although Pearl and Hermes Reef and Lisianski Island have had no permanent human inhabitants, military activities there may have had a detrimental effect on the seals. Midway Island hosted more than 70 seals during the 1950s, but by the 1970s only a handful remained. People, often accompanied by dogs, beachcombing for Japanese glass net-floats frequently disturbed mothers with pups, forcing them to abandon the preferred basking areas. At Tern Island, French Frigate Shoals, the number of monk seals using the beaches increased tenfold within a year after the Coast Guard station there was closed. A mass die-off of seals occurred at Laysan Island in 1978, apparently caused by a toxin that originated in a bloom of dinoflagellates and became concentrated in the flesh of fish eaten by the seals. Similar die-offs due to what is essentially "red-tide poisoning" have been suggested for bottlenose dolphins in the western North Atlantic and West Indian manatees in Florida.

The downward trend in the monk seal population appears to have been reversed during the 1980s. Mean beach counts rose from 474 seals in 1983 to 586 in 1987, representing a total population of more than 1500. Much of the credit for the improved status of Hawaii's seal population belongs to the recovery program led by William Gilmartin at the Honolulu laboratory of the National Marine Fisheries Service's Southwest Fisheries Center. From 1981 through early 1991, 33 weaned female pups at Kure Atoll were placed in fenced enclosures for their first summer, then released. By June 1991, 24 of these seals were still at Kure, and 2 had moved to Midway Island. A number of them had given birth in the wild. This "headstart" program, together with stricter controls on Coast Guard activities at Kure, has improved overall female survival. Beginning in 1985, emaciated and prematurely weaned female pups from French Frigate Shoals have been captured, fattened in Honolulu for 7 to 11 months, and released at Kure Atoll as yearlings. As of early 1991, 20 seals had been handled in this way; 14 were alive at Kure; one had moved to Midway. Several had already pupped. Another aspect of the recovery effort at Laysan Island was

Efforts to stop the decline of the Hawaiian monk seals have focused on improving pup survival. Some weaned or abandoned pups are housed in pens on beaches of their natal islands, fed during their most vulnerable first months, and then released. (Laysan Island, Hawaii, October 1984: Stan Minasian, Earthviews.)

to remove adult males that had participated in collective attacks on females. The males were captured, transported to Johnston Atoll, and released.

In spite of the danger that it could pose to the monk seal's survival, development of commercial fisheries in the inner reef and reef slope waters of the northwestern Hawaiian islands has been avidly promoted. Fragments of lost or discarded fishing gear pose an insidious threat to monk seals. The propensity of newly weaned pups in particular to explore fishing debris can result in entanglement leading to debilitation or death.

All important breeding rookeries, except Kure Atoll, are within the Hawaiian Islands National Wildlife Refuge administered by the U.S. Fish and Wildlife Service. Kure Atoll is administered by the state of Hawaii as a Coast Guard Loran station. In May 1986, the National Marine Fisheries Service designated "critical habitat" for the Hawaiian monk seal, according to the terms of the U.S. Endangered Species Act. This designation encompasses all beach areas, lagoon waters, and ocean waters seaward to the 10 fathom contour around most of the northwestern Hawaiian Islands. The area was extended to the 20 fathom contour in 1988.

FURTHER READING: Gerrodette and Gilmartin (1990), Johnson et al. (1982), Kenyon (1972, 1981a), Kenyon and Rice (1959), Rice (1960), Westlake and Gilmartin (1990).

Part III

SIRENIANS

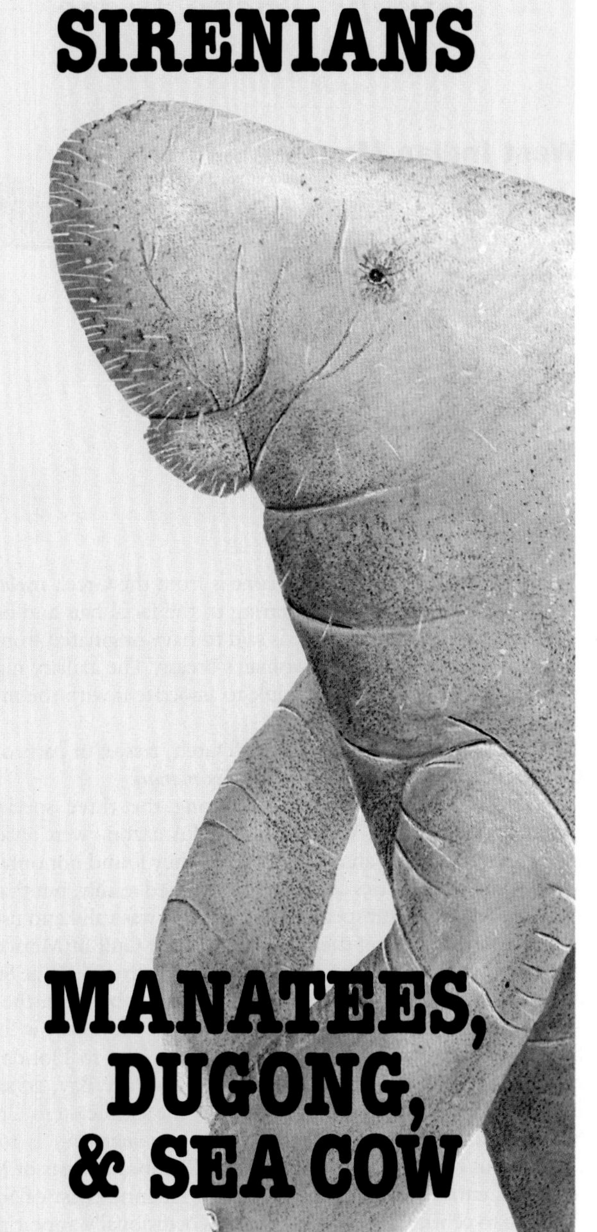

MANATEES, DUGONG, & SEA COW

Manatees

Family: Trichechidae

West Indian Manatee

Trichechus manatus
Linnaeus, 1758

NOMENCLATURE: The generic name is from the Greek *trichos* for "hair" and *ekhō* for "have," referring to the facial hair and bristly moustache. The specific name is said to have originated from the Carib *manati,* referring to a woman's breast. The axillary nipples of the manatee have caused people to associate it with the mythical mermaid.

The manatee is called sea cow (Dutch, *zeekoe*) in parts of the Caribbean region. Its Spanish name is *manatí.*

The systematic relationships among the three species of manatee–West Indian, West African, and Amazon–were reviewed recently by Domning and Hayek (1986). They found not only that skulls of the three species could be distinguished readily, but that the recognition of two subspecies of *T. manatus* was valid and useful. The cool winters along the U.S. coast of the Gulf of Mexico, in combination with the deep water and strong currents of the Straits of Florida, apparently create an effective barrier between the two subspecies. The Florida manatee (*T. manatus latirostris*) is found from Louisiana (and possibly eastern Texas) east to Florida and north seasonally to the Carolinas and Chesapeake Bay, generally inhabiting the coastal and inland waters of the southeastern United States. The Antillean manatee (*T. manatus manatus*) is found throughout the West Indies, along the Caribbean coasts of Mexico and Central America, and along the Atlantic coast of South America to central Brazil. The manatees occasionally appearing in

Texas waters are most likely from the Antillean rather than the Florida group.

DESCRIPTION: Adults average 3 to 3.5 m and weigh 500 kg, but large individuals can be close to 4 m long and weigh more than 1600 kg. Newborns may be 1 m or more long and weigh nearly 30 kg. Males and females are similar in size and appearance.

The head is small in proportion to the body, and there is no obvious neck crease. The eyes are small and deeply set; the ears have no external flaps. There is a pair of semicircular nostrils at the angle of the muzzle. When they are open during routine surfacing, the portion of face and head visible above the surface can resemble a bowling ball. The distinctive and prominent muzzle is covered with colorless bristles. Viewed head on, the pads of the upper lip hang down on both sides, covering the lower lip. These have been likened to the pendulous upper lips of a bloodhound. The lip pads are supple and can maneuver independently, grasping plants and directing them to the grinding teeth toward the rear of the mouth. The tongue and the inside of the lips are rough and coarse to the touch.

The two long pectoral flippers have several deep creases near the base. These flippers are wider distally and are squared off with rounded corners. The flexibility of the forelimbs allows a manatee to "walk" along the substrate, manipulate food, and caress companions. Nails are present on the dorsal surface near the end of the flippers. A single teat is present as a fold of skin behind the axilla of each flipper. Manatees have no dorsal fin or ridge on the back. The spatulate tail is somewhat reminiscent of a beaver's.

The finely wrinkled skin is sparsely covered with short, colorless hairs. The body is uniformly gray to gray brown and lightens somewhat with age. Some animals are blotched with areas that are lighter or darker than the dominant tone. The basic color is interrupted frequently by algal growths, barnacles, assorted encrustations, and scars.

There are five to seven low-crowned cheek teeth (reminiscent of those of pigs and people) in each jaw, situated toward the rear of the mouth. These are continuously replaced from the rear as the front teeth are shed.

DISTRIBUTION: The West Indian manatee occurs as far south as Bahia, Brazil, and north regularly to the Carolinas and, rarely, Chesapeake Bay. In addition to warm water, important factors affecting its current and historic distribution are access to fresh water and an adequate food supply.

The Florida manatee is mainly a summer migrant north of Florida, although a few individuals may remain year-round in Cumberland Sound, southeastern Georgia, where they depend on factory warm-water outfalls to survive the colder winter months. The northernmost record in North America is a sighting in the Potomac River in August 1980. Two years earlier, a manatee had been seen near Norfolk, Virginia. Manatees normally range no farther north along the west coast of Florida than the Suwannee River. Centers of abundance on Florida's east coast from March through October are the St. Johns River and the Indian and Banana rivers south to Biscayne Bay. Those on the west coast are the Suwannee, Crystal, and Homosassa rivers; the Charlotte Harbor, Matlacha Pass, and San Carlos Bay region; and the Everglades. During the colder months, manatees congregate near natural and artificial sources of warm water, including the artesian springs at the headwaters of the Crystal and Homosassa rivers and coastal power-generating plants.

The Antillean manatee has a much greater overall range. Most of the islands of the Greater Antilles with suitable vegetation and sources of fresh water have (or did have) populations of manatees. Manatees formerly occurred in parts of the Lesser Antilles, although probably in relatively low numbers. The manatees that reach the Bahamas may be vagrants from Florida.

Manatees probably were present historically along virtually the entire Gulf of Mexico and Caribbean coasts of Mexico. Now they are confined mainly to the southeast coast from Nautla, Veracruz, to the Belize border, especially the wetlands of Tabasco and Chiapas, the bays and coastal springs of Quintana Roo, and the rivers near Alvarado, Veracruz. Small numbers are present in the Soto la Marina and Palmas rivers of the state of Tamaulipas in northeastern Mexico.

In Belize, relatively large numbers inhabit Four Mile, Southern, and Placentia lagoons, the lower Belize and New rivers, and the vicinity of the cays off Belize City. In Honduras, manatees occur in the lakes and rivers of the Mosquitia region, the rivers east of Trujillo, and the rivers and lagoons west of La Ceiba. Although Nicaragua is considered to have some of the most extensive and ideal habitat for manatees in Central America, little is known about their distribution in that country. In Costa Rica, manatees are still thought to be present in the broad coastal plain of the northeast coast and possibly in a few areas of the southeast coast. Some also survive in the large, slow-moving rivers and protected lagoons of Panama. The rivers of the Bocas del Toro province and Gatun Lake and its associated waters bordering the Panama Canal have resident populations of manatees.

The Panama Canal has allowed faunal traffic between the Atlantic and Pacific oceans. In 1963, ten manatees, nine of them West Indian from elsewhere in Panama and one of them Amazonian from Peru, were introduced to the Chagres River to help control aquatic weeds. By the early 1980s, manatees were being sighted as far west in the adjacent Panama Canal as Miraflores Lake, only one lock away from the Pacific Ocean. In 1984, it was confirmed that

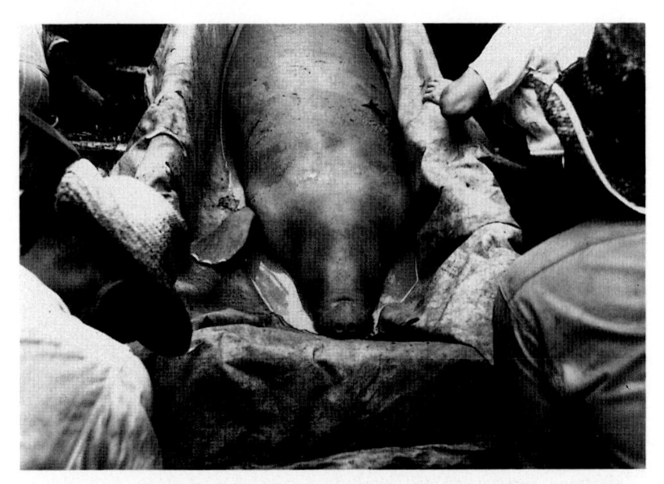

One of several Antillean manatees translocated from the northeast coast of Panama to the Panama canal complex. It was hoped that the animals would control the growth of troublesome aquatic vegetation. (September 1964: Frank S. Todd.)

manatees had entered the Pacific Ocean through the canal. Manatees or manatee-like animals have been absent from the eastern Pacific Ocean for several million years. It will be interesting to see if the dispersal continues and a population becomes established in the Pacific.

Although greatly depleted in Colombia, manatees are still present in the upper portions of the Magdalena River and in the Atrato River and its confluence with Candelaria Bay. Eastern Venezuela has a large population of manatees along the Gulf of Paria and throughout the Orinoco delta, the middle Orinoco, and the Orinoco tributaries. A smaller population inhabits the Lake Maracaibo region. The absence of manatees along more than 1500 km of the Caribbean coast of Venezuela represents a major hiatus in their overall distribution, separating the populations in Central America from those on the Atlantic coast of South America. Manatees are often sighted in the mouths of rivers and canals along much of the coast of Guyana. They are probably still present in remote areas of Surinam, including Nanni Creek and the Coesewijne, Tibiti, and Cottica rivers; some have been isolated in the drainage canals of plantations there. The Antillean manatee ranges south to the mouths of the Amazon River, where it is replaced in freshwater habitats upstream by the Amazon manatee. Historical sources suggest that the Antillean manatee once ranged as far south along the Brazilian coast as about 20° S. Thus, its full range may have extended to the limits of the 24° C mean annual isotherm in both the Northern and Southern hemispheres.

Courting manatees very near shore in the Banana River, near Cape Canaveral, Florida, 7 August 1969. (R. McDonald, courtesy of David and Melba Caldwell.)

NATURAL HISTORY: West Indian manatees are not particularly gregarious, although large groups form near sources of warm water during periods of cold weather. Mothers and calves probably maintain contact and recognize each other by calling, and possibly also by vision and smell. Young manatees generally stay in the same general area as their mother for at least several years after weaning.

Behavior interpreted to be mating has been observed in all seasons, and there is no clear peak season for calving. A female in estrus is often accompanied by several males for as long as 2 weeks, and she may mate with several of them. Reproductive maturity is usually reached at ages of more than 5 years, although some 4-year-old females are sexually mature. Intervals between births are at least 2 years and possibly as much as 5 in some cases. Twinning is rare. Gestation lasts 12 to 14 months, and calves are normally nursed for 1 to 2 years. Most young manatees are weaned when they reach a length of 2.1 to 2.6 m.

West Indian manatees eat mainly submerged vegetation, but they also eat floating and emergent plants. They sometimes partially beach themselves to reach desired food items, such as forbs and grasses growing on the banks of tidal rivers and creeks. In Florida freshwater habitats, hydrilla and water hyacinth are consumed, along with a great variety of other plants. Manatees regularly eat marsh grass in the salt-marsh areas of northeastern Florida and southeastern Georgia. Animals in the St. Johns River forage on live oak acorns during the winter. Thomas J. O'Shea (1986) observed that manatees in Blue Spring Run often concentrate their

A young manatee is suckled by its mother. (Crystal River, Florida, March 1984: Doug Perrine.)

foraging activity in the circular nests of male blue tilapia beneath overhanging oaks. These nests often serve as "collecting bowls" for acorns. O'Shea also noticed that flocks of common grackles feeding on acorns in the oak trees spill large numbers, causing periodic "pulses of acorn availability" underwater. In Venezuela, manatees reportedly feed on 18 species of plants, including mangroves, grasses, and sedges.

In marine waters, sea grasses and epiphytic algae are consumed. These manatees feed on sea grasses either by grazing only the blades or by rooting to consume the entire plant. Numerous invertebrates are ingested incidentally, and they may constitute an important supplement to the manatee's mainly vegetarian diet. In Jamaica, manatees are known to rob gill nets, stripping the flesh from ensnared fish and leaving the intact skeletons. Keepers of captive manatees have reported that they like mullet, and vitamins are administered in smelt. Horse chow is often provided to supplement their diet.

The longest submergence on record for a manatee in Florida lasted 24 minutes. Although diving time varies with activity, 4 minutes has been suggested as an average in Florida waters. Manatees may swim for short bursts at fairly high speeds, but their normal cruising speed is only 4 to 10 km per hour.

In areas where they are hunted, manatees behave secretively. For example, they are essentially nocturnal in Honduras where they are regularly hunted. In contrast, some individuals in Florida exhibit curiosity and approach people, including divers. John E. Reynolds (1981) had the unusual opportunity of studying a group of 50 wild manatees that were confined for about a decade to an artificial lake in southern Florida. Contrary to the usual impression that they are asocial and undemonstrative, Reynolds found the manatees in Blue Lagoon Lake to be mildly sociable and even playful. They rode currents downstream of flood-control structures ("body-surfing") and played "follow-the-leader" near barges.

Manatees apparently depend at least partially on sound to communicate. They seem to hear well. The whistle-squeaks and chirps recorded from Florida manatees have been mainly in the range of 2.5 to 5 kHz, but their entire vocal repertoire ranges between at least 600 Hz and 16 kHz. Call rates increase as social activities and interactions intensify; little calling takes place while the animals rest or feed.

In the southeastern United States, at the northern limit of their range, manatees move up rivers and canals in winter, seeking warmth provided by natural springs and industrial effluents. Although they have been observed in water as cold as 14° to 16°C, West Indian manatees appear not to be able to tolerate prolonged exposure to water colder than 20°C. Observations of distinctively marked individuals in Florida have shown that some manatees migrate north in summer and south in winter. Long-distance movements between areas more than 850 km apart have been documented. West Indian manatees are capable of living for prolonged periods in either fresh or salty water, although those living in a predominantly marine habitat frequently drink from sources of fresh water: springs, sewage outfalls, even hoses at piers.

Little is known about natural predation. Large sharks, crocodiles, alligators, and killer whales are potential predators. However, manatee parts have not been found and reported as stomach contents from any of these predators. At least in the southeastern United States, cold stress takes some toll in winter, and in southwestern Florida in 1982, at least 37 manatees died in less than 3 months, probably due to "red-tide poisoning."

The first manatee known to have been conceived and born in captivity was born at the Miami Seaquarium in 1975. Since then, facilities in Amsterdam, Nürnberg, and Beijing have bred this species. Captive manatees eat 55 to 75 kg of vegetation daily. One individual has lived in captivity in Florida for more than 30 years. Manatees are probably capable of living for 50 to 60 years.

HISTORY OF EXPLOITATION: Manatee meat is often likened to pork and is said to be delicious, so it is not surprising that humans have hunted and trapped them throughout their range. Many Amerindian archaeological sites have yielded manatee bones. Middens dating from A.D. 400 to 700 on Moko Cay near Belize City contained the largest number of manatee bones of any site searched thus far in the Caribbean. From as early as the sixteenth and seventeenth centuries, the Antillean manatee was hunted in Brazil by European settlers as well as Natives.

By the early sixteenth century, Spanish colonists in Central America had developed a crossbow technique for killing manatees, and a similar technique was used in Cuba. Manatees were exploited commercially in Surinam, Guyana, and Brazil from as early as the seventeenth century. The trade in manatee meat in the Guayanas ended by the late 1800s. In Brazil, some manatee meat was still being sold illegally as recently as the 1980s.

Residents of La Mosquitia in Honduras continue to hunt

A young West Indian manatee cradled in the arms of its handler. The bristly muzzle and chin, tough hairless hide, and docility are essential features of these tropical herbivores. (Early 1970s: Wometco Miami Seaquarium.)

manatees with harpoons thrown from dugout canoes. Large numbers were killed in Venezuela to supply meat markets during the mid-twentieth century, and some hunting for meat still occurs. Many Latin Americans attribute special qualities to manatee products. The meat, oil, skin, and ear bones are considered to have medicinal value.

CONSERVATION STATUS: West Indian manatees are legally protected in most countries where they occur: in Guyana since 1956, Honduras since 1959, Brazil since 1967, Venezuela since 1978, and Panama since 1967. However, poaching has continued in many countries, since there is little enforcement. In Brazil, commercial catches of manatees were even reported in government statistics through the 1980s, and manatee ribs, polished and carved, are still sold openly in the city market at Santo Domingo, Dominican Republic.

Although the status of manatees in Guyana has not been investigated since the 1960s or early 1970s, they are considered relatively abundant and well protected there. They are also still present in considerable numbers, although perhaps less well protected, in Surinam. In Venezuela, where the manatee population declined substantially during the middle of the twentieth century, the law protecting manatees has been enforced, and the waters of eastern Venezuela and the Orinoco River system are considered among the most promising areas for the future of the species. Some manatees are present in Colombia's Isla de Salamanca National Park. They are still widely distributed, abundant, and reasonably well protected in Belize. Guatemala recently established a manatee reserve (Biotopo Para la Conservacion del Manatí Chocon-Machacas), the first of its kind in Central or South America. However, manatees are relatively rare in Guatemalan waters, and illegal hunting and meat sales continue in parts of that country. In northeastern Costa Rica, manatees are present in Tortuguero National Park, and this may be an important protected area for them. Manatees generally are not hunted in Trinidad but are occasionally captured in fishing nets. Local residents claim that they are still common in the large, freshwater Nariva Swamp on the Atlantic coast.

In Mexico, the Emiliano Zapata region, which includes the Usumacinta River drainage, is said to contain a relatively large and well-protected manatee population. In northern Mexico, by contrast, the species is either very scarce or extirpated because of overexploitation. Manatees are still hunted illegally in Campeche and southern Quintana Roo. The numbers in northern Quintana Roo have been greatly depleted. Harpooning and gill netting of manatees continues throughout much of Honduras. In Jamaica, manatees taken accidentally in fishing nets are usually butchered and sold illegally.

The low densities of manatees around most of the Greater Antilles and the Bahamas are probably due in part to natural factors.

However, hunting and habitat destruction certainly have reduced the populations. Cuba is the only island where manatees are still thought to be reasonably common. Some effort has been made at enforcing the ban on hunting in Cuba, but poaching and accidental netting remain as threats.

A Florida law passed in 1893 prohibited the capture of manatees. Although poaching now rarely occurs in U.S. waters, mortality from vessel collisions, net entanglement, and crushing in flood-control structures is increasingly significant. In Florida, where many censuses have been attempted, there are at least 1400 to 1500 manatees, but the documented kill was 206 animals in 1990.

More than 200 manatees were counted in the Crystal and Homosassa rivers in January 1987, and more than 300 were counted at the Fort Myers power plant in January 1985. Off Florida's east coast, 128 manatees were counted in the Indian River in February 1986, and 292 were counted in the Banana River in April 1986. The best places in the world for viewing manatees are in Florida. At Blue Springs State Park on the St. Johns River of northeastern Florida and at the headwaters of the Crystal River on the gulf coast, the water is warm and clear. Winter is the best season for seeing manatees in both areas. They are also common, but less easily watched, in Merritt Island National Wildlife Refuge and Everglades National Park. Harassment by tourists and divers has become a serious problem in some areas, so attempts to view manatees at close range should be made with care and only in designated areas.

Among the many continuing threats to manatees over much of their range are illegal hunting, entanglement in fishing gear, collisions with powered vessels, crushing in flood- or salinity-control structures, and various environmental modifications. In Florida, changes to the environment caused by residential and recreational development have created acute problems for manatees. In 1990, Florida was the fourth most populous state in the United States. The number of registered boats in the state increased from 100,000 in the early 1960s to more than 675,000 in 1990. Some 300,000 additional boats enter Florida seasonally each year. And the number of boats using Florida's waterways is expected to double by the beginning of the twenty-first century. Reliance on warm-water springs and power-plant outfalls forces manatees to congregate in heavily used areas during cold snaps. The Everglades and the Big Bend area of northwestern Florida are the only parts of the state where major stretches of undisturbed manatee habitat remain.

In Puerto Rico, manatees congregate in coves along the southern shore of Roosevelt Roads Naval Station, feeding on the rich sea grass beds and drinking fresh water from the effluent of a nearby sewer plant. This military installation, where human activities are restricted, may provide a *de facto* sanctuary for a portion of Puerto Rico's small manatee population.

There has been considerable discussion about the potential

Power boats and fishing gear threaten manatees throughout their range. The light scar on this animal's left flipper was probably caused by a net or line. (Homosassa Springs, Florida, 5 February 1989: Howard Hall.)

use of manatees for controlling excessive plant growth in impoundments, and limited experimentation has taken place. It was reported in the mid-1970s that a few manatees in Guyana had, for decades, kept the ponds in the Georgetown Water Works clear of weeds. In Surinam, seven male manatees had kept a 10 km stretch of an irrigation canal clear of weeds.

FURTHER READING: Colmenero-R. and Zárate (1990), Domning (1981, 1982a, 1982b, 1984–91), Domning and Hayek (1986), Hartman (1979), Lefebvre et al. (1989), O'Shea (1986), O'Shea et al. (1988, 1991), Powell (1978), Powell and Rathbun (1984), Rathbun et al. (1982, 1983), Reynolds (1981), Reynolds and Wilcox (1986), Shane (1983), Sue et al. (1990).

Amazon Manatee

Trichechus inunguis
(Natterer, 1883)

NOMENCLATURE: The specific name comes from the Latin *in* for "without" and *unguis* for "nails," referring to the lack of nails on the flippers.

In the upper Amazon, the manatee is normally called *vaca marina,* meaning sea cow. In Brazil, it is known as *peixe-boi,* or ox-fish. The term *peixe-boi* is applied to both species of manatee in Brazil, although researchers identify the freshwater species as *peixe-boi da Amazônia.*

DESCRIPTION: The maximum documented length of an Amazon manatee is slightly less than 3 m; this is the smallest of the three extant species. It is nevertheless the largest mammal on the South American continent, with some individuals weighing as much as 400 to 450 kg. Birth length is 85 to 105 cm, and newborns weigh 10 to 15 kg.

The Amazon manatee's body is somewhat less cylindrical and more fusiform than those of the African and West Indian manatees. This is probably because it has fewer ribs and thus a shorter rib cage.

The skin has a strikingly different texture from that of its congeners: smooth, slick, and unwrinkled–rubberlike. Newborns have a heavily wrinkled outer layer of skin that peels off in the first few weeks after birth.

Most Amazon manatees have a conspicuous white or bright pink ventral patch. It is usually centered on the chest and abdomen but can extend forward onto the throat or backward onto the tail. Some individuals are uniformly gray or black, with no ventral markings.

As suggested by its scientific name, the Amazon manatee has no nails on its flippers. It differs in this respect from African and West Indian manatees, and this difference is diagnostic. Specimens from near the mouths of the Amazon River can be readily distinguished by the presence (*T. manatus*) or absence (*T. inunguis*) of nails, and less reliably by the Amazon manatee's white ventral markings.

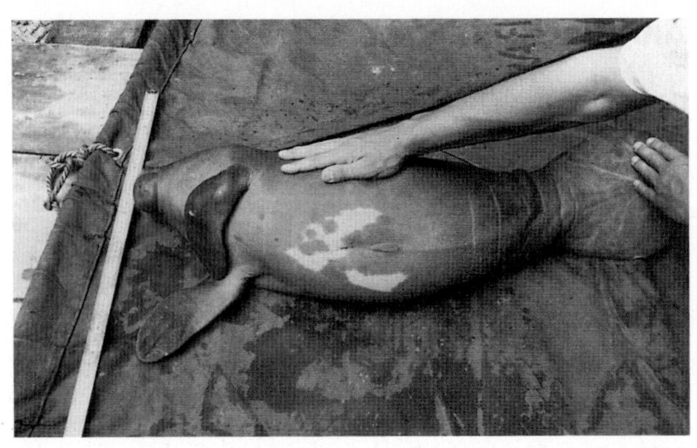

The splashes of white on this young Amazon manatee's belly are typical of the species. Also, the flippers lack nails entirely. (Manaus, Brazil, May 1977: Daryl P. Domning.)

The adult tooth row contains seven to eight fully erupted teeth, which are continuously replaced horizontally throughout the animal's life.

Distribution: This manatee's distribution in Brazil extends from Marajó Island throughout the Amazon drainage, including lakes, major tributaries, and associated floodplains. It is present in all three primary water types–white, black, and clear. Reports that it may reach the Essequibo and Rupununi rivers of Guyana and the Takatu River on the Brazil-Guyana border during flood periods are unconfirmed. Manatees are excluded from areas affected by fast currents, rapids, or waterfalls. The ranges of the Amazon and West Indian manatees overlap at the mouths of the Amazon, but such overlap has not been well documented. The large differences in karyotype between the two species would, in any case, make interbreeding unlikely.

Amazon manatees are also found in the upper Amazon basin in eastern Peru, including the drainages of at least eight rivers (the Napo, Tigre, Marañón, Samiria, Pacaya, Ucayali, Huallaga, and Purus) and in some of the blackwater lagoons of Napo Province in eastern Ecuador. In southeastern Colombia, they inhabit the Amazon, Putumayo, Caquetá, and lower Apaporis systems.

Many accounts indicate that Amazon manatees inhabit the Orinoco River basin as well as the Amazon. However, the documentation for this claim is weak, and it is likely that only the West Indian manatee occurs in the Orinoco system.

Natural History: Although large aggregations were reported in the past, and the animals certainly must have been densely distributed in some areas to support the large amount of exploitation, groups of more than a few individuals are rarely seen today. There is no firm evidence that this species moves into the marine environment; it is essentially confined to fresh water. Observation in the wild is hampered by the animals' remaining completely submerged in opaque waters, with only their nostrils showing when they breathe. It is feasible to radio track manatees, but unfortunately this technique has been applied only to a very limited extent in Amazonia.

Water levels in the central Amazon basin vary by 10 to 15 m during a typical year. Manatees have adapted behaviorally and physiologically to survive and flourish in this dynamic environment. When water levels are at their lowest in November and December, the manatees are restricted to the deeper channels and pools of major rivers and the deeper parts of large lakes. During the flood season, which peaks in June, they are widely dispersed. Amazon manatees share much of their extensive range with the Amazon River dolphin, or boto, as well as with the South American estuarine dolphin *Sotalia fluviatilis*.

The late Robin C. Best (1983) studied a group of 500 to 1000 manatees that were confined to a large Brazilian lake during the prolonged dry season of 1979–80. He noticed, as the water level dropped in late August and early September, that aquatic macrophytes became stranded on the beaches and that grasses and floating aquatic plants became unavailable to the manatees. By the end of September, Best could find no floating manatee feces in the lake, although these had been common and easy to find in August. By the following March, it appeared that, apart from a 2 week period in late December and early January, the manatees in the lake had been denied access to food for nearly 7 months (September to March). Nevertheless, most apparently survived. Best examined four large males killed by fishermen in February and found the manatees to be emaciated, nearing the end of their fat reserves. The manatees in the lake may have eaten dead vegetative matter on the bottom of the lake, but this would have contributed little to their energy supplies. Local fishermen claim that in years with prolonged dry seasons they often find dead manatees whose bowels are obstructed by quantities of clay. Best concluded that Amazon manatees are able to fast for several months, probably by greatly reducing their activity and living off their thick layer of fat. In normal years, the fasting period is only 3 to 4 months long.

The Amazon manatee usually gives birth between December and July, with a probable peak in February through March. The birthing season coincides with the season of rising water levels in the Amazon basin. This presumably ensures that the mother manatee has the best possible chance of keeping herself and her calf well nourished during the critical (and energetically costly) months of lactation.

Amazon manatees appear to prefer floodplain lakes with abundant floating vegetation. Like West Indian manatees, they seem to prefer plants that can be consumed as low in the water column as possible. They usually pull floating or emergent vegetation below the surface before consuming it. In Brazil, aquatic vascular plants, including grasses (family Gramineae, or Poaceae) and herbs (family Pontederiaceae), are the basic diet; floating palm fruits may also be eaten from time to time. Manatees in some areas graze on inundated terrestrial sedges. In Ecuador, they are said to eat water hyacinths and various aquatic grasses.

A captive individual proved capable of remaining submerged for 14 minutes, but it is likely that wild manatees, particularly during fasts when their metabolic rate is lower than normal, dive for considerably longer.

Natural predators are said to include jaguars and caimans, but there is little firm evidence of such predation.

Butterball, a male Amazon manatee, lived for more than 17 years in San Francisco's Steinhart Aquarium. He made sounds with a fundamental frequency between 6 and 8 kHz. The physiology of captive manatees has been studied extensively at the National Institute of Amazonian Research in Manaus, Brazil. In spring 1991, the institute had 11 captive manatees.

HISTORY OF EXPLOITATION: The manatees of Amazonia were exploited on a large scale from as early as the mid-1600s. As many as 20 shiploads of manatee meat are said to have been exported annually during the seventeenth century (some or much of this may have been from West Indian manatees). During 2 years in the 1780s, the meat and lard of some 1500 manatees were produced by the royal fishery near Santarém. Hides were used at this time to make whips. A large manatee can yield up to 100 kg of fat. The oil was valued as an illuminant, and it was used widely for cooking as well. Sometimes manatee oil was mixed with pitch to make a caulking for boats. Large amounts of mixira (manatee meat preserved in its own fat) as well as some dried and salted manatee meat continued to be exported through the early twentieth century.

In addition to the exported products, manatee meat was consumed in substantial quantities by Natives and settlers of Amazonia. Additionally, manatee hides were used to make stiff, resilient shields, and the Tikuna of the middle Amazon used manatee scapulae as spatulas for cooking.

The most destructive exploitation occurred in comparatively recent times. Between 1935 and 1954, tens of thousands of manatee hides were exported from Amazonia. An estimated 80,000 to 140,000 manatees were killed during this 20 year period, not including those killed for subsistence in remote regions. More than 19,000 manatees were taken in the state of Amazonas alone from 1938 to 1942. The hides were used for making high-pressure valves, machine belts, hoses, gaskets, glue, and cords for cotton looms. This hide industry collapsed abruptly in 1954, for reasons not entirely understood. Nevertheless, manatees continued to be killed in great numbers. Daryl P. Domning (1982), who reviewed the history of the Amazon manatee's exploitation, concluded that the heaviest mortality on record, taking into account both domestic and commercial consumption of meat, occurred in the late 1950s. He estimated that more than 6500 manatees were killed in 1959.

Even though manatee hunting was totally banned in Brazil by 1973, hunting for meat has continued. In addition to harpooning them from canoes, fishermen in the lower Amazon take manatees by at least two other methods: by harpooning them from specially constructed stationary platforms built offshore and by trapping them behind a wooden fence erected in tide-washed grazing areas. There are three seasonal peaks in manatee hunting in the Brazilian sector of the Amazon basin: one in July and August as the animals migrate from feeding areas into the deeper lakes, another in November and December when the animals are restricted to the deep portions of lakes and rivers, and a third in January through March when the manatees can be found feeding in the shallow, recently inundated grass meadows and along the shores of the lakes where luxuriant vegetation has again become accessible. Prolonged dry seasons can be especially disastrous. For example, in 1963, the driest year on record, hundreds of manatees were killed in the large lakes of the central Amazon region.

Local Natives were capturing manatees in the Napo River of Ecuador when an Italian explorer and naturalist visited their land during the 1840s. They continued doing so until the early 1970s. At least one commercial hunter specializes in manatee hunting on both sides of the Peru-Ecuador border; he sells the salted meat to local Ecuadorian and Peruvian military bases. The animals are harpooned from a dugout canoe. This hunter claimed in 1983 to have killed approximately 100 manatees in his career; he had taken at least 7 to 10 that year.

CONSERVATION STATUS: Although there are no good estimates of past or present population size in any area, in most parts of their range, Amazon manatees are thought to be seriously depleted. The combination of commercial and subsistence hunting has certainly had a severe impact on many local populations.

Manatees have been legally protected in Brazil since 1967. However, there has been little effective enforcement, and poaching has continued. They are not legally protected in Ecuador, although the Siona Indians (themselves depleted to only some 50 individuals) observe a self-imposed moratorium on hunting them in the upper Cuyabeno River. Hunting by military personnel, settlers, and commercial meat-hunters threatens the survival of the manatee in Ecuador and Peru.

Four of the 30 or more Amazon manatees that were transported 1000 km from Brazil's Japurá River to the Curuá-Una hydroelectric reservoir near Santarém in the early 1980s. It was hoped that this project would both provide the manatees with a safe home and prevent the reservoir from becoming choked with unwanted vegetation. (WWF/Robin Best.)

Brazil's ambitious plans for opening up the Amazon region to development and tapping its enormous hydroelectric potential do not bode well for manatees and other wildlife. As of 1988, the intention was to flood some 60,000 square kilometers, an area the size of Holland. The huge Balbina dam alone, completed in 1988 to supply power to Manaus, was expected to destroy some 2300 km^2 of rain forest once its reservoir completely filled. Plans for introducing manatees to this and other artificial lakes as a way of controlling the spread of water hyacinth may help offset the loss of pristine manatee habitat and the ill effects of other human activities. An attempt was made to establish a breeding population of manatees in a hydroelectric reservoir near Santarém beginning in 1980. More than 40 manatees were translocated, mainly from the Japurá River.

FURTHER READING: Best (1983, 1984), Domning (1981, 1982a, 1982b, 1984–91), Montgomery et al. (1981), Timm et al. (1986).

West African Manatee

Trichechus senegalensis
Link, 1795

NOMENCLATURE: The specific name comes from Senegal, West Africa, where this manatee reaches its northern limit of distribution; the Latin suffix *-ensis* denotes locality or country.

The French word for manatee is *lamantin*. In German it is called *seekuh* or *manati*.

There is a great variety of local tribal names for the manatee. In Nigeria, "mammy water" connotes a humanlike mythical river animal, regarded with dread.

DESCRIPTION: Adults are generally 2.5 to 3.4 m (maximum, 4 m) long and weigh 400 to 500 kg. At birth, they are about 1 m long.

In external appearance, the West African manatee is very similar to the West Indian manatee. The two species probably share a more recent common ancestor than either does with the Amazon manatee. Like the West Indian manatee, the West African manatee has nails on the dorsal surface of its flippers.

DISTRIBUTION: The West African manatee inhabits coastal waters, estuaries, rivers, swamplands, and coastal lagoons from the Senegal River (approximately 16° N) in the north to the Cuanza River of Angola (approximately 17° S) in the south. It occurs as far up the Niger River as 2000 km from the sea. Manatees are not known to occur in Lake Chad. Although present in the mouth of the Congo River, they apparently always have been blocked by cataracts from the upper Congo. They occur around islands in the Bijagos Archipelago, which extends from the coast of Guinea-Bissau to about 65 km offshore.

NATURAL HISTORY: In comparison with the West Indian manatee, this species has been little studied. Most of its habitat is turbid,

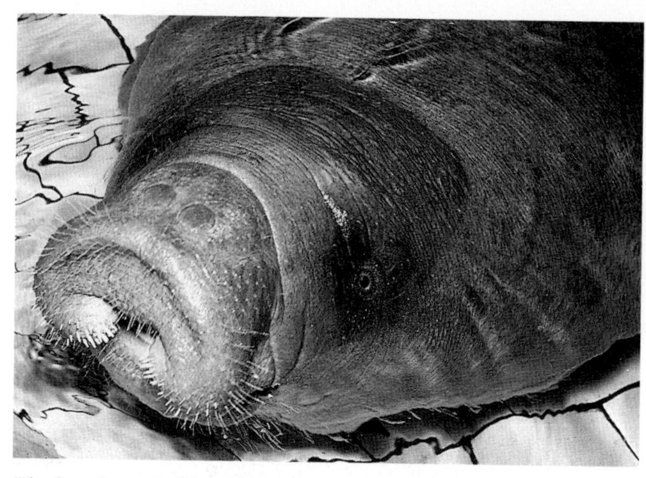

The face of a male West African manatee. (Antwerp Zoo, Belgium, 1957: Royal Zoological Society of Antwerp.)

and there are few if any areas where underwater observation is even remotely possible. James A. Powell recently made the first detailed study of the West African manatee's behavior and ecology. He discovered that the manatees in Ivory Coast feed mainly on emergent vegetation. They often come close to villages, where they eat bits of cassava discarded into the water by the villagers. Local manatee traps are often baited with fresh cassava peels. In Sierra Leone, manatees are widely believed to eat rice, although there is no direct evidence to support this claim. Fishermen in Sierra Leone complain that manatees take fish from gill nets (as do West Indian manatees in Jamaica). West African manatees certainly eat mangrove leaves that hang over the water's edge, and mangrove fruits are used to bait manatee traps in northern Sierra Leone.

Powell's radio-tracking work has shown West African manatees to be shy and secretive. They are most active at night and often rest on the bottom during much of the day. Groups of up to six individuals have been seen. The manatees in Powell's study area in Ivory Coast generally remain in the coastal lagoons, entering rivers only occasionally. In contrast, manatees in Nigeria's lower Benue River system must move seasonally into and out of lakes, oxbows, and pools of larger rivers, according to seasonal changes in water level.

A 1.38 m long, 43 kg calf caught in Ivory Coast had fibrous plant material in its feces, indicating that weaning had at least begun.

Nothing is known about natural predation, but sharks and crocodiles inhabit some of the same areas as West African manatees. Many of the waterways inhabited by manatees are, or were historically, also inhabited by two other large, herbivorous aquatic mammals: the common and pygmy hippopotamuses. However, since common hippos generally come out of the water and graze at night, they probably compete little, if at all, with manatees for food. Pygmy hippos are more terrestrial than aquatic and forage at night in forests.

West African manatees have been maintained in captivity in various places, sometimes for long periods. The manatee Hukunga lived at the Antwerp Zoo from 1954 to 1970.

History of Exploitation: Like other sirenians, West African manatees are highly prized for their meat. Many Africans also value them for their skin, bones, and oil. Belief in the restorative, medicinal, and aphrodisiacal properties of manatee products greatly strengthens the incentive to kill these animals.

Hunting by tribal peoples has often been marked by rituals and taboos. For example, among the Kalabari people of the Niger delta, the manatee was traditionally regarded as a sacred animal or even as a reincarnated human being. Anyone who killed a manatee was obliged to remain in his house for 3 days, rubbing himself with cam wood and yellow powder. Women of his family would sing at dawn and dusk outside the house, in hopes of appeasing the animal's wounded spirit. After the 3 day rite, a piece of manatee meat was given to each head of household to lay before the

Manatees are netted and trapped in many parts of West Africa, mainly for food but also to eliminate them from areas where they are considered nuisance animals. (Gbandakor, Malen River, Sierra Leone, April 1982: Harry Spaling.)

shrine of his forefathers. The men who catch manatees in northern Sierra Leone are reluctant to discuss their activities, not so much because such catching is illegal but because it involves secret taboos and superstitions.

Commercial exploitation on a significant scale is not known to have occurred in any part of this manatee's range.

In Sierra Leone, where manatees have long been trapped and hunted by villagers, they are considered pests because of supposed damage to rice crops and fishing nets. There and in Ivory Coast, special traps and nets are used to catch manatees. In Ivory Coast, harpoons and baited hooks are used as well. In Nigeria, villagers along much of the Niger and Benue systems have specially constructed manatee harpoons, and they also catch manatees with hooks, nets, and fish traps. On Formosa Island, Senegalese fishermen harpoon manatees from platforms built near freshwater seeps.

Conservation Status: Although legally protected for several decades in most countries where they regularly occur, West African manatees are still hunted, trapped, and netted for food over much of their range. Unfortunately, few well-guarded parks or reserves contain substantial numbers of manatees. A small lake in the Pandam region of the lower Benue River system in Nigeria was declared a state manatee reserve in the early 1970s, but it probably serves only as seasonal habitat for a small number of animals. Hydroelectric dams in Nigeria have fragmented the manatee population in the Niger River, a formerly well-populated river system. Similar problems have and will arise in many other parts of this manatee's range. Such developments can benefit manatee populations by making the animals harder to hunt in the large lakes formed by the dams and by expanding available habitat. On balance, though, the damming of major river systems is likely to be contributing to the general decline of the species. The development of waterways for ship navigation has helped fragment Ivory Coast's manatee population into five or six isolated remnants. Along the Senegal River, and probably elsewhere, the draining of mangrove swamps has eliminated some manatee habitat. A few animals in such circumstances have been captured and relocated to the Djoudj National Park, where they are protected.

Like other sirenians, West African manatees are susceptible to capture in nets and traps set for fish.

Further Reading: Domning (1982b, 1984–91), Hatt (1934), Husar (1978), Reeves et al. (1988), Roth and Waitkuwait (1986).

Dugong and Sea Cow

Family: Dugongidae

Dugong

Dugong dugon
Lacépède, 1799

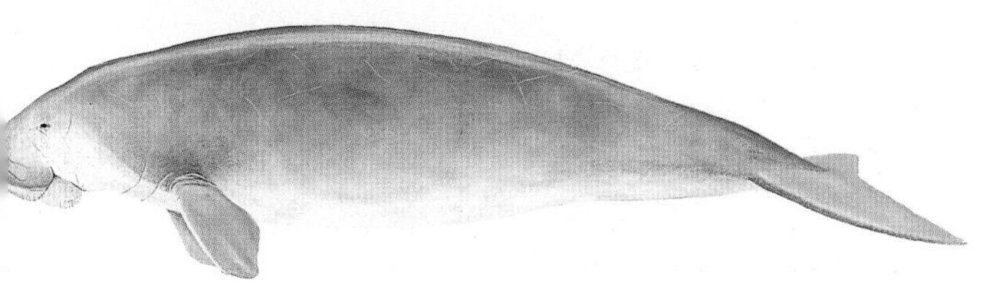

NOMENCLATURE: The term dugong comes from the Malayan and Javanese word *duyong.*

A few of the more interesting local names given to dugongs throughout their extensive range are: *arus-al-bahr,* Arabic for "bride of the sea," or mermaid; *kandal pandi,* Tamil (Sri Lanka) for "sea pig"; and *ri,* used in New Ireland, Papua New Guinea.

DESCRIPTION: Most adults are 2.4 to 3.0 m long and weigh 250 to 420 kg. The maximum reliably measured length was just over 3.3 m. Dugongs exhibit no obvious sexual dimorphism apart from the short tusks, which usually erupt only in adult males. Size at birth is slightly more than 1 m and 20 to 35 kg. There appears to be some tendency for females to grow slightly larger than males.

The head is squarish with a bulbous chin. Nostrils are situated on the crown, or summit, of the snout and are semicircular like those of manatees. The muzzle is directed ventrally and terminates in a large, horseshoe-shaped facial disk in front of the mouth. The disk, and especially the jowls on either side of the upper lip, is covered with bristles and hairs.

The spindle-shaped body appears smooth, but there are short, fair hairs spaced a few centimeters apart, and small barnacles may colonize the skin. There is no dorsal fin. The paddlelike flippers (about 15 percent as long as the body) have no nails. They are not generally used for propulsion but are flexible and useful for sculling,

Dugongs in a drained tank at Jaya Ancol Oceanarium in Djakarta, Indonesia. Although some dugongs have fared well in captivity, there has been no reproduction thus far. (May 1983: Tas'an.)

sweeping, and balancing on the sea floor. The flippers are normally pressed against the body during high-speed swimming. The body tapers to a narrow peduncle and ends in a large tail that provides the primary means of locomotion. The trailing edges of the whalelike flukes can be straight or concave, and they are sometimes notched.

Dugongs are pale at birth and darken to pale gray-brown early in life. With age, the skin continues to darken until adults are a deep slate gray to bronze, lighter below than above. Older animals frequently have blotches of unpigmented skin on the back. Linear scars are usually present, especially on the back.

Adult males and some old females have a pair of upper incisors that protrude through the upper lip enough to be easily visible. These tusks erupt at 12 to 15 years of age in males. Six pairs of grinding cheek teeth are present in each jaw quadrant in young animals; only two or three pairs are present in older individuals.

DISTRIBUTION: The historic range of the dugong included most coastal areas of the tropical and subtropical Indo-West Pacific with shallow sea grass meadows, from eastern Africa (north from Delagoa Bay, Mozambique) and Madagascar north and east, including the Red Sea and Arabian (Persian) Gulf, to the Pacific coast of Asia as far north as the margins of the South China and East China seas, thence south and east to New Guinea, Australia, the Solomon Islands, Vanuatu, and New Caledonia.

Many relict populations are now separated by vast distances, and the aggregate range of the species has shrunk. In Australia, dugongs are distributed around the coast, anticlockwise, from Moreton Bay in the east to Shark Bay in the west. A young female recently appeared on the south coast of New South Wales, well outside the normal range of the species. In the northern Indian Ocean, historically important dugong areas were the Gulf of Mannar and Palk Bay between India and Sri Lanka, the Gulf of Kutch off northwestern India, and the Andaman and Nicobar islands. Dugongs also inhabit(ed) much of Indonesia, northern Madagascar, and the Aldabra Archipelago; individuals once ranged even farther north in the Seychelles. Although their historic range included much of Melanesia and the Caroline Islands, it apparently did not include most of Micronesia or the Ellice and Fiji islands.

NATURAL HISTORY: Dugongs can form aggregations of 100 or more animals, for example, along portions of the Australian coast and in the Arabian Gulf. However, in most areas they are circumspect, difficult to approach, and seldom encountered in groups of more than a few individuals. The size of groups probably depends, at least to some extent, on the size of patches of suitable habitat. It also depends on how a given observer chooses to interpret spatial relationships among individuals. Little is known about social affiliations and interactions. It is unclear whether large herds are social groupings or simply adventitious aggregations in prime feeding areas. Dugongs and green turtles frequent the same sea grass beds.

Their tropical distribution allows dugongs to breed in all seasons, although there may be seasonal peaks. A male rut may occur during May through October in northern Australia, where the peak calving season is August through January. This calving peak coincides with the period of sea grass regeneration after the winter decline in biomass. Observations of mating behavior in Moreton and Shark bays have been interpreted as evidence that dugongs lek. There is also clear evidence that males fight for access to estrous females.

Dugongs probably seek shallow areas for calving. The gestation period is 13 to 14 months. Calves are probably nursed for more than a year (up to 18 months) and can remain with their mothers for at least 2 years, even though they begin eating sea grass soon after birth. While being suckled, the calf generally lies horizontally beside the mother. The mother often continues grazing or moving while nursing the calf. Individual nursing bouts can last up to 90 seconds. In aerial surveys of the Great Barrier Reef and Torres Strait areas, calves were estimated to comprise slightly more than 14 percent of the dugong populations.

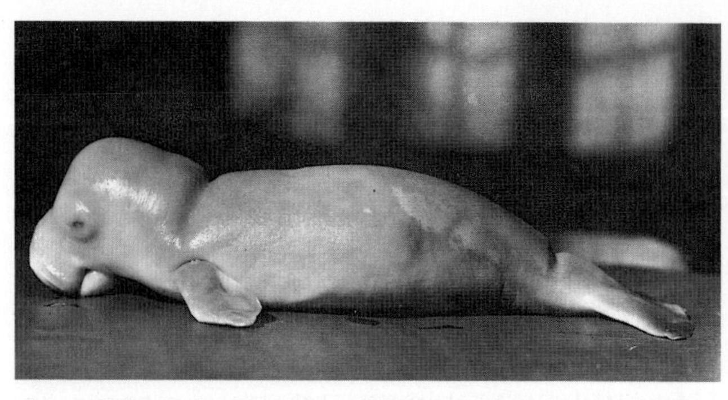

This small dugong fetus, preserved in formalin and housed in a Sri Lankan museum, already shows the essential features of its species. (Colombo, Sri Lanka, 10 August 1985: Stephen Leatherwood.)

Sexual maturity is not reached until at least 9 to 10 years of age in both sexes. Some females do not reproduce until they are 15 to 17 years old. Adult females give birth at estimated mean intervals of 3 to 7 years. Under the best of circumstances, dugong populations probably cannot increase at a rate higher than about 5 percent per year. Dugongs may live for up to 70 years.

Movements into and out of shallow coastal waters are apparently related to changes in tides, weather, and food availability. In areas where they are hunted, dugongs may enter shallow areas to feed only at night. At least short-range seasonal movements occur in response to shifts in food availability and changing weather patterns. For example, in Shark Bay, dugongs concentrate near the eastern shore in summer and near the western shore in winter, apparently responding to changes in water temperature by seeking warmer areas. The distance between the two centers of seasonal abundance is some 140 km. Evidence for longer migrations is equivocal. However, a dugong carrying a satellite transmitter journeyed between two bays 200 km apart three times in 9 weeks. During recent surveys in Torres Strait and the Great Barrier Reef lagoon off northern Australia, dugongs were found up to 58 km offshore and in inter-reef areas up to 37 m deep. A mother and calf were recently observed on Ashmore Reef, some 400 km off the northern coast of Western Australia and approximately 140 km from Timor. Dugongs almost never enter fresh water but may move several kilometers up tidal creeks and inlets.

Most dugongs are seen in water 2 to 6 m deep, although they also feed in offshore sea grass beds as deep as 20 m. In Australia, dugongs appear to prefer water 18°C or warmer, and they favor calm, sheltered inshore or reef areas rather than exposed coasts. Submergence time is usually 0.5 to 4.0 minutes during feeding, longer when traveling or disturbed. In Shark Bay, dugongs grazing the canopy of *Amphibolis antarctica* have shorter dive times than those in Shoalwater Bay grubbing the bottom for *Zostera*. The longest timed dive in captivity lasted 8 minutes, 26 seconds; in the wild, 6 minutes, 40 seconds. The dugong's dense, heavy bones have been said to function as a built-in "weight belt," helping offset the animal's buoyancy as it feeds on or near the bottom in saltwater. (Manatees have even heavier skeletons than dugongs.) Their average traveling speed is not more than 10 km per hour, although spurts of up to 22 km per hour may be possible. Dugongs normally show little of their body when surfacing, but at times the head is raised far enough above the surface to expose the eyes and ears. Locomotion is achieved primarily by the powerful tail. The snout is used to grub whole plants from the bottom, and this activity often leaves shallow trenches in the bottom sediments. The snout is also used to strip clusters of leaves from the branching stems of taller species, the roots of which are not eaten.

Sea grasses (families Potamogetonaceae and Hydrocharitaceae) are dietary staples. Dugongs exploit opportunities to graze during high tide on sea grass beds that are exposed at low tide. When these beds are inadequate, dugongs may supplement their diet with marine algae. They generally feed day and night.

Hunters attest to the dugong's acute hearing, but little has been quantified. Recorded sounds have generally been birdlike chirps in the frequency range of 1 to 8 kHz.

Although virtually defenseless, dugongs do not appear to suffer from heavy predation. Sharks, large marine crocodiles, and killer whales are confirmed predators. Hurricanes (cyclones) also pose a threat. For example, at least 27 animals were left stranded on the mudflats of the McArthur River delta, northern Australia, after cyclone Kathy had crossed the Gulf of Carpentaria in March 1984. Herculean efforts by scientists, rangers, and local residents led to the rescue and successful return to the sea of 23 of them.

More than 30 dugongs have been brought into captivity since 1959. A high proportion of these were sucklings when caught, and their survival has been poor. There is no record of breeding in captivity. Two dugongs were maintained at an Indian fisheries institute for 11 years. In 1955, a young male speared by fishermen in Palau was transported live (on a rubber air mattress, wrapped in wet blankets!) to San Francisco's Steinhart Aquarium. Not long after his arrival, however, Eugenie (as he was misnamed) died from the spear wounds inflicted during his capture.

HISTORY OF EXPLOITATION: The dugong's meat tastes like veal, beef, or pork; therefore, the long history of subsistence hunting throughout most of its range is not surprising. An average adult produces 100 to 150 kg of meat and 19 to 36 liters of oil. The Orang Laut (sea gypsies) hunted dugongs in the Strait of Malacca during the early twentieth century and sold the flesh to Malays and Chinese, who regarded it as a delicacy. A commercial fishery for oil began in Queensland during the middle of the nineteenth century. Dugong oil is believed by some to have medicinal and cosmetic properties. The hide has been used for making leather sandals in some countries. Bones and tusks have been used for carving such things as knife handles. Cigarette holders made from tusks and ribs were being sold in parts of Indonesia as recently as 1990. At the Aru Islands, this use apparently remains an incentive for killing dugongs. In Palau, the atlas vertebrae have been used to make wristlets that, at least in the past, could be worn only by chiefs of villages or municipalities. In the Philippines, dugong teeth are used as pendants on necklaces. Dugong "tears" may still be sold as aphrodisiacs in parts of Indonesia, as is powdered dugong bone in the Philippines.

Several methods have been used to catch dugongs. For example, fishermen near António Enes, Mozambique, use special large-mesh nets made traditionally of coconut fibers and more recently of nylon. The nets, buoyed by wooden floats and anchored with lumps of coral, are set over sea grass beds. Some dugongs are taken in tidal traps made of wooden fences set near the mangrove fringe in the Aru Islands. Aborigines of the Wellesley Islands in the southeastern Gulf of Carpentaria traditionally took dugongs by netting and spearing them from rafts. These capture methods have given way to the use of a *wap,* a traditional harpoon with a detachable spearhead, thrown from a motor-driven aluminum dinghy. In Torres Strait, dugongs were traditionally harpooned from portable hunting platforms built over sea grass beds and from canoes. The hunter was taken to a platform and left there overnight. He simply waited for a dugong to come within spearing distance. Use of these platforms ended after World War II. Double-outrigger canoes specially adapted by the Kiwai enabled them to spend more time on the reefs and to carry more dugongs (as many as four at a time) home from the hunting grounds. The introduction and proliferation of motorized craft has made hunting easier and increased the range of the hunters. Today, people from the Papuan side of Torres Strait harpoon dugongs from motor- or sail-powered canoes and also net them. Some of the men who own trawlers use these vessels as "mother-ships" from which smaller craft are deployed on the reefs. Australian islanders generally hunt from outboard-powered dinghies, often in association with crayfishing (lobstering) boats. The deliberate netting of dugongs became commonplace during the 1970s in Torres Strait. All dugong hunters in Torres Strait catch green turtles as well. In southern India, the demand for dugong meat is particularly strong at Kilakarai. Large-mesh gill nets are set in shallow waters for the express purpose of capturing dugongs. Also, fishing with explosives began in the Palk Bay region in 1981, and soon afterward the fishermen began using sticks of dynamite to kill dugongs.

The importance of dugongs to human communities throughout the Indo-Pacific has been affected by cultural and economic changes as well as by the animals' reduced availability. Dugongs were hunted traditionally in the Arabian Gulf. One archaeological site at Abu Dhabi is known for its 4000-year-old collection of dugong bones. However, little direct hunting of dugongs is known to occur in the Arabian Gulf and Red Sea today. Those taken in gill nets continue to be sold in the local market at Abu Dhabi. In much of Indonesia, dugong hunting continues. In the Aru Islands, the introduction of shark gill nets by Taiwanese and Japanese fishermen allowed the islanders to begin netting dugongs deliberately on a large scale. The annual catch during the late 1970s was estimated at 1000 dugongs; this seems improbably high. A team visiting the islands in 1990 found that dugong meat was still popular but that the annual catch had declined.

Three groups in Torres Strait traditionally depended on dugongs: the Kiwai of the Western Province of Papua New Guinea, the Torres Strait Islanders, and the Aborigines of eastern Australia. Beginning in 1957, the Kiwai supplied a commercial meat market in Daru, the capital of Western Province. With this new commercial outlet for meat, the Kiwai catch rose from approximately 25 to 75 dugongs per year. Although an estimated 500 to 1000 dugongs were taken annually in Torres Strait during the late 1970s, the catch declined substantially during the 1980s. There are at least two reasons: a ban on the sale of dugong meat in Daru in 1984 and a ban on crayfish boats carrying dugong meat. Perhaps even more significant was the boom in the crayfish industry in the late 1980s. The Torres Strait Islanders were too busy making money selling crayfish to bother with dugong hunting!

The subsistence hunting done along the east coast of Cape York appears to be within sustainable limits. In New Caledonia, dugongs can still be killed under permit to supply meat for festivals. From 1976 to 1984, 16 were killed (9 by harpoon and 7 by netting) for such occasions.

CONSERVATION STATUS: The following summary of the dugong's status is given by region, beginning on the east coast of Africa and working clockwise. Although there are clear grounds for regarding the species as vulnerable to extinction, the dugong's conservation status in many parts of its range is inadequately known.

In the early 1970s, substantial numbers of dugongs were present in Canal de Quilla, Mozambique. However, they were being exploited in a net fishery at the time. No recent information from this area has come to our attention. Although the dugong could still be described as plentiful in the Lamu district of Kenya during the early 1960s, all current evidence suggests that it is scarce there today. Surveys along the remainder of the Kenyan coast in the mid- to late 1980s resulted in only a handful of sightings, over the grass beds south and west of Wasini and Kisite islands. A herd of some 30 dugongs was seen in 1980 in the western Gulf of Aden near the Djibouti-Somalia border. Although there is some incidental mortality of dugongs in this region, direct hunting is not known to occur. Dugongs became legally protected in Djibouti in 1980.

Aerial surveys in the mid- to late 1980s dramatically changed earlier perceptions of dugong abundance in the Middle East. There may be some 4000 in the Red Sea, including 1800 in Saudi Red Sea waters and 200 in waters of the Yemen Arab Republic. Only 50 to 70 animals were believed to be present in the western Arabian Gulf in the early 1980s, but subsequent surveys resulted in an estimate of slightly more than 7300 dugongs. This discovery allayed fears that the large Nowruz oil spill in 1983 had reduced the gulf population to a dangerously low level (at least 37 dugongs died coincident with the spill). At least 14 dugongs are known to have died in Saudi and Bahrain waters between February and early May 1991, following the huge oil spill from the Gulf War in January 1991. Scientists are not certain of what caused these deaths. The deliberate killing of dugongs was banned in Bahrain in 1986. The deterioration of habitat throughout the Arabian Gulf, due to oil pollution, land reclamation, and military activities, threatens the dugong's existence there. Incidental mortality in fishing gear is also a serious threat to dugongs in both the Red Sea and Arabian Gulf.

The formerly large dugong population in the waters between India and Sri Lanka has declined greatly in recent years. Among the factors believed responsible are the intensive trawl fishery for prawns in Palk Bay, heavy mechanized boat traffic in the coastal waters of the Gulf of Mannar, continued poaching, and incidental killing in fish nets.

There are at least two areas of Indonesia–South Sulawesi and the Aru Islands–where dugongs were common through the 1970s. However, the Indonesian law against hunting these animals was not being enforced, direct and incidental mortality was increasing, and the local dugong populations may have been (and continue to be) overexploited there. Dugongs have been severely depleted along the northern coast of Borneo. Individuals still appear occasionally in parts of Malaysia, but the population in this region generally is considered much reduced.

In the Philippines, dugongs have probably been extirpated from the Central Visayas, but small numbers are still present around Palawan and in portions of the Sulu Sea. Some fishermen in the Philippines intentionally kill dugongs with dynamite, and most fishermen eagerly butcher any dugongs taken in fishing nets. The prospects for their survival in the Philippines appear poor.

In southern China, dugongs were abundant along the coasts of Guangxi and western Guangdong; at least 120 were caught there during 1958 to 1960. Catches have declined since, possibly because the population is depleted. Groups of up to 20 animals have been seen along the coast of Hepu County in recent years, however, and dugongs are reportedly present in substantial numbers in the harbors of Dui-da and DongXing Zhushan in Bei-bu Gulf. The species reportedly is badly depleted off Taiwan and the Ryukyu Archipelago.

A few tens of dugongs have been seen during recent surveys near Palau, but continued poaching is a serious threat to this small, probably isolated population. Palauan fishermen have used dynamite obtained from construction sites to kill dugongs.

In Australian waters, dugongs are legally protected except for subsistence hunting by Aborigines and Torres Strait Islanders living in native communities. There is some poaching, and substantial numbers continue to be taken in fishing nets (especially barramundi nets) and antishark nets set to protect bathers in parts of Australia. Dugong entanglement in commercial gill nets set in Hervey Bay, southeastern Queensland, has been publicized in recent years. One of the most promising areas in the world for the dugong's long-term survival is the 31,288 km² Great Barrier Reef Marine Park. Recent population estimates for park waters total more than 12,000 animals; 600 were seen in recent years congregated near the mouth of the Starcke River in the Cairns section. Aboriginal hunting in this area is limited to two communities that together take some 40 dugongs per year, and gill netting operations are limited.

There are also relatively large populations of dugongs in other parts of Australia. The waters of Princess Charlotte Bay, Magnetic Island and Cleveland Bay near Townsville, and Shoalwater Bay are particularly good dugong areas on the Queensland coast. Several thousand dugongs inhabit Moreton and Hervey bays north of Brisbane. Large, probably stable populations are present on the north coast in the Gulf of Carpentaria and on the northwest coast near Broome. On the west coast there are at least 10,000 dugongs in Shark Bay and another 2000 in the Exmouth Bay–Ningaloo Reef area north of Shark Bay.

Overexploitation has reduced the dugong population in Torres Strait, where some 12,500 dugongs were estimated to be present in November 1987. A management and education program developed by the Kiwai during the late 1970s proved ineffective in the face of pressure to expand the crayfish and barramundi fisheries, human population growth in Western Province of Papua New Guinea, and the continued local demand for dugong meat. As indicated above, this situation may have changed due to the current crayfishing boom, but the situation in this important dugong stronghold could again deteriorate.

Dugongs have been described as relatively abundant in New Caledonia lagoon. In Vanuatu, they are widespread, and small-scale hunting does not appear to be a serious threat.

Dugongs may be susceptible to capture stress, which means that even the animals that escape from nets or hunters die later from the stress of being temporarily captured or pursued.

It is encouraging that there are far more dugongs than previously supposed in northern Australia. A conservative management policy is already in effect in Queensland, where traditional hunting by the people of Hopevale and Lockhart River is considered to be within sustainable limits. Sanctuaries for dugongs should be established promptly in other areas where there are viable populations. Conservation status should not be based on abundance per se, but on population trends. These have not been established and will not be for at least a decade because of the problems involved in sampling a population with a low rate of increase. Meanwhile, management should be conservative. In many parts of their range, dugongs require complete protection if they are to survive. At the same time, the sea grass beds on which they depend must be kept intact and unpolluted.

FURTHER READING: Anderson, P. K. (1979, 1981, 1986), Anderson, P. K., and Birtles (1978), Domning (1984–91), Hudson (1986), Leatherwood and Reeves (1989), Marsh (1981, 1988), Marsh et al. (1978, 1984), Marsh and Saalfeld (1989, 1990), Nishiwaki and Marsh (1985), Preen (1989), Wang and Sun (1986).

Steller's Sea Cow

Hydrodamalis gigas
(Zimmerman, 1780)

NOMENCLATURE: The generic name is from the Greek *hydro* for "water" and *damalis* for "a young cow"; the specific name is from the Greek *gigas* for "giant."

DESCRIPTION: There are no stuffed specimens in museums, and even purported shreds of sea cow skin are of doubtful identity. What we have of this extinct species are some collections of bones, a set of the palatal masticating plates or pads, and a few written accounts describing the animal. G. W. Steller, a German naturalist, discovered the sea cow while shipwrecked with a Russian exploring party on Bering Island in 1741. Steller remained on the island for 10 months, and he made the only detailed written record of the habits and appearance of the species.

Steller's sea cows reached lengths of at least 7.4 m and possibly weights of 10 metric tons. The head was small in proportion to the rotund body, and the tail was bilobed like that of the dugong. They had very tough skin, like the bark of an old oak tree according to Steller. They were generally dark brown, sometimes spotted or streaked with white. Steller described the forelimbs as follows:

> There are no traces of fingers, nor are there any nails or hoofs; . . . the ends of the arms are something like claws, or rather like a horse's hoof; but a horse's hoof is sharper and more pointed, and so better suited to digging (Stejneger, 1936).

Steller's sea cow had no functional teeth.

DISTRIBUTION: In the Recent period, Steller's sea cow apparently was limited to the nearshore waters of the Commander Islands. During the Pliocene and into the late Pleistocene, however, this sea cow, or close relatives, also lived along the rim of the North Pacific from Japan to Baja California.

NATURAL HISTORY: As toothless, strictly algivorous herbivores, Steller's sea cows apparently relied entirely on kelp that grew on the shallow banks around Bering and Copper islands. Steller described four kinds of "seaweed" that the animals ate selectively. Their appetites seemed large: "Where they have been staying even for a single day there may be seen immense heaps of roots and stems." They spent most of their time feeding, according to Steller. They grazed head-down, with their backs exposed above the surface. Gulls would land on their backs to feed on what Steller called "the vermin infesting the skin." These were probably crustaceans similar to "whale lice." The average submergence time of the feeding sea cows was 4 to 5 minutes.

Steller described the animals as living in shallow, sandy areas along shore, particularly "the mouths of the gullies and brooks, the rushing fresh water of which always attracts them in herds." They would approach shore so closely during the flood tide that a person could prod them with a pole or spear, or even touch them. There is some uncertainty about whether these sea cows were capable of diving. If they were not, their feeding seaward of the intertidal zone would have been restricted to the shallow edges of this zone and to the very top of the offshore kelp canopy.

Steller found the sea cows solicitous toward their young. When traveling in groups, calves were always kept in the middle, surrounded by larger individuals. He also observed that other members of a herd would try to help an injured or tethered animal escape from its captors.

Steller guessed, judging by the number of young calves present in autumn, that this was the peak period of births. However, he believed that some reproduction occurred year-round. He saw what he interpreted to be mating in spring and concluded that the gestation period was greater than 12 months. Litter size appeared to be one. It would be surprising if all of Steller's speculations about the sea cow's reproductive cycle were correct, especially the one concerning timing of births. Giving birth in the fall seems an unlikely strategy for a high-latitude herbivore utterly dependent upon sea plants for energy.

It is not hard to imagine that the life of these sea cows was a perilous one, even before their discovery by hungry humans. Steller claimed that they were often smothered by ice in winter or dashed against cliffs in storm surges. During winter they became "so emaciated that not only the ridge of the backbone but every rib shows." It seems unlikely that they would have escaped the predatory attentions of killer whales.

HISTORY OF EXPLOITATION: Scientists have speculated that the ancient maritime Aleuts and other prehistoric human populations of the North Pacific coasts hunted sea cows, but no archaeological evidence has been found to support this speculation. A sixteenth-century Eskimo midden at the confluence of the Noatak River and Kangiguksuk Creek in northwestern Alaska contained a partial rib that might be that of a sea cow. The site is 100 km from the ocean today.

Steller and his companions on the wrecked Russian vessel *St. Peter* caught sea cows with a large iron hook connected to a long, stout rope. Some 30 men held the end of the rope on shore while a crew of 5 or 6 rowed offshore and implanted the hook in their prey. As the tethered beast was dragged ashore, the men in the boat fastened another rope to it and impaled it repeatedly with bayonets, knives, and other implements. Sea cows generally were butchered while still alive. Very large, old animals were easier to kill than young ones. Calves were faster swimmers and sometimes escaped when the hook tore loose from the skin.

Steller described the sea cow's blubber, 3 to 4 inches thick over most of the body, in glowing terms: "Melted, it tastes so sweet and delicious that we lost all desire for butter. In taste it comes pretty close to the oil of sweet almonds." This delicacy had the further advantage of keeping well, even through the hottest days. The meat was similarly resistant to spoilage, and Steller likened its taste and texture (when cooked) to that of beef. The fat of calves resembled fresh lard, and their meat was like veal.

CONSERVATION STATUS: Steller's sea cows were slow-moving, fearless, and good eating. At the time of their discovery, it was estimated that 1500 to 2000 animals existed, and these, according to L. Stejneger, Steller's biographer, "were the last survivors of a once more numerous and more widely distributed species" (Stejneger 1936). From 1743 to 1763, fur hunters made regular visits to the Commander Islands, dining heartily on sea cow flesh. By 1768 the supply had run out–utterly. The sea cow was extinct. There have been occasional unsubstantiated accounts of sightings since 1768, but unfortunately they appear to have been inspired more by wishful thinking than by accurate observation.

FURTHER READING: Domning (1978), Haley (in Haley 1986), Stejneger (1887, 1936).

Part IV

MUSTELIDAE

OTTERS

Sea Otter

Enhydra lutris
(Linnaeus, 1758)

NOMENCLATURE: The generic name is from the Greek *enhydris,* for "otter." The specific name is from the Latin *lutra,* for "otter." Three subspecies are recognized, and they are discussed in the Distribution section.

DESCRIPTION: Sea otters are some 60 cm long and weigh 2 to 2.3 kg at birth. Males grow larger than females, reaching a maximum length of close to 150 cm and weights of up to 45 kg. Females grow no larger than about 140 cm and 33 kg. Average adult weights are 27 to 38 kg for males and 16 to 27 kg for females. The sea otter is the largest member of the family Mustelidae, which includes weasels, skunks, badgers, and freshwater otters. It differs from other mustelids in not having functional anal scent glands.

Karl Kenyon, whose fieldwork in Alaska during the 1950s and 1960s culminated in his authoritative monograph on the sea otter (1969), identified the following important physical characteristics of this animal: a coat of sparse guard hairs and dense underlying fur; flattened hind feet (flippers) for propulsion; retractile claws on the front feet only; a loose flap or pouch of skin under each foreleg and extending across the chest, used to hold or store food items; rounded, flattened posterior cheek teeth with no cutting cusps; a horizontally flattened tail; and an external ear resembling those of otariid seals more than those of river otters. Adult males have a thicker, more massive head and neck than females.

The fluffy natal pelage differs in color and structure from the dense fur of adults. Pups do not lose their light buff natal pelage completely until they are some 13 weeks old, although it is obscured within a few weeks after birth by yellowish guard hairs. The adult pelage is evenly light or dark brown to black. The sparse guard hairs can be dark or silvery white. The head becomes whiter with

A California sea otter rafting in the kelp off Monterey, California. The grizzled head suggests that this is an adult male. (August 1989: Mark Conlin.)

age, and some grizzling often occurs on other parts of the body. The molt is continuous, although more fur fibers are shed during summer than winter. Given the differences in water and air temperatures experienced in the two environments, it is somewhat surprising that the pelage characteristics of Alaska and California otters do not differ significantly.

DISTRIBUTION: The genus *Enhydra* arose in and has remained confined to the North Pacific. Sea otters of the genera *Enhydriodon* and *Enhydrithereum* were widespread in the Late Miocene and Pliocene. They are best represented in the fossil record from the North Atlantic.

Before the arrival of Europeans, sea otters lived along most of the rim of the North Pacific from the northern Japanese archipelago to central Baja California, including the Kuril, Commander, Aleutian, and Pribilof islands. After having been reduced to near-extinction by commercial killing, the species has reoccupied most of its historic range in Alaska and the U.S.S.R. Otters assigned to the subspecies *E. lutris lutris* occur in the western Pacific from the Commander Islands southward through the Kuril. Those classified as *E. lutris kenyoni* inhabit the Aleutian Islands and formerly occurred north to the Pribilofs and south to Oregon. The subspecies *E. lutris nereis* occurs in California and formerly ranged as far south as Morro Hermoso, Baja California.

The small relict populations at the San Benito Islands (Baja California) and the Queen Charlotte Islands (British Columbia) were extirpated by 1919 and 1929, respectively, and sea otters remain absent from these areas. Small translocated populations are present in British Columbia and Washington, but the populations translocated to Oregon and the Pribilof Islands have not become established. The population in central California is distributed mainly from Santa Cruz to Pismo Beach. Otters occasionally show up farther south, among the Channel Islands, along the mainland coast of the United States, and rarely off the Pacific coast of northern Baja California. More than 100 sea otters were translocated to San Nicolas Island in the southern California Channel Islands between 1987 and 1990, but only about a dozen remained by late 1990. Many have returned to central California, some are known to have died or been killed, and others are unaccounted for.

Sea otters are fairly tame in areas where they are not deliberately hunted or harassed. For example, they can be observed closely, at virtually any time, along the Cannery Row wharf in Monterey Bay.

NATURAL HISTORY: Sea otters are social, but males are segregated from females and young during much of the year. Adult males are territorial during the summer and fall breeding season, when they associate with groups of females. Although they are completely at home in the water, sea otters in some areas haul out on land, usually remaining within a few meters of the shoreline. In California, most haul-out sites are rocks located at least 25 m from shore.

Three sea otters in an uncharacteristic terrestrial setting—a snowy beach. (Bering Island, western Bering Sea, January 1978: Sasha Zorin.)

Often the same rocks are used by harbor seals. While hauled out and motionless, otters can be very difficult to detect. Their dark coats serve as good camouflage in mats of algae. In California, their often light-colored heads blend well with the white and tan granite and feldspar rocks or the clumps of goose barnacles. Kenyon reported seeing sea otters in Alaska sleeping in grass as far as 50 to 75 m from the water, but this was on remote islands where human disturbance was minimal. In California, one otter was known to haul out regularly on a tidally exposed sandspit under a wharf in the Monterey harbor. Usually no more than 6 otters are seen hauled out together in California, but on one occasion 18 adults and 4 pups were seen out of the water together at Cypress Point near Monterey. A translocated male hauled out regularly on a sandy beach at San Nicolas Island among California sea lions, northern elephant seals, and harbor seals. As many as 50 males can be seen hauled out and sleeping together on favorite, undisturbed beaches in Alaska. Sea otters are slow and clumsy walkers.

Sea otters have a strong homing instinct that makes translocation difficult. Otters released on San Nicolas Island, nearly 70 miles from the California mainland, managed to cross the intervening deep water and then navigate the 200 or so miles north along the coast, arriving back at the sites where they were originally captured.

Male sea otters become sexually mature at 5 to 6 years of age. Females generally bear their first pup at 4 years of age, although some may not pup until 6 to 8 years old.

A single pup is normally born after a gestation period of some 8 months. Delayed implantation occurs in the sea otter, but apparently the length of the delay can vary. Pups can be born either on land or in water. In all except the harshest parts of their range, sea otters give birth in all months. The peak pupping season varies by area.

Pups remain fairly passive during the first month of life, spending considerable time on their mother's chest. They begin swimming actively only at 4 to 5 weeks of age. Even after they swim, pups continue to try to climb onto their mother whenever she returns to the surface from a feeding dive. During their tenth week, pups make the transition from surface swimming with their belly down to swimming on their backs, as adult sea otters do. They begin diving at some 6 weeks of age but do not become proficient at it for another 6 to 8 weeks. They begin taking solid food in their sixth week. During their sixth month, pups forage on their own and open hard-shelled items with the mouth or a rock, but they also continue to accept solid food from their mothers. By an age of 6 months, sea otters are usually capable of full independence. However, some pups remain with their mothers for as long as a year. Occasionally, a young sea otter may rejoin its mother after a few months of separation, remaining with her and receiving supplemental food for perhaps another half-year.

A mother sea otter floating, asleep, with her sleeping pup clasped to her chest. (Amchitka Island, Aleutians, 27 January 1957: Karl W. Kenyon.)

Grooming is a vital aspect of a sea otter's behavior because a clean pelage is essential to retention of trapped air, the animal's primary insulation against heat loss. Pups begin grooming in their first month of life, and by 8 to 14 weeks of age they are nearly self-sufficient in this regard.

Most adult females probably give birth annually or at intervals of less than 2 years. They generally come into estrus within a few days or weeks after weaning their pups. Mating is a hazardous experience for females; they are lucky to escape with no more than a torn nose. The male grasps the female's head or jaws, including the nose, with his own jaws, while she lies rigidly on her back on the surface of the water. The male presses his belly against the female's back and clasps her with his paws. Thus, copulation occurs with both animals positioned upside down near the water surface.

Sea otters feed primarily in water less than 20 fathoms deep, although they are known to forage in depths of up to 55 fathoms. They prey on nearshore inhabitants of rocky and soft bottoms, sometimes in submerged intertidal areas. Occasionally they clamber over rocks and enter tide pools to feed on urchins, and there is a published observation of an otter foraging on mussels while hauled out on rocks. Their most important prey are mollusks, echinoderms, and crustaceans, but in some parts of the Aleutian, Commander, and Kuril islands the otters also eat fish regularly. Sea otters forage at night as well as during the daytime. Adult California sea otters frequently use rocks or hard shells as tools to break open hard-shelled prey. Their teeth, often worn and broken, are adapted for crushing hard-bodied prey.

Although as a species the sea otter is a feeding generalist, individual otters seem to have strong dietary preferences and, in fact, behave as feeding specialists. A recent study in Monterey Bay revealed that the diet of most otters is dominated by one to three prey items, and that these clear-cut preferences are maintained for at least several years.

Sea otters have a major impact on the bottom communities that sustain them. From a human viewpoint, a given impact can be seen as positive or negative, depending on one's prejudices. In Alaska, predation by otters limits the intensity of kelp grazing by herbivorous sea urchins, essentially forestalling the development of "urchin barrens" within kelp forests. Healthy kelp beds support associated fish populations, which in turn become a part of the otters' diet and, of course, also gladden fishermen. On the other hand, though, otters often clash with human shellfish harvesters. There is no doubt that sea otters have either eliminated or hastened the decline of abalone fisheries in parts of California. They also have dramatically affected fisheries for crabs and clams in both California and Alaska. In the Elkhorn Slough, California, recent studies found that the deep-burrowing clams *Tresus nuttallii* and *Saxidomus nuttalli* constituted 61 percent of the prey of several dozen sea otters; the shallow-burrowing Pismo clam was eaten as well but was less preferred.

James A. Estes and Glenn A. Van Blaricom (1985) summarized a predictable sequence of events occurring as sea otters expand their range in California. First, groups of males, which tend to occupy the periphery of the population's range, arrive. They generally forage on a few high-energy, locally abundant prey species. As such prey declines, these male bands leave, to be replaced by breeding females, territorial males, and dependent pups. The local density of otters generally declines greatly (often by a factor of 10), reducing the pressure on prey stocks. Thus, while the short-term consequences of the sea otter's recolonization of parts of its former range may be clear, the longer-term effects on commercially desirable shellfish stocks are less certain.

Sea otters commonly live to ages of 10 to 15 years, and one captive male remained sexually potent until at least 19 years of age. Longevity records for wild sea otters are 23 years for females and 18 years for males.

The most uncertain time for a sea otter probably comes immediately after independence, when natural mortality can approach 50 percent. Ice and weather conditions can affect survival in northern regions; shark predation is a more serious threat to southern sea otters. Coyotes have been seen to capture recently weaned juveniles in Prince William Sound. Most of the pup carcasses found in the nests of bald eagles at Amchitka Island are thought to have been scavenged, although eagles do take live pups at least occasionally. The otters that haul out regularly on Kamchatka Penninsula are preyed upon by the locally abundant brown bears.

History of Exploitation: Sea otters were exploited by aboriginal maritime peoples in many areas. In upper and lower California, Spaniards bartered pelts from Indians during the early eighteenth century. It was not until the middle of the century, however, that the commercial value of sea otter pelts became widely recognized. Fur hunters from Siberia hurried eastward to the Commander and Aleutian islands to exploit the large populations of seals and otters discovered by shipwrecked members of the Bering expedition in 1741–42. Later, otter pelts obtained from Vancouver Island by Captain James Cook's expedition appeared in Asian markets, leading to a frenzy of fur collecting along the west coast of North America. By the end of the nineteenth century, the remaining stocks of sea otters were so small that they had no further economic importance. Kenyon estimates that about half a million sea otters were harvested between 1740 and 1911.

By the 1950s, it had become clear that some sea otter populations in Alaska had recovered. After Alaskan statehood in 1959, the Department of Game initiated a program of experimental harvesting. From 1962 to 1971, state agents took 2933 sea otters to obtain zoological data and test the fur market. The pelts sold at auction for an average price of $134.

Conservation Status: After 170 years of intensive commercial exploitation, sea otters in Alaska were given complete legal protection in 1899. They were given additional protection under the North Pacific Fur Seal Act of 1910. Although by the turn of the century sea otters had been extirpated over much of their range, a substantial recovery in both range and overall abundance has occurred since then. This recovery was facilitated by the translocation of some 400 otters into southeastern Alaska during the 1960s. The Near Islands, westernmost of the Aleutians, provide an example of the otters' ability to expand their range. Although extirpated from these islands during the nineteenth century, otters were back in very small numbers by the mid-1960s. They have continued to increase in numbers by approximately 15 percent per year and have repopulated much of the suitable habitat in this island group.

Estimates in the mid- to late 1980s suggested that there were 6000 to 7000 sea otters in the Kuril Islands, 2000 to 2500 in the Commander Islands, and 2500 to 3000 off Kamchatka, giving a total western Pacific population of 10,500 to 12,500. Alaska had 100,000 to 150,000, British Columbia 350, Washington 211 (counted in July 1989), and California 1400. From an original 402 animals released in southeastern Alaska in 1965 and 1968, the regional total had grown to more than 3500 otters in five populations by 1987. Sea otters are absent today at the Pribilof Islands, although they were numerous there in 1786 when the islands were discovered. They also remain absent from a number of other former strongholds, including Bering Island, the Queen Charlotte Islands, and Baja California.

The sea otter population in Alaska is regarded as healthy and growing. However, hunting resumed in recent years, particularly in southeastern Alaska near Sitka. The Marine Mammal Protection Act exempts Natives from the ban on hunting, so long as their harvest is nonwasteful and for subsistence or for making and selling authentic handicrafts and clothing. One hunter is known to have taken 45 sea otters from the Yakobi-Chichagof population in 1985. Although this population is large (more than 600 animals) and expanding, others that have been exploited recently by legal Native hunting and illegal killing by people concerned about competition for shellfish resources are much smaller and may be declining. An Alaska Sea Otter Commission was established in the late 1980s to oversee and regulate hunting.

The California population was designated as threatened under the U.S. Endangered Species Act in 1977. The U.S. Fish and Wildlife Service began a program in 1987 to reestablish a breeding population of sea otters at San Nicolas Island, California. With less than 2000 otters distributed along the mainland of central California and no prospect of immigration from northern populations, game management officials saw the reintroduction as a way of ensuring that a catastrophic event such as an oil spill would not put the entire California population at risk. As of November 1989, 135 otters had been translocated from central California to the island, but less than 10 percent of them remained alive at San Nicolas. At the time of this writing, the success of the translocation program had not yet been assured.

In May 1991, a group of about 10 sea otters, including a pup, was discovered at San Miguel Island, about 70 miles north of San Nicolas. Under an agreement between the U.S. Fish and Wildlife Service and the California Department of Fish and Game, any sea otters found in southern California outside the San Nicolas Island translocation zone must be captured and moved to central California or San Nicolas Island. In June 1991, efforts began to capture the otters that had settled at San Miguel Island.

A variety of human activities, besides deliberate hunting, are known to harm sea otters. The shock waves caused by underground nuclear tests caused mass mortality of sea otters in Alaska. Significant numbers of otters die from entanglement in gill and trammel nets; also, some are killed accidentally in purse seines and crab pots. In California, net entanglement is considered the chief reason for the sea otter population's failure to grow from the early 1970s to the mid-1980s. Legal restrictions on fishing activities have been in force since 1982 in parts of the state's Sea Otter Refuge,

Sea otters have mated and given birth in captivity. This mother was pregnant when she became oiled during the Exxon Valdez *spill in Prince William Sound. Brought to southern California and rehabilitated, she gave birh in spring 1990 to a healthy pup. (Sea World, San Diego, California, October 1990: Brent S. Stewart.)*

an area comprising some 100 miles of coastline from just south of Monterey Bay to just south of San Simeon. Pups sometimes die from collisions with boat propellers. Unusually high mortality at Kodiak Island in summer 1987 was attributed partly to paralytic shellfish poisoning that the otters may have contracted from eating butter clams.

Oil contamination is a serious threat to sea otters. Their fur is finely tuned for heat conservation. Maintenance of a clean, waterproof coat is essential to an individual otter's survival. Exposure to oil is disastrous for sea otters. Since they are so small in size for a marine mammal, have no blubber layer, and carry little fat in their body tissues, sea otters need to eat an estimated 23 to 37 percent of their body weight in food each day simply to sustain normal metabolic functions. They depend on the air trapped within their fur to prevent heat loss. Crude oil fouling of more than 20 to 30 percent of an otter's body surface is enough to cause hypothermia-related death. Although experiments conducted in the wild and in captivity had demonstrated their extreme vulnerability to oiling, the major oil spill in Prince William Sound in March 1989 laid any conceivable doubts to rest. The 250,000 barrels of crude oil spilled from the *Exxon Valdez* coated beaches and destroyed marine life along Kenai Peninsula, Kodiak Island, and Alaska Peninsula as well as in Prince William Sound. More than 1000 carcasses of sea otters were found; 348 ailing otters were captured for treatment. Of these, somewhat more than 190 were rehabilitated and released, while 114 died in captivity. It is expected to take decades for the sea otter populations in the affected areas to recover.

Further Reading: Chanin (1985), Estes (1990), Estes and Van Blaricom (1985), Kenyon (1969, 1981b), Kvitek et al. (1988), Ralls and Siniff (1990), Riedman and Estes (1990), Rotterman and Simon-Jackson (1988), Van Blaricom and Estes (1988), Wilson et al. (1991), Woodhouse et al. (1977).

Marine Otter

Lutra felina
(Molina, 1782)

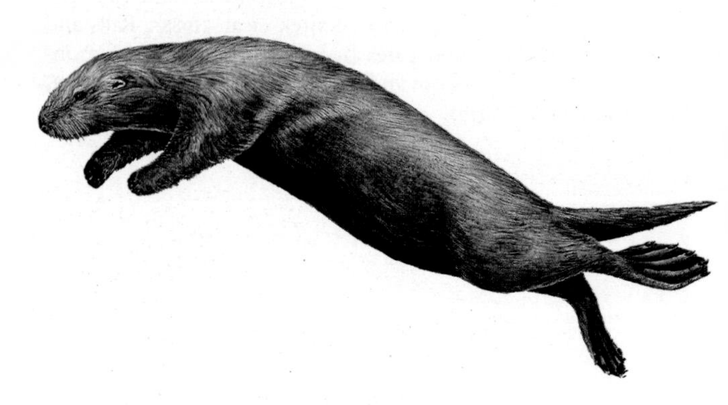

NOMENCLATURE: The name *Lutra* is Latin for "otter"; *felina* is from the Latin *felis,* for "cat." Although known in English as marine otter, this species is called *chungungo, gata de mar, gato marino, huallaque, nutria de mar,* and *chinchmen* in different coastal regions of Chile and Peru. In some translated works it is mistakenly called the sea otter.

DESCRIPTION: This otter is the smallest in the genus *Lutra.* The maximum body length, head to tip of tail, is about 1.15 m, and adults weigh only 4 to 4.5 kg. The tail is relatively short.

The guard hairs are dark brown overall, lightened at the tips. The underfur is brownish gray. The chin, cheeks, and throat may be somewhat lighter than the rest of the foreparts, and the ventral region is generally lighter than the dorsal. Individuals in Peru and northern Chile are said to be lighter than those farther south, and the northern population is considered by some scientists to be a subspecies.

DISTRIBUTION: As the only exclusively marine species in the genus *Lutra,* the marine otter is confined to the coastal waters of western South America from at least Chimbote, central Peru, in the north to Cape Horn and the Estrecho de le Maire (Strait of Le Maire) in the south. It may occur near Isla de los Estados (Staten Island), off the southern tip of Argentina, but it is apparently absent from the eastern Strait of Magellan, the Atlantic coast of Tierra del Fuego, and the north side of Navarino Island.

There are reports of marine otters in rivers as far upstream as 650 m above sea level. However, a problem in authenticating such records is that the freshwater otter, *Lutra provocax* (known locally in Chile as *huillin*), occurs from central Chile south to Cape

The marine otter is ecologically intermediate between the sea otter, a truly marine species, and otters that live year-round in riverine or lacustrine environments. (At Curiñaco, near Valdivia, Chile, February 1991; Gonzalo Medína Vogel, CODEFF.)

Horn and north into Argentina. This situation invites confusion and necessitates care in verifying sighting and specimen records.

NATURAL HISTORY: The marine otter is principally diurnal, although it can also be active at night and at dusk and dawn. It is considered timid and secretive. Its preferred habitat is said to be rocky, exposed coasts where there is abundant prey as well as cover in the form of caves and crevices for denning. At least in southern Chile, marine otters are invariably associated with the brown alga, *Durvillea antarctica.*

Gestation probably lasts 60 to 65 days, but there is a period of delayed implantation. The average litter size is probably two, but some litters may have as many as four cubs. An 86 cm (total length) female had twin fetuses. The female has four mammary glands. There is some evidence suggesting monogamy in this species, with biparental care of the young. However, this would be inconsistent with what is known about the social relations of most freshwater otters. The evidence for territoriality is inconclusive. Fights have been observed between two adult otters hauled out on islets or rocks. Such fights are accompanied by high-pitched squeaking.

The diet is varied, including many kinds of fish, freshwater prawns, and various marine crustaceans and mollusks. South of Beagle Channel, marine otters are thought to feed mainly on fish, supplemented by crustaceans and mollusks. A recent study of the otters around Chiloe Island revealed that crustaceans and mollusks form the main part of their diet. Sea urchins and marine worms also may be taken in some areas. Pairs of otters have been seen fishing cooperatively and sharing the large fish caught.

HISTORY OF EXPLOITATION: Marine otters have been hunted in Chile, especially south of Chiloe Island, for their valuable pelts, and in Peru because of the widespread belief among fishermen that they damage commercial shellfish stocks.

CONSERVATION STATUS: Although the marine otter is listed as endangered and legally protected in Peru and Chile, there is much poaching, especially in the parts of Chile where appreciable numbers are still available. It is generally assumed that the aggregate population is severely depleted and that it is now divided into several essentially isolated subpopulations: one in Peru and three in Chile, north, central-south, and south. The largest remaining numbers probably are in the remote southern channels of Chile. In the 1970s, estimated densities of 10 marine otters per kilometer of rocky coastline were reported for the northwestern sector of Chiloe Island; 2.5 per kilometer, for a portion of Coquimbo province. The population south of 48°50′ S was estimated to be 4700 to 9400 in the late 1980s. This was based on an estimate of available habitat and a supposed density of 1 to 2 otters per kilometer of coastline.

Marine otters are present in at least two national parks–Pan de Azucar in the Atacamá region of northern Chile and Chiloe in southern Chile.

FURTHER READING: Brownell (1978), Ostfeld et al. (1989), Sielfeld (1988a, 1988b), Sielfeld K. (1983).

Part V

URSIDAE

POLAR BEAR

Polar Bear

Ursus maritimus
Phipps, 1774

NOMENCLATURE: The generic name is the Latin word for "bear"; *maritimus* refers to the maritime habitat of this species. For many years, the polar bear was classified in a separate genus, *Thalarctos*. However, it is now assigned to *Ursus* along with several other widely distributed modern bears. Polar bears are also called white bears. *Nanuk* or *nanook* is the name used by the Eskimos and Inuit.

DESCRIPTION: Males grow larger than females. The average adult size is 2 to 2.5 m (tip of nose to tip of tail) and 300 to 800 kg for males, and 1.8 to 2 m and 150 to 300 kg for females. Cubs weigh less than a kilogram at birth and usually have grown to 10 to 15 kg by the time they emerge from the maternity den at 3 to 4 months of age. Yearlings weigh 45 to 80 kg; 2-year-olds, 70 to 140 kg. The adult weight is attained by 5 years of age in females and 8 to 10 years in males.

The polar bear's general appearance is familiar to everyone. Of all the world's bears, including eight well-defined species in the carnivore family Ursidae, the polar bear is the largest (with the possible exception of the largest brown bears). Its body lacks the shoulder hump characteristic of brown and black bears. The neck is longer and the head smaller than those of other ursids. The ears are relatively small. Polar bears are believed to have evolved fairly recently from a stock of Siberian brown bears that became isolated during the mid-Pleistocene. The crossing of brown bears and polar bears has produced fertile hybrids.

Polar bears are not always creamy white. They can appear white, yellow, gray, or even brownish according to the season and lighting conditions. Immediately after molting, the fur is nearly pure white. In summer, it often takes on a yellowish cast, probably

A mother polar bear and her yearling cubs near Churchill on the west coast of Hudson Bay. During the ice-free season, this area becomes an important congregating site for polar bears, and a tour business has developed to take advantage of the opportunities for close approaches. (November 1981: Fred Bruemmer.)

due to oxidation by the sun. The nose, lips, and foot pads are black.

Polar bears have five toes on each foot, with nonretractable claws 5 to 7 cm long in adults. The large, oarlike forepaws are adapted for swimming. Females normally have four pectoral mammary glands.

Polar bears have 38 to 42 teeth, including 3 unspecialized incisors in each row and 1 elongated, conical, slightly hooked canine on either side in the upper and lower jaws. The last cheek teeth are reduced in size compared with those of other modern bears.

Distribution: The polar bear has a circumpolar distribution. Some shrinkage of its range has occurred historically, mainly as it has been excluded by humans from southern areas such as Newfoundland and the Gulf of St. Lawrence. Polar bears range far to the north of the American, European, and Asian continents, reaching at least 84° N. They have been seen within 300 km of the North Pole.

Polar bears occasionally wander over enormous distances. One bear tagged in Svalbard was killed a year later in southwestern Greenland, some 3200 km from the tagging site. In general, however, tag returns have demonstrated that individual bears tend to remain in the same general area from one year to the next. The world population is best thought of as including several essentially discrete stocks, with only limited interchange among them. In the U.S.S.R., three stocks are recognized, with the bears in the Laptev Sea and its adjacent islands having little interchange with the populations in the eastern and western Eurasian Arctic. It is likely that if intensive mark-recapture studies were done in Eurasia, as have been done in North America, more subpopulations would be defined. There may be some movement between Svalbard and the vicinities of Franz Josef Land and Novaya Zemlya, or between Svalbard and Greenland, but the Svalbard stock is fairly discrete, as is the east Greenland stock. Alaska has two polar bear stocks, one northern and one western. There is some movement of bears between Wrangel Island and the north coast of Alaska, and the bears in the western Canadian Arctic and northern Alaska belong to one population. Canada has a number of different stocks, one or more of them shared with west and north Greenland.

Mother polar bears require specific conditions for their maternity dens, and denning areas help to separate and define the various stocks. In Canada, known major denning areas are on the west shore of Hudson Bay (the Manitoba and Ontario coasts), Southampton Island, the west coast of Banks Island, Simpson Peninsula, southeastern Baffin Island, and Gateshead Island in the central Arctic. Denning is also widespread throughout much of the Canadian Arctic archipelago, although in lower densities than in the areas listed above. Some denning probably occurs on the western shores of the Belcher Islands, in northern Labrador, on Akpatok Island in Ungava Bay, and on the north coast of the Ungava Peninsula. Polar bears are not abundant in Ungava Bay and northern Labrador, probably at least in part because of hunting pressure. They were common historically in southern Labrador and on the island of Newfoundland, but they are rare south of Nain, Labrador, today. The Churchill area on the Manitoba coast of Hudson Bay is the best-known site for watching polar bears in North America. Each spring some 100 to 150 cubs emerge from the maternity dens in this area. Farther south in Hudson and James bays, some polar bears wander as far as 150 km inland from the coast, entering coniferous forest and at times living more like brown or black bears than like other polar bears.

High-density denning areas in Soviet territory are on Wrangel Island in the Chukchi Sea and Novaya Zemlya and Franz Josef Land in the Barents Sea. The Norwegian-controlled Svalbard archipelago contains important denning habitat, especially on northern Nordaustlandet and Kong Karls Land.

Bears that are not in dens during the winter (only the pregnant females den) tend to concentrate in areas of unstable ice where they can prey on immature ringed seals. Generally speaking, the polar bear's preferred habitat is periodically active ice, where wind or currents cause shifting and fracturing that results in refreezing. The lanes and patches of recently refrozen ice provide excellent opportunities for seal hunting. These conditions occur at the floe edge or flaw zone (the interface between land-fast ice and drifting pack ice), at the mouths of bays, and along tidally active coasts.

Natural History: The ringed seal is the staple of the polar bear's diet. The bears prey on all age classes of ringed seals; they take large numbers of pups in spring by digging into the birth lairs. When hunting in unstable pack ice and along the floe edge, polar bears mainly catch juvenile ringed seals. Although many fewer bearded seals are taken, the relatively high energy payoff of killing one of these large seals could mean that they form an important part of the bears' diet in some areas. Harp and hooded seals are particularly vulnerable to predation on the spring pupping grounds, where at times polar bears will kill far more pups than can be consumed. Walruses are formidable adversaries and have been known to injure or even kill polar bears, but some walruses, particularly pups, succumb to predation. Polar bears stalk seals by crawling across the ice and by swimming between ice floes. For detecting and approaching prey, they rely mainly on their keen sense of smell. The eyesight of polar bears is probably about as good as that of humans. Most hunts on old winter ice are "still hunts," in which the bear lies or stands by a seal breathing hole, waiting patiently for its prey. When the seal appears, the bear hits it with a paw, then grabs it with its teeth and drags the carcass onto the ice. When "stalk hunting," the bear keeps low, creeps forward, and attempts to hide behind any irregularity on the ice surface. If it manages to close within 15 to 30 m of a basking seal, there is a good chance that the bear's final rush will enable it to catch the seal.

Polar bears also kill and consume narwhals and belugas when these small whales become trapped in pools or cracks in the ice, or when the whales are traveling through narrow lead systems. On one occasion in the northern Bering Sea, polar bears killed at least 40 entrapped belugas and dragged them onto the ice. Polar bears often leave considerable portions of a carcass uneaten, and this provides opportunities for scavenging birds and arctic foxes. Scavenging from polar bear kills may be an important supplement to the diet of coastal arctic fox populations.

In spite of their well-earned reputation as supreme predators, polar bears are not above scavenging. They often congregate to feed on whale carcasses. For example, at least 75 different bears gathered to take advantage of three bowhead whale carcasses at Barter Island, Alaska, in October and November 1981. After several hundred walruses had been killed by Norwegian hunters on a small island in southeastern Svalbard in 1852, 60 or more polar bears congregated to feast on the abandoned carcasses. Their attraction to bait has, of course, made polar bears especially vulnerable to trapping and hunting. Also, the bears' tendency to congregate at dump sites, as near Churchill, greatly increases the frequency of encounters with people, which in turn leads to the destruction of "problem bears."

Ian Stirling (1988) has commented on how clean polar bears like to keep themselves. After every meal, they either wash in water or rub and roll in the snow to get the grease off. Even the badly soiled bears at the Churchill dump make a point of walking to the bay and bathing periodically. Stirling, who has studied several thousand wild polar bears in his career, claims to have found very few of them to be dirty, in spite of the fact that their diet is dominated by greasy seal blubber.

While on land during the late summer and fall, polar bears may eat some grass, berries, and seaweed along the beach. They also occasionally eat eggs, birds, and small mammals when the absence of sea ice makes it difficult for them to procure seals. In winter, the bears sometimes eat kelp, which they pull out of seal breathing holes onto the ice; they also have been seen diving for kelp. Some polar bears in Hudson Bay regularly dive under flocks of ducks and catch them by surprise. There is little evidence that polar bears catch fish, even though arctic char are available in shallow rivers throughout the Arctic. Old historical records suggest that polar bears once congregated along rivers in Labrador to take advantage of the spawning runs of Atlantic salmon, much as brown bears do today in parts of Alaska. Polar bears are capable of prolonged fasting, both while on land in summer and when in maternity dens in winter. At such times, they are in a physiological state similar to that of a hibernating black bear.

The availability of ringed seals strongly influences the movements of polar bears. A dramatic example was provided between 1974 and 1975 when, following a major climate-induced reduction in the population of ringed seals in the eastern Beaufort Sea, tagged bears moved west from Canada into Alaska. Correspondingly, there was no documented movement of tagged bears from Alaska into Canada in 1974 and 1975, the years of low ringed seal abundance in Canadian waters.

Male polar bears can become sexually mature at 3 years of age, but few mate successfully until about 6 years old. Females can mature at 3.5 years of age, but most do not become pregnant until they are 4 to 6 years old, giving birth for the first time at 5 to 7 years of age.

Mating occurs between late March and mid-May. Implantation is thought to be delayed, resulting in a total gestation period of 195 to 265 days. Pregnant females enter maternity dens dug below snowdrifts in November and December and give birth to one, two, or occasionally three cubs in December or January. The cubs, hairless, blind, and relatively helpless at birth, develop and mature in the maternity den until they are capable of padding across the snow and ice with their mothers. They usually emerge together in spring. Timing of emergence from the den is influenced by snow conditions and climate. Bears in higher latitudes tend to stay in the den longer than those in lower latitudes. For instance, in the Hudson and James bay areas, the peak of emergence is around 20 March; whereas in the Resolute Bay area, it is around 5 to 10 April. Most bears emerge from dens in Alaska in April.

The reasons for a mother polar bear's choosing one area over another for denning are not obvious. The ready availability of prey (especially ringed seals) at the time when they emerge from the den with newborn cubs may be a prime consideration. Most dens are within 8 km of the coast; rarely are they more than 48 km offshore. The southwest coast of Hudson Bay presents an exceptional situation. There, pregnant females commonly overwinter in earth dens 20 to 100 km from the coast. Individual mothers tend to return repeatedly to the same general area for denning. Dens can be detected by searching land areas for mounds of excavated snow, called "porches" by scientists. The configuration of maternity dens varies, from single-chambered structures with a short tunnel to elaborate multichambered structures with several long tunnels. Mothers with cubs sometimes dig temporary dens in the snow to shelter from storms. Polar bears also sometimes dig temporary dens in the earth during the summer and fall. Bears of all ages are known to den at least occasionlly, but maternity dens are of greatest concern from a conservation standpoint. Apparently unlike female polar bears elsewhere, those in Alaska often den at sea. Maternity dens have been found in the active offshore pack ice of the Beaufort Sea, as much as 550 km north of the Alaskan coast.

Immature polar bears remain with their mothers until they are approximately 2.5 years old; in some instances weaning may not be complete until the cubs are almost 3 years of age. The longer the period of dependency, the better the cub's chances of surviving to adulthood. From her first pregnancy until her death, a female polar bear is almost never unencumbered by cubs. Those adult females that are alone are usually pregnant. The mean breeding interval is between 3 and 4 years. Polar bears live for 20 to 30 years. Females can be reproductively active until at least 21 years of age; males to 19.

Polar bears are powerful swimmers and walkers. They have been known to swim 100 km across open water. According to Stirling, a subadult can probably run 25 to 30 miles per hour in a short sprint, for example, when charging a seal. The polar bear is kept warm by a thick layer of subcutaneous fat (5 to 10 cm in adults) and its impressive coat of hair. It uses approximately twice as much oxygen while walking as do most mammals. Excess heat is dissipated by conduction through the ears, nose, face, inguinal region, foot pads, and shoulders, and by panting.

Apart from humans, polar bears appear to have no enemies. Some mortality results from accidents, starvation, and injuries sustained while hunting (for example, those inflicted by walruses). Avalanches and snowslides probably kill some bears while in their dens during winter and early spring. Males sometimes fight during the mating season, but they probably kill one another only rarely. Peaceful, even playful, aggregations of adult males on the west coast of Hudson Bay during the ice-free season have been explained by the lack of competition for food and the absence of breeding activity at this time of the year. Females with cubs shun the company of adult males, as the latter pose a serious threat to the cubs. Wolves have been known to kill and eat cubs as the cubs walked from their maternity den to the sea ice. In northern Alaska and the western Canadian Arctic, brown (grizzly) bears range onto the coastal lowlands and are seen occasionally on the sea ice. Also in this area, polar bears have been seen chasing caribou on land.

HISTORY OF EXPLOITATION: Polar bears have been hunted and trapped by humans for thousands of years. Native peoples have valued the bear's meat and hide for subsistence, but in more recent times the hide's commercial value has become of paramount interest. With the closure of hunting in some areas and tighter management in others, the price of polar bear rugs soared during the early 1970s. A good pelt was said to be worth as much as $10,000 during 1973–74. Sport hunters from all over the world have gone north to kill polar bears as trophies. Commercial whalers and sealers have killed bears for a combination of sport, profit, and in some cases self-defense, as have many adventurers, explorers, and researchers. The orphaned cubs of killed females have often been taken alive and kept briefly as pets or sold to zoos.

Consumption of polar bear meat and organs carries some risks. Many bears are hosts for *Trichinella spiralis,* the parasite that causes deadly trichinosis in humans. Bears can acquire the parasite from eating infected seals or dogs. Polar bears are otherwise remarkably free of parasites.

In Svalbard before 1970, Norwegian trappers and hunters wintering on the islands killed bears, seals, and foxes for their hides and fat. Also, weather-station crews took some 100 bears each year

An angry bear lunges at a sled dog on Jones Sound near Ellesmere Island, eastern Canadian Arctic. Held at bay by dogs, the bear was shot and killed by Inuit hunters. (May 1967: Fred Bruemmer.)

for hides. Most were killed with baited set-gun traps, which were unselective and wasteful, killing mothers (and thus orphaning cubs) and wounding several bears at a time. In addition, beginning in 1952, sport hunters from Tromsø cruised through the pack ice southeast of Spitsbergen, shooting bears from the ship's deck or chasing them on foot across large ice pans. This sport hunting accounted for some 30 to 40 bears per year. The annual kill at Svalbard in postwar years was more than 300 bears. Polar bears can be killed at Svalbard today only in self-defense. The reported kill from 1976 to 1985 was 38 bears. Two more were killed at a Norwegian weather station on Jan Mayen Island.

In Alaska, polar bears are hunted by coastal residents from Nome to Kaktovik and by the people of St. Lawrence Island and the Diomede Islands. Most of the killing is done during November through January. Polar bear hunting in Alaska through the 1940s was done mainly by Natives for subsistence and the sale of hides. Annual harvests between 1925 and 1953 have been estimated as 117 bears, based on fur-export records. In the late 1940s, sport hunting with the use of aircraft began; this continued until 1972 when it was banned by the state. With the aid of aircraft, the sport hunters took bears as far as 450 km offshore from Alaska. The annual kill increased to 160 during 1954 to 1960 and then to 260 during 1960 to 1972. In the latter period, only about 13 percent of the kill (34 bears) was taken by Natives. The kill during the rest of the 1970s was not well monitored. Although sport hunting had stopped, Native hunting continued and probably increased. From July 1980, when monitoring resumed, to June 1985, 676 polar bears were reported killed in Alaska, for an average of 135 per year.

In Greenland, polar bear hunting is particularly important in the northern districts. The Polar Eskimos of the Thule district are known for their traditional polar bear pants and boots (*kamiks*). These are still worn by hunters who consider them irreplaceable as protection from the harsh environment. Approximately 100 to 125 polar bears are killed each year in Greenland.

CONSERVATION STATUS: The aggregate world population of polar bears is more than 20,000. Stirling believes that the total may be as many as 40,000. Most stocks are believed to be stable or increasing slowly under protection or managed harvests.

The population of polar bears between Point Barrow, Alaska, and Cape Bathurst in the western Canadian Arctic was estimated as 1300 to 2500 in 1972–83. This Beaufort Sea population receives little immigration from surrounding areas. The most important denning area on land in Alaska is the Arctic National Wildlife Refuge, whose rich underground supply of oil and gas makes it a prime target for industrial development. At least 1100 polar bears are believed to inhabit the central Canadian High Arctic. There are a few thousand in Hudson and James bays and perhaps 2000

Evidence of successful hunting—the hides of seven recently killed polar bears outside a hut at Resolute, Cornwallis Island, Canadian Arctic. (May 1967: Fred Bruemmer.)

in the Lancaster Sound region and Baffin Bay. An estimate of 5000 has been suggested for the Barents and Greenland seas. Soviet investigators estimated 3600 for the Soviet Arctic, based on aerial surveys. Polar bears are notoriously difficult to census, and many parts of their range have yet to be surveyed adequately.

Polar bear exploitation has been managed under the International Agreement on the Conservation of Polar Bears and Their Habitat, which was ratified in 1973 and took effect in 1976. All nations with jurisdiction over polar bear habitat—Norway, the U.S.S.R., the United States, Canada, and Denmark (Greenland)—are parties to this agreement. Polar bears had already been given complete protection from commercial hunting in the U.S.S.R. in 1955. This protection continues, and bears can be taken there only in compliance with a decision of the responsible government agency. Several bears are killed each year when they attack people or enter human settlements. The Norwegian government limited the number of permits issued for polar bear hunting on Svalbard beginning in 1970. Then, in 1973, a 5-year hunting ban was introduced, and it has yet to be lifted.

Less extreme measures have been taken in Canada and the United States, where polar bear hunting by Native peoples remains an important traditional activity. Canada's Northwest Territories government began setting kill quotas, by settlement, in 1968. In recent years, about 620 polar bears have been killed annually by Inuit hunters in the Northwest Territories. Polar bear hunting has been illegal in the province of Newfoundland, including Labrador, since the end of 1970, although this did not stop the killing of four bears that came ashore on the island of Newfoundland in March 1973.

Alaska established an annual limit of three polar bears per person for residents in 1971, and in the same year placed a limit on the number of permits issued for sport hunting. Polar bear management in Alaska was removed from state jurisdiction in 1972 with the passage of the federal Marine Mammal Protection Act. This meant that sport hunting was banned, but hunting by Native Alaskans (for commercial sale of pelts) was no longer regulated in any way. Females and cubs once again could be hunted legally, and there was no requirement to report kills. An agreement reached in early 1988 between the Inuvialuit Game Council in Inuvik, Northwest Territories, and the North Slope Borough Fish and Game Management Committee in Barrow, Alaska, calls for protection of denning bears and family groups, closed seasons, and the setting of sustainable limits on the harvest.

There is much reason for concern about the polar bear's future, particularly in areas like Alaska where the human population is growing and where industrial development is proceeding rapidly. It is difficult for these powerful, aggressive animals to coexist with humans. In Canada, where the hunting of polar bears has been strictly controlled by a quota system for some time, 20 to 40 bears are killed each year to protect human life and property. Various means have been used to reduce or at least manage encounters between bears and people. Electric fences, trip wires, and laser beams that set off alarms when crossed can warn people of an approaching bear. Noisemakers and projectiles such as rubber bullets can help scare off bears.

Several important denning areas have been set aside as polar bear reserves. In Canada, Ontario's Polar Bear Provincial Park protects from development more than 15,000 km² along the south coast of Hudson Bay. Human activities are strictly regulated to protect polar bear denning sites in the U.S.S.R.'s Wrangel Island Republic Reserve. Several islands in Svalbard, accounting for more than 21,000 km² and comprising the most important denning habitat in the archipelago, are encompassed within the Northeast and Southeast Svalbard Nature Reservations. The huge (972,000 km²) Northeast Greenland National Park contains much of the east Greenland polar bear population.

The polar bear can be regarded, at least for the moment, as a species in no immediate danger of extinction, thanks largely to intensive research and to conservation measures based on the results of that research.

FURTHER READING: Amstrup and DeMaster (1988), Amstrup et al. (1986), Burns et al. (1985), DeMaster and Stirling (1981), Kurtén (1964), Larsen (1978), Lentfer (1982), Lønø (1970), Lowry, L.F., et al. (1987), Manning (1971), Stirling (1986, 1988), Stirling and Archibald (1977).

Note: Also see the series *Bears–Their Biology and Management,* published irregularly since 1972 by the International Union for

Conservation of Nature and Natural Resources (IUCN, or World Conservation Union), Gland, Switzerland. The most recent contribution in this series—*Polar Bears: Proceedings of the Tenth Working Meeting of the IUCN/SSC Polar Bear Specialist Group*—was published in 1991 and is available from the IUCN Publications Unit, 219c Huntingdon Road, Cambridge CB3 0DL, United Kingdom.

Bibliography

THIS READING LIST IS MEANT only to guide the reader toward a broad, voluminous literature. It includes but a small sample of the sources consulted during this book's preparation. A number of scientific journals were particularly useful, including *Arctic, Biological Conservation, Canadian Journal of Fisheries and Aquatic Sciences, Canadian Journal of Zoology, Journal of Mammalogy, Journal of Wildlife Management, Journal of Zoology, Marine Mammal Science, Polar Biology,* and many others. The abstracts of presentations at various scientific conferences, especially those of the international Society for Marine Mammalogy and the group of experts on aquatic mammals of South America, were also helpful. Our reading list has some obvious biases. We have emphasized materials in English that are readily available at a good university library. Although several classic studies are cited, we have usually cited recent titles rather than older ones. Also, we have included review papers, which themselves often contain lengthy bibliographies. Short notes on natural history, range, research methods, taxonomy, and the like have generally not been listed. We expect that readers wishing to explore a topic in more depth will use this list only to begin their search.

The list of Further Reading at the end of each species account does not necessarily include all the titles from the Bibliography that contain information on the species. For example, if one were interested in the Galápagos fur seal, several of the more general references, such as Bonner (1982, 1989a), Food and Agriculture Organization of the United Nations (1982), King (1983), the relevant account in Ridgway and Harrison (1981a), and certainly Gentry and Kooyman (1986), should be consulted in addition to those listed on page 67. Considered together, the bibliographies in these sources provide a good entry to the literature on the species. A valuable annotated bibliography on pinnipeds is available from the International Council for the Exploration of the Sea, Palaegade 2-4, DK-1261 Copenhagen, Denmark.

No effort was made to update the bibliography past mid-1991, when the copy-edited manuscript was returned to Sierra Club Books.

AARDE, R. J. VAN. 1980. Fluctuations in the population of southern elephant seals *Mirounga leonina* at Kerguelen Island. *South African Journal of Zoology* 15:99–106.

ABBOTT, I. 1979. The past and present distribution and status of sea lions and fur seals in Western Australia. *Records of the Western Australian Museum* 7:375–90.

AGUAYO L., A. 1978. The present status of the antarctic fur seal *Arctocephalus gazella* at South Shetland Islands. *Polar Record* 19:167–76.

ALLEN, J. A. 1880. *History of North American pinnipeds. A monograph of the walruses, sea-lions, sea-bears and seals of North America.* Washington, DC: U.S. Geological and Geographical Survey of the Territories, Misc. Pub. 12.

ALLEN, J. A. 1887. The West Indian seal (*Monachus tropicalis* Gray). *Bulletin of the American Museum of Natural History* II(1):1–34.

AMSTRUP, S. C., AND D. P. DEMASTER. 1988. Polar bear *Ursus maritimus*. In J. W. Lentfer, ed., *Selected marine mammals of Alaska*. Washington, DC: Marine Mammal Commission, pp. 39–56.

AMSTRUP, S. C., I. STIRLING, AND J. W. LENTFER. 1986. Past and present status of polar bears in Alaska. *Wildlife Society Bulletin* 14:241–54.

ANDERSON, P. K. 1979. Dugong behavior: On being a marine mammalian grazer. *The Biologist* 61: 113–44.

ANDERSON, P. K. 1981. The behavior of the dugong (*Dugong dugon*) in relation to conservation and management. *Bulletin of Marine Science* 31:640–47.

ANDERSON, P. K. 1986. Dugongs of Shark Bay, Australia–Seasonal migration, water temperature, and forage. *National Geographic Research* 2:473–90.

ANDERSON, P. K., AND A. BIRTLES. 1978. Behaviour and ecology of the dugong, *Dugong dugon* (Sirenia): Observations in Shoalwater and Cleveland bays, Queensland. *Australian Wildlife Research* 5:1–23.

ANDERSON, S. 1988. *The grey seal*. Bucks, England: Shire Natural History Series 26.

ANDERSON, S. S., J. R. BAKER, J. H. PRIME, AND A. BAIRD. 1979. Mortality in grey seal pups: Incidence and causes. *Journal of Zoology, London* 189:407–17.

ANTONELIS, G. A., JR., M. S. LOWRY, D. P. DEMASTER, AND C. H. FISCUS. 1987. Assessing northern elephant seal feeding habits by stomach lavage. *Marine Mammal Science* 3:308–22.

ANTONELIS, G. A., B. S. STEWART, AND W. F. PERRYMAN. 1990. Foraging characteristics of female northern fur seals (*Callorhinus ursinus*) and California sea lions (*Zalophus californianus*). *Canadian Journal of Zoology* 68:150–58.

BARLOW, G. W. 1972. A paternal role for bulls of the Galapagos Islands sea lion. *Evolution* 26:307–10.

BARNES, L. G., D. P. DOMNING, AND C. E. RAY. 1985. Status of studies on fossil marine mammals. *Marine Mammal Science* 1:15–53.

BARTHOLOMEW, G. A., JR. 1952. Reproductive and social behavior of the northern elephant seal. *University of California Publications in Zoology* 47:369–472.

BARTHOLOMEW, G. A., AND L. G. HOEL. 1953. Reproductive behavior of the Alaska fur seal. *Mammalia* 34:417–36.

BEDDINGTON, J. R., R.J.H. BEVERTON, AND D. M. LAVIGNE, EDS. 1985. *Marine mammals and fisheries*. London: Allen & Unwin.

BEENTJES, M. P. 1989. Haul-out patterns, site fidelity and activity budgets of male Hooker's sea lions (*Phocarctos hookeri*) on the New Zealand mainland. *Marine Mammal Science* 5:281–97.

BEENTJES, M. P. 1990. Comparative terrestrial locomotion of the Hooker's sea lion (*Phocarctos hookeri*) and the New Zealand fur seal (*Arctocephalus forsteri*): Evolutionary and ecological implications. *Zoological Journal of the Linnean Society* 98:307–25.

BENGTSON, J. L., AND R. M. LAWS. 1985. Trends in crabeater seal age at maturity: An insight into antarctic marine interactions. In W. R. Siegfried, P. R. Condy, and R. M. Laws, eds., *Antarctic nutrient cycles and food webs*. Berlin: Springer-Verlag, pp. 669–75.

BENGTSON, J. L., AND D. B. SINIFF. 1981. Reproductive aspects of female crabeater seals (*Lobodon carcinophagus*) along the Antarctic Peninsula. *Canadian Journal of Zoology* 59:92–102.

BENGTSON, J. L., AND B. S. STEWART. 1992. Diving and haulout behavior of crabeater seals in the Weddell Sea, Antarctica, during March 1986. *Polar Biology.* In press.

BENJAMINSEN, T. 1973. Age determination and the growth and age distribution from cementum growth layers of bearded seals at Svalbard. *Fiskeridirektoratets Skrifter Serie Havundersøkelser* 16:159–70.

BEST, R. C. 1983. Apparent dry-season fasting in Amazonian manatees (Mammalia: Sirenia). *Biotropica* 15:61–64.

BEST, R. C. 1984. The aquatic mammals and reptiles of the Amazon. In H. Sioli, ed., *The Amazon. Limnology and landscape ecology of a mighty tropical river and its basin.* Dordrecht, Netherlands: Junk, pp. 371–412.

BESTER, M. N. 1980. The southern elephant seal *Mirounga leonina* at Gough Island. *South African Journal of Zoology* 15:235–39.

BESTER, M. N. 1981. Seasonal changes in the population composition of the fur seal *Arctocephalus tropicalis* at Gough Island. *South African Journal of Wildlife Research* 11:49–55.

BESTER, M. N. 1982. Distribution, habitat selection and colony types of the Amsterdam Island fur seal *Arctocephalus tropicalis* at Gough Island. *Journal of Zoology, London* 196:217–31.

BESTER, M. N. 1987. Subantarctic fur seal, *Arctocephalus tropicalis,* at Gough Island (Tristan da Cunha Group). *NOAA Technical Report NMFS* 51:57–60.

BIGG, M. A. 1969. The harbour seal in British Columbia. *Bulletin of the Fisheries Research Board of Canada* 172.

BIGG, M. A. 1988a. Status of the California sea lion, *Zalophus californianus,* in Canada. *Canadian Field-Naturalist* 102:307–14.

BIGG, M. A. 1988b. Status of the Steller sea lion, *Eumetopias jubatus,* in Canada. *Canadian Field-Naturalist* 102:315–36.

BONESS, D. J., W. D. BOWEN, AND O. T. OFTEDAL. 1988. Evidence of polygyny from spatial patterns of hooded seals (*Cystophora cristata*). *Canadian Journal of Zoology* 66:703–6.

BONNER, W. N. 1968. The fur seal of South Georgia. *British Antarctic Survey Scientific Reports* 56.

BONNER, W. N. 1972. The grey seal and common seal in European waters. *Oceanography and Marine Biology Annual Review* 10:461–507.

BONNER, W. N. 1981. Grey seal *Halichoerus grypus* Fabricius, 1791. In S. H. Ridgway and R. J. Harrison, eds., *Handbook of marine mammals.* Vol. 2, *Seals.* London: Academic Press, pp. 111–44.

BONNER, W. N. 1982. *Seals and man. A study of interactions.* Seattle: University of Washington Press.

BONNER, W. N. 1984. Seals of the Galapagos Islands. *Biological Journal of the Linnean Society* 21:177–84.

BONNER, W. N. 1985. Impact of fur seals on the terrestrial environment at South Georgia. In E. R. Siegfried, P. R. Condy, and R. M. Laws, eds., *Antarctic nutrient cycles and food webs.* Berlin: Springer-Verlag, pp. 641–46.

BONNER, W. N. 1989a. *The natural history of seals.* New York: Facts on File.

BONNER, W. N. 1989b. Seals and man—A changing relationship. *Biological Journal of the Linnean Society* 38:53–60.

BONNER, W. N., AND R. M. LAWS. 1985. Marine mammals. In W. Fischer and J. C. Hureau, eds., *FAO identification sheets for fishery purposes. Southern Ocean, Vol.* 2. Rome: Food and Agriculture Organization of the United Nations, pp. 401–45.

BONNOT, P. 1951. The sea lions, seals and sea otter of the California coast. *California Fish and Game* 37:371–89.

BORN, E. W. 1984. Status of the Atlantic walrus *Odobenus rosmarus rosmarus* in the Svalbard area. *Polar Research* n.s. 2:27–45.

BOULVA, J., AND I. A. MCLAREN. 1979. Biology of the harbor seal, *Phoca vitulina,* in eastern Canada. *Bulletin of the Fisheries Research Board of Canada* 200.

BOWEN, W. D. 1985. Harp seal feeding and interactions with commercial fisheries in the north-west Atlantic. In J. R. Beddington, R.J.H. Beverton, and D. M. Lavigne, eds., *Marine mammals and fisheries*. London: Allen & Unwin, pp. 135–52.

BOWEN, W. D., R. A. MYERS, AND K. HAY. 1987. Abundance estimation of a dispersed, dynamic population: Hooded seals (*Cystophora cristata*) in the northwest Atlantic. *Canadian Journal of Fisheries and Aquatic Sciences* 44:282–95.

BOWEN, W. D., O. T. OFTEDAL, AND D. J. BONESS. 1985. Birth to weaning in 4 days: Remarkable growth in the hooded seal, *Cystophora cristata. Canadian Journal of Zoology* 63:2841–46.

BOYD, I. L. 1991. Environmental and physiological factors controlling the reproductive cycles in pinnipeds. *Canadian Journal of Zoology* 69:1135–48.

BOYD, I. L., AND T. S. MCCANN. 1989. Pre-natal investment in reproduction by female antarctic fur seals. *Behavioral Ecology and Sociobiology* 24:377–85.

BRAHAM, H. W., R. D. EVERITT, AND D. J. RUGH. 1980. Northern sea lion population decline in the eastern Aleutian Islands. *Journal of Wildlife Management* 44:25–33.

BROTHERS, N., AND D. PEMBERTON. 1990. Status of Australian and New Zealand fur seals at Maatsuyker Island, southwestern Tasmania. *Australian Wildlife Research* 17:563–69.

BROWNELL, R. L., JR. 1978. Ecology and conservation of the marine otter, *Lutra felina*. In N. Duplaix, ed., *Otters. Proceedings of the first working meeting of the IUCN Otter Specialist Group, Paramaribo, Suriname, 27–29 March 1977*. Morges, Switzerland: International Union for Conservation of Nature, pp. 104–6.

BRUEMMER, F. 1983. Sea lion shenanigans. *Natural History* 92(7):32–41.

BRUEMMER, F. 1989. *World of the polar bear*. Toronto, Ontario: Key Porter Books.

BRUEMMER, F. 1988. A fate unsealed. *Natural History* 97 (11):58–65.

BURNS, J. J. 1981a. Bearded seal *Erignathus barbatus* Erxleben, 1777. In S. H. Ridgway and R. J. Harrison, eds., *Handbook of marine mammals*. Vol. 2, *Seals*. London: Academic Press, pp. 145–70.

BURNS, J. J. 1981b. Ribbon seal *Phoca fasciata* Zimmermann, 1783. In S. H. Ridgway and R. J. Harrison, eds., *Handbook of marine mammals*. Vol. 2, *Seals*. London: Academic Press, pp. 89–109.

BURNS, J. J., K. J. FROST, AND L. F. LOWRY, EDS. 1985. *Marine mammals species accounts*. Alaska Department of Fish and Game, Game Technical Bulletin 7.

BURT, W. H., ED. 1971. *Antarctic Pinnipedia*. Washington, DC: American Geophysical Union, Antarctic Research Series 18.

BUSCH, B. C. 1985. *The war against the seals. A history of the North American seal fishery*. Kingston, Ontario, and Montreal: McGill-Queen's University Press.

CAMPAGNA, C. 1985. The breeding cycle of the southern sea lion, *Otaria byronia. Marine Mammal Science* 1:210–18.

CAMPAGNA, C., AND B. J. LE BOEUF. 1988a. Reproductive behaviour of southern sea lions. *Behaviour* 104:233–61.

CAMPAGNA, C., AND B. J. LE BOEUF. 1988b. Thermoregulatory behaviour of southern sea lions and its effect on mating strategies. *Behaviour* 107:72–90.

CAMPAGNA, C., B. J. LE BOEUF, AND H. L. CAPPOZZO. 1988a. Group raids: A mating strategy of male southern sea lions. *Behaviour* 105:224–49.

CAMPAGNA, C., B. J. LE BOEUF, AND H. L. CAPPOZZO. 1988b. Pup abduction and infanticide in southern sea lions. *Behaviour* 107:44–60.

CAMPBELL, R. R. 1987. Status of the hooded seal, *Cystophora cristata,* in Canada. *Canadian Field-Naturalist* 101:253–65.

CARRICK, R., AND S. E. INGHAM. 1960. Ecological studies of the southern elephant seal, *Mirounga leonina* (L.), at Macquarie Island and Heard Island. *Mammalia* 24:325–42.

CARRICK, R., AND S. E. INGHAM. 1962. Studies on the southern elephant seal, *Mirounga leonina* (L.). II. Canine tooth structure in relation to function and age determination. *CSIRO Wildlife Research* 7:102–18.

CAWTHORN, M. W., M. C. CRAWLEY, R. H. MATTLIN, AND G. J. WILSON. 1985. Research on pinnipeds in New Zealand. *Wildlife Research Liaison Group Research Review* 7.

CHANIN, P. 1985. *The natural history of otters.* New York: Facts on File.

CHAPMAN, J. A., AND G. A. FELDHAMER, EDS. 1982. *Wild mammals of North America. Biology, management, and economics.* Baltimore: Johns Hopkins University Press.

CLARK, T. W. 1975. *Arctocephalus galapagoensis. Mammalian Species* 64: 1–2.

CLARKE, M. R., AND F. TRILLMICH. 1980. Cephalopods in the diet of fur seals of the Galapagos Islands. *Journal of Zoology, London* 190:211–15.

COLMENERO-R., L. C., AND B. E. ZÁRATE. 1990. Distribution, status and conservation of the West Indian manatee in Quintana Roo, México. *Biological Conservation* 52:27–35.

CONDY, P. R. 1978. Distribution, abundance, and annual cycle of fur seals (*Arctocephalus* spp.) on the Prince Edward Islands. *South African Journal of Wildlife Research* 8:159–68.

CONDY, P. R. 1979. Annual cycle of the southern elephant seal *Mirounga leonina* at Marion Island. *South African Journal of Zoology* 14:95–102.

CONDY, P. R. 1981. Annual food consumption, and seasonal fluctuations in biomass of seals at Marion Island. *Mammalia* 45:21–30.

COOPER, C. F., AND B. S. STEWART. 1983. Demography of northern elephant seals, 1911–1982. *Science* 219:969–71.

CORBET, G. B., AND S. HARRIS, EDS. 1991. *Handbook of British mammals,* 3d ed. Oxford: Mammal Society, Blackwell Scientific Publications.

CRAWFORD, R.J.M., J.H.M. DAVID, A. J. WILLIAMS, AND B. M. DYER. 1989. Competition for space: Recolonising seals displace endangered, endemic seabirds off Namibia. *Biological Conservation* 48:59–72.

CROXALL, J. P., AND R. L. GENTRY, EDS. 1987. *Status, biology, and ecology of fur seals. Proceedings of an international symposium and workshop, Cambridge, England, 23–27 April 1984.* NOAA Technical Report NMFS 51.

CROXALL, J. P., AND L. HIBY. 1983. Fecundity, survival and site fidelity in Weddell seals, *Leptonychotes weddelli. Journal of Applied Ecology* 20:19–32.

CROXALL, J. P., S. RODWELL, AND I. L. BOYD. 1990. Entanglement in man-made debris of antarctic fur seals at Bird Island, South Georgia. *Marine Mammal Science* 6:221–23.

DAMPIER, W. 1729. *A new voyage around the world.* Vol. 1, 7th ed. London: James and John Knapton.

DAVID, J. H. 1989. Seals. In A.I.L. Payne and R.J.M. Crawford, eds., *Oceans of life off South Africa*. Vlaeberg, South Africa: Vlaeberg Publishers, pp. 228–302.

DAVIES, J. L. 1957. The geography of the gray seal. *Journal of Mammalogy* 38:297–310.

DELONG, R. L., G. A. ANTONELIS, C. W. OLIVER, B. S. STEWART, M. C. LOWRY, AND P. K. YOCHEM. 1991. Effects of the 1982–83 El Niño on several population parameters and diet of California sea lions on the California Channel Islands. In F. Trillmich and K. Ono, eds., *Pinnipeds and El Niño: Responses to environmental stress*. Berlin: Springer-Verlag, pp. 166–72.

DELONG, R. L., AND B. S. STEWART. 1991. Diving patterns of northern elephant seal bulls. *Marine Mammal Science* 7:369–84.

DEMASTER, D. P., AND I. STIRLING. 1981. *Ursus maritimus*. *Mammalian Species* 145:1–7.

DIERAUF, L. A., ED. 1990. CRC handbook of marine mammal medicine: Health, disease, and rehabilitation. Boston, Massachusetts: CRC Press.

DIETZ, R., M.-P. HEIDE-JØRGENSEN, AND T. HÄRKÖNEN. 1989. Mass deaths of harbor seals (*Phoca vitulina*) in Europe. *Ambio* 18(5):258–64.

DOMNING, D. P. 1978. *Sirenian evolution in the North Pacific Ocean*. Berkeley: University of California Publications in Geological Sciences 118.

DOMNING, D. P. 1981. Distribution and status of manatees *Trichechus* spp. near the mouth of the Amazon River, Brazil. *Biological Conservation* 19:85–97.

DOMNING, D. P. 1982a. Commercial exploitation of manatees *Trichechus* in Brazil c. 1785–1973. *Biological Conservation* 22:101–26.

DOMNING, D. P. 1982b. Evolution of manatees: A speculative history. *Journal of Paleontology* 56:599–619.

DOMNING, D. P., ED. 1984–91. *Sirenews. Newsletter of the IUCN/SSC Sirenia Specialist Group,* Nos. 1–16. Washington, DC: Howard University, Department of Anatomy.

DOMNING, D. P., AND L. C. HAYEK. 1986. Interspecific and intraspecific morphological variation in manatees (Sirenia: *Trichechus*). *Marine Mammal Science* 2:87–144.

EBERHARDT, L. L. 1981. Population dynamics of the Pribilof fur seals. In C. W. Fowler and T. D. Smith, eds., *Dynamics of large mammal populations*. New York: Wiley, pp. 197–220.

ELIASON, J. J., T. C. JOHANOS, AND M. A. WEBBER. 1990. Parturition in the Hawaiian monk seal (*Monachus schauinslandi*). *Marine Mammal Science* 6:146–50.

ELSNER, R., D. WARTZOK, N. B. SONAFRANK, AND B. P. KELLY. 1989. Behavioral and physiological reactions of arctic seals during under-ice pilotage. *Canadian Journal of Zoology* 67: 2506–13.

ESTES, J. A. 1990. Growth and equilibrium in sea otter populations. *Journal of Animal Ecology* 59:385–401.

ESTES, J. A., AND G. R. VAN BLARICOM. 1985. Sea otters and shellfisheries. In J. R. Beddington, R.J.H. Beverton, and D. M. Lavigne, eds., *Marine mammals and fisheries*. London: Allen & Unwin, pp. 187–235.

FAY, F. H. 1981. Walrus *Odobenus rosmarus* (Linnaeus, 1758). In S. H. Ridgway and R. J. Harrison, eds., *Handbook of marine mammals*. Vol. 1, *The walrus, sea lions, fur seals and sea otter*. London: Academic Press, pp. 1–23.

FAY, F. H. 1982. Ecology and biology of the Pacific walrus, *Odobenus rosmarus divergens* Illiger. *North American Fauna* 74:1–279.

FAY, F. H. 1985. *Odobenus rosmarus*. *Mammalian Species* 238:1–7.

Fay, F. H., and G. A. Fedoseev, eds. 1984. *Soviet-American cooperative research on marine mammals.* Vol. 1– *Pinnipeds.* Seattle, Washington: NOAA Technical Report NMFS 12.

Fay, F. H., B. P. Kelly, and B. A. Fay, eds. 1990. *The ecology and management of walrus populations. Report of an international workshop,* 26–30 March 1990, Seattle. Washington, DC: Marine Mammal Commission.

Fay, F. H., B. P. Kelly, and J. L. Sease. 1989. Managing the exploitation of Pacific walruses: A tragedy of delayed response and poor communication. *Marine Mammal Science* 5:1–16.

Fay, F. H., G. C. Ray, and A. A. Kibal'chich. 1984. Time and location of mating and associated behavior of the Pacific walrus, *Odobenus rosmarus divergens* Illiger. In F. H. Fay and G. A. Fedoseev, eds., *Soviet-American cooperative research on marine mammals.* Vol. 1–*Pinnipeds.* Seattle, Washington: NOAA Technical Report NMFS 12, pp. 89–99.

Fedak, M. A., and S. S. Anderson. 1982. The energetics of lactation: Accurate measurements from a large wild mammal, the grey seal (*Halichoerus grypus*). *Journal of Zoology, London* 198:473–79.

Feldkamp, S. D., R. L. DeLong, and G. A. Antonelis. 1989. Diving patterns of California sea lions, *Zalophus californianus. Canadian Journal of Zoology* 67:872–83.

Finley, K. J., M.S.W. Bradstreet, and G. W. Miller. 1990. Summer feeding ecology of harp seals (*Phoca groenlandica*) in relation to arctic cod (*Boreogadus saida*) in the Canadian High Arctic. *Polar Biology* 10:609–18.

Finley, K. J., and C. R. Evans. 1983. Summer diet of the bearded seal (*Erignathus barbatus*) in the Canadian High Arctic. *Arctic* 36:82–89

Finley, K. J., G. W. Miller, R. A. Davis, and W. R. Koski. 1983. A distinctive large breeding population of ringed seals (*Phoca hispida*) inhabiting the Baffin Bay pack ice. *Arctic* 36(2): 162–73.

Fleischer, L. A. 1987. Guadalupe fur seal, *Arctocephalus townsendi. NOAA Technical Report NMFS* 51:43–48.

Food and Agriculture Organization of the United Nations. 1982. *Mammals in the seas.* Vol. 4, *Small cetaceans, seals, sirenians and otters.* FAO Fisheries Series no. 5. Rome: Food and Agriculture Organization of the United Nations.

Fowler, C. W. 1987. Marine debris and northern fur seals: A case study. *Marine Pollution Bulletin* 18:326–35.

Fowler, C. W. 1990. Density dependence in northern fur seals (*Callorhinus ursinus*). *Marine Mammal Science* 6:171–95.

Francour, P., D. Marchessaux, A. Argiolas, P. Campredon, and G. Vuignier. 1990. La population de phoque moine *(Monachus monachus)* de Mauritanie. *Revue d'Écologie (La Terre et la Vie)* 45:55–64.

Frost, K. J., and L. F. Lowry. 1980. Feeding of ribbon seals (*Phoca fasciata*) in the Bering Sea in spring. *Canadian Journal of Zoology* 58:1601–7.

Frost, K. J., and L. F. Lowry. 1981. Ringed, Baikal and Caspian seals *Phoca hispida* Schreber, 1775; *Phoca sibirica* Gmelin, 1788 and *Phoca caspica* Gmelin, 1788. In S. H. Ridgway and R. J. Harrison, eds., *Handbook of marine mammals.* Vol. 2, *Seals:* London: Academic Press, pp. 29–53.

Gales, N. J., M. Adams, and H. R. Burton. 1989. Genetic relatedness of two populations of the southern elephant seal, *Mirounga leonina. Marine Mammal Science* 5:57–67.

Gaskin, D. E. 1972. *Whales, dolphins and seals. With special reference to the New Zealand region.* London: Heinemann.

Gentry, R. L., and J. H. Johnson. 1981. Predation by sea lions on northern fur seal neonates. *Mammalia* 45:423–30.
Gentry, R. L., and G. L. Kooyman, eds. 1986. *Fur seals. Maternal strategies on land and at sea.* Princeton, N.J.: Princeton University Press.
Geraci, J. R., and D. J. St. Aubin, eds. 1990. *Sea mammals and oil: Confronting the risks.* San Diego, California: Academic Press.
Gerrodette, T., and W. G. Gilmartin. 1990. Demographic consequences of changed pupping and hauling sites of the Hawaiian monk seal. *Conservation Biology* 4:423–30.
Goebel, M. E., J. L. Bengtson, R. L. DeLong, R. L. Gentry, and T. R. Loughlin. 1991. Diving patterns and foraging locations of female northern fur seals. *U.S. Fishery Bulletin* 89:171–79.
Grachev, M. A., et al. 1989. Distemper virus in Baikal seals. *Nature* 338:209.
Guerra C., C., and D. Torres N. 1987. Presence of the South American fur seal, *Arctocephalus australis,* in northern Chile. NOAA Technical Report NMFS 51:169–75.
Haley, D., ed. 1986. *Marine mammals of eastern North Pacific and arctic waters.* 2d ed., rev. Seattle: Pacific Search Press.
Hamilton, J. E. 1934. The southern sea lion, *Otaria byronia* (deBlainville). *Discovery Reports* 8:269–318.
Hamilton, J. E. 1939. A second report on the southern sea lion, *Otaria byronia* (deBlainville). *Discovery Reports* 19:121–64.
Hammill, M. O., and T. G. Smith. 1991. The role of predation in the ecology of the ringed seal in Barrow Strait, Northwest Territories, Canada. *Marine Mammal Science* 7:123–35.
Harestad, A. S., and H. D. Fisher. 1975. Social behavior in a non-pupping colony of Steller sea lions (*Eumetopias jubata*). *Canadian Journal of Zoology* 53:1596–1613.
Härkönen, T. 1987. Seasonal and regional variations in the feeding habits of the harbour seal, *Phoca vitulina,* in the Skagerrak and the Kattegat. *Journal of Zoology, London* 213:535–43.
Härkönen, T., and M.-P. Heide-Jørgensen. 1990a. Comparative life histories of east Atlantic and other harbour seal populations. *Ophelia* 32:211–35.
Härkönen, T., and M.-P. Heide-Jørgensen. 1990b. Density and distribution of the ringed seal in the Bothnian Bay. *Holarctic Ecology* 13:122–29.
Hartman, D. S. 1979. *Ecology and behavior of the manatee (Trichechus manatus) in Florida.* American Society of Mammalogists, Special Pub. no. 5.
Harwood, J., ed. 1987. *Population biology of the Mediterranean monk seal in Greece.* Cambridge, England: Natural Environment Research Council, Sea Mammal Research Unit.
Harwood, J., et al. 1989. New approaches for field studies of mammals: Experiences with marine mammals. *Biological Journal of the Linnean Society* 38:103–11.
Harwood, J., and J.J.D. Greenwood. 1985. Competition between British grey seals and fisheries. In J. R. Beddington, R.J.H. Beverton, and D. M. Lavigne, eds., *Marine mammals and fisheries.* London: Allen & Unwin, pp. 153–69.
Harwood, J., and J. H. Prime. 1978. Some factors affecting the size of British grey seal populations. *Journal of Applied Ecology* 15:401–11.
Hatt, R. T. 1934. A manatee collected by the American Museum Congo

Expedition, with observations on the Recent manatees. *Bulletin of the American Museum of Natural History* 66(4):533–66.
HEIDE-JØRGENSEN, M.-P., B. S. STEWART, AND S. LEATHERWOOD. 1991. Satellite tracking of ringed seals off Northwest Greenland. *Holarctic Ecology* (in press).
HEIMARK, R. J., AND G. M. HEIMARK. 1986. Southern elephant seal pupping at Palmer Station, Antarctica. *Journal of Mammalogy* 67:189–90.
HELLE, E. 1980. Age structure and sex ratio of the ringed seal *Phoca (Pusa) hispida* Schreber population in the Bothnian Bay, northern Baltic Sea. *Zeitschrift für Säugetierkunde* 45:310–17.
HELLE, E., H. HYVÄRINEN, AND T. SIPILÄ. 1984. Breeding habitat and lair structure of the Saimaa ringed seal *Phoca hispida saimensis* Nordq. in Finland. *Acta Zoologica Fennica* 172:125–27.
HINDELL, M. A. 1991. Some life-history parameters of a declining population of southern elephant seals, *Mirounga leonina*. *Journal of Animal Ecology* 60:119–34.
HOOK, O., AND A. G. JOHNELS. 1972. The breeding and distribution of the grey seal (*Halichoerus grypus* Fab.) in the Baltic Sea, with observations on other seals of the area. *Proceedings of the Royal Society of London* B 182:37–58.
HUBBS, C. L., AND K. S. NORRIS. 1971. Original teeming abundance, supposed extinction, and survival of the Juan Fernández fur seal. In W. H. Burt, ed., *Antarctic Pinnipedia*. Washington, DC: American Geophysical Union, Antarctic Research Series 18, pp. 35–52.
HUBER, H. R. 1987. Natality and weaning success in relation to age of first reproduction in northern elephant seals. *Canadian Journal of Zoology* 65:1311–16.
HUDSON, B.E.T. 1986. Dugongs and people. *Oceanus* 29(2):100–6.
HUNTLEY, A. C., D. P. COSTA, G.A.J. WORTHY, AND M. A. CASTELLINI, EDS. 1987. *Approaches to marine mammal energetics*. Society for Marine Mammalogy, Special Pub. 1.
HUSAR, S. L. 1978. *Trichechus senegalensis*. *Mammalian Species* 89:1–3.
JOHNSON, A. M., R. L. DELONG, C. H. FISCUS, AND K. W. KENYON. 1982. Population status of the Hawaiian monk seal (*Monachus schauinslandi*), 1978. *Journal of Mammalogy* 63:415–21.
JONES, E. 1981. Age in relation to breeding status of the male southern elephant seal, *Mirounga leonina* (L.), at Macquarie Island. *Australian Wildlife Research* 8:327–34.
KAPEL, F. O. 1975. Recent research on seals and seal hunting in Greenland. *Rapports et procès-verbaux des réunions. Conseil international pour l'exploration de la mer* 169:462–78.
KELLY, B. P. 1988a. Bearded seal *Erignathus barbatus*. In J. W. Lentfer, ed., *Selected marine mammals of Alaska*. Washington, DC: Marine Mammal Commission, pp. 77–93.
KELLY, B. P. 1988b. Ribbon seal *Phoca fasciata*. In J. W. Lentfer, ed., *Selected marine mammals of Alaska*. Washington, DC: Marine Mammal Commission, pp. 95–106.
KELLY, B. P. 1988c. Ringed seal *Phoca hispida*. In J. W. Lentfer, ed., *Selected marine mammals of Alaska*. Washington, DC: Marine Mammal Commission, pp. 57–75.
KELLY, B. P., AND L. T. QUAKENBUSH. 1990. Spatiotemporal use of lairs by ringed seals (*Phoca hispida*). *Canadian Journal of Zoology* 68:2503–12.
KENYON, K. W. 1962. History of the Steller sea lion at the Pribilof Islands, Alaska. *Journal of Mammalogy* 43:68–75.

Kenyon, K. W. 1969. The sea otter in the eastern Pacific Ocean. *North American Fauna* 68:1–352.

Kenyon, K. W. 1972. Man versus the monk seal. *Journal of Mammalogy* 53(4):687–96.

Kenyon, K. W. 1977. Caribbean monk seal extinct. *Journal of Mammalogy* 58:97–98.

Kenyon, K. W. 1981a. Monk seals *Monachus* Fleming, 1822. In S. H. Ridgway and R. J. Harrison, eds., *Handbook of marine mammals*. Vol. 2, *Seals*. London: Academic Press, pp. 195–220.

Kenyon, K. W. 1981b. Sea otter *Enhydra lutris (*Linnaeus, 1758*)*. In S. H. Ridgway and R. J. Harrison, eds., *Handbook of marine mammals*. Vol. 1, *The walrus, sea lions, fur seals and sea otter*. London: Academic Press, pp. 209–23.

Kenyon, K. W., and D. W. Rice. 1959. Life history of the Hawaiian monk seal. *Pacific Science* 13:215–52.

Kerley, G.I.H. 1983. Relative population sizes, trends, and hybridization of *Arctocephalus tropicalis* and *A. gazella* at the Prince Edward Islands, Southern Ocean. *South African Journal of Zoology* 18:388–92.

King, J. E. 1983. *Seals of the world*. 2d ed. Oxford, England: Oxford University Press.

Kingsley, M.C.S. 1990. Status of the ringed seal, *Phoca hispida,* in Canada. *Canadian Field-Naturalist* 104:138–45.

Klages, N.T.W., and V. G. Cockcroft. 1990. Feeding behaviour of a captive crabeater seal. *Polar Biology* 10:403–4.

Kooyman, G. L. 1981a. Leopard seal *Hydrurga leptonyx* Blainville, 1820. In S. H. Ridgway and R. J. Harrison, eds., *Handbook of marine mammals*. Vol. 2, *Seals*. London: Academic Press, pp. 261–74.

Kooyman, G. L. 1981b. *Weddell seal: Consummate diver*. Cambridge, England: Cambridge University Press.

Kooyman, G. L. 1989. *Diverse divers*. Berlin: Springer-Verlag.

Kovacs, K. M. 1987. Maternal behaviour and early behavioural ontogeny of grey seals (*Halichoerus grypus*) on the Isle of May, U.K. *Journal of Zoology, London* 213:697–715.

Kovacs, K. M. 1990. Mating strategies in male hooded seals (*Cystophora cristata*)? *Canadian Journal of Zoology* 68:2499–2502.

Kovacs, K. M., and D. M. Lavigne. 1985. Neonatal growth and organ allometry of Northwest Atlantic harp seals (*Phoca groenlandica*). *Canadian Journal of Zoology* 63:2793–99.

Kovacs, K. M., and D. M. Lavigne. 1986. *Cystophora cristata*. *Mammalian Species* 258:1–9.

Kurtén, B. 1964. The evolution of the polar bear, *Ursus maritimus* Phipps. *Acta Zoologica Fennica* 108:1–30.

Kvitek, R. G., A. K. Fukayama, B. S. Anderson, and B. K. Grimm. 1988. Sea otter foraging on deep-burrowing bivalves in a California coastal lagoon. *Marine Biology* 98:157–67.

Larsen, T. 1978. *The world of the polar bear*. London: Hamlyn.

Lavigne, D. M. 1978. The harp seal controversy reconsidered. *Queen's Quarterly* 85(3):377–88.

Lavigne, D. M., and K. M. Kovacs. 1988. *Harps and hoods. Ice-breeding seals of the northwest Atlantic*. Waterloo, Ontario, Canada: University of Waterloo Press.

Laws, R. M. 1960. The southern elephant seal (*Mirounga leonina* Linn.) at South Georgia. *Norsk Hvalfangst-tidende* 49:466–76, 520–42.

LAWS, R. M. 1984. Seals. In R. M. Laws, ed., *Antarctic ecology,* Vol. 2. London: Academic Press, pp. 621–715.

LEATHERWOOD, S., AND R. R. REEVES, EDS. 1989. *Marine mammal research and conservation in Sri Lanka* 1985–1986. Nairobi, Kenya: United Nations Environment Programme, Marine Mammal Technical Report 1.

LE BOEUF, B. J. 1972. Sexual behavior in the northern elephant seal *Mirounga angustirostris. Behaviour* 41:1–26.

LE BOEUF, B., K. W. KENYON, AND B. VILLA-RAMIREZ. 1986. The Caribbean monk seal is extinct. *Marine Mammal Science* 2:70–72.

LE BOEUF, B. J., Y. NAITO, A. C. HUNTLEY, AND T. ASAGA. 1989. Prolonged, continuous, deep diving by northern elephant seals. *Canadian Journal of Zoology* 67:2514–19.

LEFEBVRE, L. W., T. J. O'SHEA, G. B. RATHBUN, AND R. C. BEST. 1989. Distribution, status, and biogeography of the West Indian manatee. In C. A. Woods, ed., *Biogeography of the West Indies.* Gainesville, FL: Sandhill Crane Press, pp. 567–610.

LENTFER, J. W. 1982. Polar bear *Ursus maritimus.* In J. A. Chapman and G. A. Feldhamer, eds., *Wild mammals of North America. Biology, management, and economics.* Baltimore: Johns Hopkins Univ. Press, pp. 557–66.

LENTFER, J. W., ED. 1988. *Selected marine mammals of Alaska. Species accounts with research and management recommendations.* Washington, DC: Marine Mammal Commission.

LING, J. K. 1987. New Zealand fur seal, *Arctocephalus forsteri,* (Lesson), in South Australia. *NOAA Technical Report NMFS* 51:53–55.

LIPINSKI, M. R., AND J.H.M. DAVID. 1990. Cephalopods in the diet of the South African fur seal (*Arctocephalus pusillus pusillus*). *Journal of Zoology, London* 221:359–74.

LLOYD, D. S., C. P. MCROY, AND R. H. DAY. 1981. Discovery of northern fur seals (*Callorhinus ursinus*) breeding on Bogoslof Island, southeastern Bering Sea. *Arctic* 34:318–20.

LOUGHLIN, T. R., M. A. PEREZ, AND R. L. MERRICK. 1987. *Eumetopias jubatus. Mammalian Species* 283:1–7.

LOWRY, L. F., J. J. BURNS, AND R. R. NELSON. 1987. Polar bear, *Ursus maritimus,* predation on belugas, *Delphinapterus leucas,* in the Bering and Chukchi seas. *Canadian Field-Naturalist* 101:141–46.

LOWRY, L. F., AND F. H. FAY. 1984. Seal eating by walruses in the Bering and Chukchi seas. *Polar Biology* 3:11–18.

LOWRY, L. F., K. J. FROST, AND J. J. BURNS. 1980a. Feeding of bearded seals in the Bering and Chukchi seas and trophic interaction with Pacific walruses. *Arctic* 33:330–42.

LOWRY, L. F., K. J. FROST, AND J. J. BURNS. 1980b. Variability in the diet of ringed seals, *Phoca hispida,* in Alaska. *Canadian Journal of Fisheries and Aquatic Sciences* 37:2254–61.

LOWRY, L. F., J. W. TESTA, AND W. CALVERT. 1988. Notes on winter feeding of crabeater and leopard seals near the Antarctic Peninsula. *Polar Biology* 8:475–78.

LOWRY, M. S., B. S. STEWART, C. B. HEATH, P. K. YOCHEM, AND J. M. FRANCIS. 1991. Seasonal and annual variability in the diet of California sea lions *Zalophus californianus* at San Nicolas Island, California, 1981–86. *U.S. Fishery Bulletin* 89:331–36.

LUNN, N. J. 1985. Observations of nonaggressive behavior between polar bear family groups. *Canadian Journal of Zoology* 64:2035–37.

LØNØ, O. 1970. The polar bear in the Svalbard area. *Norsk Polarinstitut Skrifter* 149:1–103.

MAJLUF, P. 1987. South American fur seal, *Arctocephalus australis,* in Peru. *NOAA Technical Report NMFS* 51:33–35.

MAJLUF, P., AND F. TRILLMICH. 1981. Distribution and abundance of sea lions (*Otaria byronia*) and fur seals (*Arctocephalus australis*) in Peru. *Zeitschrift für Säugetierkunde* 46:384–93.

MALOUF, A. H. 1986. *Seals and sealing in Canada. Report of the Royal Commission.* 3 vols. Ottawa, Ontario, Canada: Supply and Services Canada.

MANNING, T. H. 1971. Geographical variation in the polar bear *Ursus maritimus* Phipps. *Canadian Wildlife Service Report Series* 13:1–27.

MANSFIELD, A. W. 1988. *The grey seal.* Ottawa, Ontario, Canada: Communications Directorate, Department of Fisheries and Oceans.

MANSFIELD, A. W., AND B. BECK. 1977. *The grey seal in eastern Canada.* Sainte-Anne-de-Bellevue, Québec, Canada: Fisheries and Marine Service Technical Report 704.

MARCHESSAUX, D. 1989. Distribution et statut des populations du phoque moine *Monachus monachus* (Hermann, 1779). *Mammalia* 53:621–42.

MARKUSSEN, N. H., R. RYG, AND N. A. ØRITSLAND. 1990. Energy requirements for maintenance and growth of captive harbour seals, *Phoca vitulina. Canadian Journal of Zoology* 68:423–26.

MARLOW, B. J. 1968. The sea-lions of Dangerous Reef. *Australian Natural History* 16 (2):39–44.

MARLOW, B. J. 1975. The comparative behaviour of the Australasian sea lions *Neophoca cinerea* and *Phocarctos hookeri* (Pinnipedia: Otariidae). *Mammalia* 39:159–230.

MARLOW, B. J., AND J. E. KING. 1974. Sea lions and fur seals of Australia and New Zealand–The growth of knowledge. *Australian Mammalogy* 1:117–35.

MARSH, H., ed. 1981. *The dugong: Proceedings of a seminar/workshop held at James Cook University,* 8–13 May 1979. Townsville, Australia: James Cook University of North Queensland, Department of Zoology.

MARSH, H. 1988. An ecological basis for dugong conservation in Australia. In M. L. Augee, ed., *Marine mammals of Australasia.* Sydney, Australia: Royal Zoological Society of New South Wales, pp. 9–21.

MARSH, H., G. E. HEINSOHN, AND L. M. MARSH. 1984. Breeding cycle, life history and population dynamics of the dugong, *Dugong dugon* (Sirenia: Dugongidae). *Australian Journal of Zoology* 32:767–88.

MARSH, H., AND W. K. SAALFELD. 1989. Distribution and abundance of dugongs in the Northern Great Barrier Reef Marine Park. *Australian Wildlife Research* 16:429–40.

MARSH, H., AND W. K. SAALFELD. 1990. The distribution and abundance of dugongs in the Great Barrier Reef Marine Park south of Cape Bedford. *Australian Wildlife Research* 17:511–24.

MARSH, H., A. V. SPAIN, AND G. E. HEINSOHN. 1978. Physiology of the dugong. *Comparative Biochemistry and Physiology,* 61A:159–68.

MATHISEN, O. A., R. T. BAADE, AND R. J. LOPP. 1962. Breeding habits, growth and stomach contents of the Steller sea lion in Alaska. *Journal of Mammalogy* 43:469–77.

MATTLIN, R. H. 1987. New Zealand fur seal, *Arctocephalus forsteri,* within the New Zealand region. *NOAA Technical Report NMFS* 51:49–51.

MCCANN, T. S. 1980a. Population structure and social organization of southern elephant seals, *Mirounga leonina* (L.). *Biological Journal of the Linnean Society* 14:133–50.

McCann, T. S. 1980b. Territoriality and breeding behaviour of adult male antarctic fur seal, *Arctocephalus gazella*. *Journal of Zoology, London* 192: 295–310.
McCann, T. S. 1981. Aggression and sexual activity of male southern elephant seals, *Mirounga leonina*. *Journal of Zoology, London* 195:295–310.
McCann, T. S. 1982. Aggressive and maternal activities of female southern elephant seals (*Mirounga leonina*). *Animal Behaviour* 30:268–76.
McCann, T. S. 1983. Activity budgets of southern elephant seals, *Mirounga leonina*, during the breeding season. *Zeitschrift für Tierpsychologie* 61:111–26.
McCann, T. S., and P. Rothery. 1988. Population size and status of the southern elephant seal (*Mirounga leonina*) at South Georgia, 1951–1985. *Polar Biology* 8:305–9.
McLaren, I. A. 1958a. The biology of the ringed seal (*Phoca hispida* Schreber) in the eastern Canadian Arctic. *Bulletin of the Fisheries Research Board of Canada* 118.
McLaren, I. A. 1958b. Some aspects of growth and reproduction of the bearded seal, *Erignathus barbatus* (Erxleben). *Journal of the Fisheries Research Board of Canada* 15:219–27.
Merrick, R. L., T. R. Loughlin, and D. G. Calkins. 1987. Decline in abundance of the northern sea lion, *Eumetopias jubatus*, in Alaska, 1956–86. *U.S. Fishery Bulletin* 85:351–65.
Miller, E. H., 1974. A paternal role in Galapagos sea lions? *Evolution* 28:473–76.
Miller, E. H. 1975. Walrus ethology. I. The social role of tusks and applications of multidimensional scaling. *Canadian Journal of Zoology* 53:590–613.
Miller, E. H. 1976. Walrus ethology. II. Herd structure and activity budgets of summering males. *Canadian Journal of Zoology* 54:704–15.
Miller, E. H. 1985. Airborne acoustic communication in the walrus *Odobenus rosmarus*. *National Geographic Research* 1:124–45.
Miller, E. H., and D. J. Boness. 1979. Remarks on display functions of the snout of the grey seal, *Halichoerus grypus* (Fab.), with comparative notes. *Canadian Journal of Zoology* 57:140–48.
Montgomery, G. G., R. C. Best, and M. Yamakoshi. 1981. A radio-tracking study of the Amazonian manatee *Trichechus inunguis* (Mammalia: Sirenia). *Biotropica* 13:81–85.
Naito, Y., and S. Konno. 1979. The post-breeding distributions of ice-breeding harbour seal (*Phoca largha*) and ribbon seal (*Phoca fasciata*) in the southern Sea of Okhotsk. *Scientific Reports of the Whales Research Institute* 31:105–19.
Naito, Y., and M. Nishiwaki. 1972. The growth of two species of the harbour seal in the adjacent waters of Hokkaido. *Scientific Reports of the Whales Research Institute* 24:127–44.
Naito, Y., and M. Nishiwaki. 1975. Ecology and morphology of *Phoca vitulina largha* and *Phoca kurilensis* in the southern Sea of Okhotsk and northeast of Hokkaido. *Rapports et procès-verbaux des réunions. Conseil international pour l'exploration de la mer* 169:379–86.
Naito, Y., and M. Oshima. 1976. The variation in the development of pelage of the ribbon seal with reference to the systematics. *Scientific Reports of the Whales Research Institute* 28:187–97.
Nijhoff, P. 1979. Lake Baikal endangered by pollution. *Environmental Conservation* 6:111–15.
Nishiwaki, M., and H. Marsh. 1985. Dugong *Dugong dugon* (Müller,

1776). In S. H. Ridgway and R. J. Harrison, eds., *Handbook of marine mammals*. Vol. 3, *The sirenians and baleen whales*. London: Academic Press, pp. 1–31.

NORRIS, K. S., AND W. A. WATKINS. 1971. Underwater sounds of *Arctocephalus philippii,* the Juan Fernández fur seal. In W. H. Burt, ed., *Antarctic Pinnipedia*. Washington, DC: American Geophysical Union, Antarctic Research Series 18, pp. 169–71.

ODELL, D. K. 1975. Breeding biology of the California sea lion, *Zalophus californianus. Rapports et procès-verbaux des réunions. Conseil international pour l'exploration de la mer* 169:374–78.

OLESIUK, P. F., M. A. BIGG, AND G. M. ELLIS. 1990. Recent trends in the abundance of harbour seals, *Phoca vitulina,* in British Columbia. *Canadian Journal of Fisheries and Aquatic Sciences* 47:992–1003.

ORR, R. T., AND R. C. HELM. 1989. *Marine mammals of California,* rev. ed. Berkeley: University of California Press.

ORR, R. T., AND T. C. POULTER. 1967. Some observations on reproduction, growth, and social behavior in the Steller sea lion. *Proceedings of the California Academy of Sciences,* Fourth Series 35:193–226.

O'SHEA, T. J. 1986. Mast foraging by West Indian manatees (*Trichechus manatus*). *Journal of Mammalogy* 67:183–85.

O'SHEA, T. J., M. CORREA-VIANA, M. E. LUDLOW, AND J. G. ROBINSON. 1988. Distribution, status, and traditional significance of the West Indian manatee *Trichechus manatus* in Venezuela. *Biological Conservation* 46:281–301.

O'SHEA, T. J., G. B. RATHBUN, R. K. BONDE, C. D. BUERGELT, AND D. K. ODELL. 1991. An epizootic of Florida manatees associated with a dinoflagellate bloom. *Marine Mammal Science* 7:165–79.

OSTFELD, R. S., L. EBENSPERGER, L. L. KLOSTERMAN, AND J. C. CASTILLA. 1989. Foraging, activity budget, and social behavior of the South American marine otter *Lutra felina* (Molina 1782). *National Geographic Research* 5(4):422–38.

PASCAL, M. 1985. Numerical changes in the population of elephant seals (*Mirounga leonina* L.) in the Kerguelen Archipelago during the past 30 years. In J. R. Beddington, R.J.H. Beverton, and D. M. Lavigne, eds., *Marine mammals and fisheries*. London: Allen & Unwin, pp. 170–86.

PASTUKHOV, V. D. 1989. Anthropogenic factors affecting the population of the Baikal seal. In A. V. Yablokov and M. Olsson, eds., *Influence of human activities on the Baltic ecosystem. Proceedings of the Soviet-Swedish symposium Effect of toxic substances on dynamics of seal populations, Moscow,* 14–18 *April 1986*. Leningrad: Gidrometeoizdat, pp. 42–52.

PAYNE, M. R. 1977. Growth of a fur seal population. *Philosophical Transactions of the Royal Society of London* B279:67–80.

PAYNE, M. R. 1978. Population size and age determination in the antarctic fur seal *Arctocephalus gazella*. *Mammal Review* 8:67–73.

PAYNE, M. R. 1979. Fur seals *Arctocephalus tropicalis* and *A. gazella* crossing the Antarctic Convergence at South Georgia. *Mammalia* 43:93–98.

PAYNE, P. M., AND L. A. SELZER. 1989. The distribution, abundance and selected prey of the harbor seal, *Phoca vitulina concolor,* in southern New England. *Marine Mammal Science* 5:173–92.

PEREZ, M. A., AND M. A. BIGG. 1986. Diet of northern fur seals, *Callorhinus ursinus,* off western North America. *U.S. Fishery Bulletin* 84:957–71.

PETERSON, R. S., AND G. A. BARTHOLOMEW. 1967. *The natural history and behavior of the California sea lion*. American Society of Mammalogists, Special Pub. 1.

PETERSON, R. S., C. L. HUBBS, R. L. GENTRY, AND R. L. DELONG. 1968. The Guadalupe fur seal: Habitat, behavior, population size, and field identification. *Journal of Mammalogy* 49:665–75.

PITCHER, K. W. 1981. Prey of the Steller sea lion, *Eumetopias jubatus,* in the Gulf of Alaska. *U.S. Fishery Bulletin* 79:467–72.

PITCHER, K. W., AND D. G. CALKINS. 1981. Reproductive biology of Steller sea lions in the Gulf of Alaska. *Journal of Mammalogy* 62:599–605.

PLÖTZ, J. 1986. Summer diet of Weddell seals (*Leptonychotes weddelli*) in the eastern and southern Weddell Sea, Antarctica. *Polar Biology* 6:97–102.

POPOV, L. A. 1982. Status of the main ice-living seals inhabiting inland waters and coastal marine areas of the USSR. *FAO Fisheries Series* 5(4):361–81.

POWELL, J. A., JR. 1978. Evidence of carnivory in manatees (*Trichechus manatus*). *Journal of Mammalogy* 59:442.

POWELL, J. A., AND G. B. RATHBUN. 1984. Distribution and abundance of manatees along the northern coast of the Gulf of Mexico. *Northeast Gulf Science* 7:1–28.

PREEN, A. 1989. Observations of mating behavior in dugongs (*Dugong dugon*). *Marine Mammal Science* 5:382–87.

PRIME, J. H. 1985. The current status of the grey seal *Halichoerus grypus* in Cornwall, England. *Biological Conservation* 33:81–87.

QUAKENBUSH, L. T. 1988. Spotted seal *Phoca largha*. In J. W. Lentfer, ed., *Selected marine mammals of Alaska*. Washington, DC: Marine Mammal Commission, pp. 107–24.

RALLS, K., AND D. B. SINIFF. 1990. Time budgets and activity patterns in California sea otters. *Journal of Wildlife Management* 54(2):251–69.

RATHBUN, G. B., R. K. BONDE, AND D. CLAY. 1982. The status of the West Indian manatee on the Atlantic coast north of Florida. In R. R. Odom and J. W. Guthrie, eds., *Proceedings of the symposium on nongame and endangered wildlife*. Georgia Department of Natural Resources, Game and Fish Division, Technical Bulletin WL5, pp. 152–65.

RATHBUN, G. B., J. A. POWELL, AND G. CRUZ. 1983. Status of the West Indian manatee in Honduras. *Biological Conservation* 26:301–8.

RATHBUN, G. B., J. P. REID, AND G. CAROWAN. 1990. Distribution and movement patterns of manatees (*Trichechus manatus*) in northwestern peninsular Florida. *Florida Marine Research Publications* 48:1–33.

RAY, C., W. A. WATKINS, AND J. J. BURNS. 1969. The underwater song of *Erignathus* (bearded seal). *Zoologica* 54:79–83.

RAY, C. E., F. REINER, D. E. SERGEANT, AND C. N. QUESADA. 1982. Notes on the past and present distribution of the bearded seal, *Erignathus barbatus,* around the North Atlantic Ocean. *Memórias do Museu do Mar,* Sér. Zool. 2 (23):1–32.

RAY, G. C. 1981. Ross seal *Ommatophoca rossi* Gray, 1844. In S. H. Ridgway and R. J. Harrison, eds., *Handbook of marine mammals*. Vol. 2, *Seals*. London: Academic Press, pp. 237–60.

REEVES, R. R. 1978. The Atlantic walrus (*Odobenus rosmarus rosmarus*): A literature review and status report. U.S. Fish and Wildlife Service, *Wildlife Research Report* 10:1–41.

REEVES, R. R., D. TUBOKU-METZGER, AND R. A. KAPINDI. 1988. Distribution and exploitation of manatees in Sierra Leone. *Oryx* 22:75–84.

REIJNDERS, P.J.H., M. N. DE VISSCHER, AND E. RIES, EDS. 1988. *The Mediterranean monk seal*. Gland, Switzerland: International Union for Conservation of Nature and Natural Resources.

RENOUF, D., ED. 1991. *The Behaviour of Pinnipeds*. London: Chapman and Hall.

Repenning, C. A., R. S. Peterson, and C. L. Hubbs. 1971. Contributions to the systematics of the southern fur seals, with particular reference to the Juan Fernández and Guadalupe species. In W. H. Burt, ed., *Antarctic Pinnipedia*. Washington, DC: American Geophysical Union, Antarctic Research Series 18, pp. 1–34.

Repenning, C. A., C. E. Ray, and D. Grigorescu. 1979. Pinniped biogeography. In J. Gray and A. J. Boucot, eds., *Historical biogeography, plate tectonics, and the changing environment*. Corvallis, Oregon: Oregon State University Press, pp. 357–69.

Reynolds, J. E., III. 1981. Aspects of the social behaviour and herd structure of a semi-isolated colony of West Indian manatees, *Trichechus manatus*. *Mammalia* 45:431–51.

Reynolds, J. E., III, and J. R. Wilcox. 1986. Distribution and abundance of the West Indian manatee *Trichechus manatus* around selected Florida power plants following winter cold fronts: 1984–85. *Biological Conservation* 38:103–13.

Rice, D. W. 1960. Population dynamics of the Hawaiian monk seal. *Journal of Mammalogy* 41(3):376–85.

Rice, D. W. 1973. Caribbean monk seal (*Monachus tropicalis*). In *Seals*. Morges, Switzerland: International Union for Conservation of Nature and Natural Resources, IUCN Publications New Series, Supplementary Paper 39, pp. 98–112.

Ridgway, S. H., and R. J. Harrison, eds. 1981a. *Handbook of marine mammals*. Vol. 1, *The walrus, sea lions, fur seals and sea otter*. London: Academic Press.

Ridgway, S. H., and R. J. Harrison, eds. 1981b. *Handbook of marine mammals*. Vol. 2, *Seals*. London: Academic Press.

Ridgway, R. H., and R. J. Harrison, eds. 1985. *Handbook of marine mammals*. Vol. 3, *The sirenians and baleen whales*. London: Academic Press.

Riedman, M. 1990. *The pinnipeds. Seals, sea lions, and walruses*. Berkeley: University of California Press.

Riedman, M. L., and J. A. Estes. 1990. The sea otter *(Enhydra lutris):* Behavior, ecology, and natural history. *U.S. Fish and Wildlife Service Biological Reports* 90(14), 126 pp.

Robinson, A. C., and T. E. Dennis. 1988. The status and management of seal populations in South Australia. In M. L. Augee, ed., *Marine mammals of Australasia. Field biology and captive management*. Sydney, Australia: Royal Zoological Society of New South Wales, pp. 87–110.

Ronald, K., and R. Duguy, eds. 1979. *The Mediterranean monk seal. Proceedings of the First International Conference, Rhodes, Greece, 2–5 May 1978*. Oxford, England: Pergamon Press.

Ronald, K., and R. Duguy, eds. 1984. *Les phoques moines monk seals. Proceedings of the Second International Conference, La Rochelle, France, 5–6 October 1984*. La Rochelle, France: Annales de la Société des Sciences Naturelles de la Charente-Maritime, Supplément.

Ronald, K., P. J. Healey, J. Dougan, L. J. Selley, and L. Dunn. 1983. *An annotated bibliography on the Pinnipedia Supplement I*. International Council for the Exploration of the Sea, Copenhagen, Denmark.

Ronald, K., L. M. Hanly, P. J. Healey, and L. J. Selley. 1976. *An annotated bibliography on the Pinnipedia*. International Council for the Exploration of the Sea. Charlottenlund, Denmark.

Ronald, K., and A. W. Mansfield, eds. 1975. *Biology of the seal. Proceedings of a symposium held in Guelph, 14–17 August 1972.* Charlottenlund, Denmark: Conseil International pour l'Exploration de la Mer, Rapports et Procès-verbaux des Réunions 169.

ROTH, H. H., AND E. WAITKUWAIT. 1986. Répartition et statut des grandes espèces de mammifères en Côte-d'Ivoire III. Lamantins. *Mammalia* 50:227–42.

ROTTERMAN, L. M., AND T. SIMON-JACKSON. 1988. Sea otter *Enhydra lutris.* In J. W. Lentfer, ed., *Selected marine mammals of Alaska.* Washington, DC: Marine Mammal Commission, pp. 237–75.

ROUX, J.-P. 1987. Subantarctic fur seal, *Arctocephalus tropicalis,* in French subantarctic territories. *NOAA Technical Report NMFS* 51:79–81.

ROUX, J.-P., AND A. D. HES. 1984. The seasonal haul-out cycle of the fur seal *Arctocephalus tropicalis* (Gray, 1872) on Amsterdam Island. *Mammalia* 48:377–89.

ROWLEY, J. 1929. Life history of the sea lions on the California coast. *Journal of Mammalogy* 10:1–36.

RYG, M., T. G. SMITH, AND N. A. ØRITSLAND. 1990. Seasonal changes in body mass and body composition of ringed seals (*Phoca hispida*) on Svalbard. *Canadian Journal of Zoology* 68:470–75.

SALTER, R. E. 1979. Site utilization, activity budgets, and disturbance responses of Atlantic walruses during terrestrial haul-out. *Canadian Journal of Zoology* 57:1169–80.

SAKURAI, Y., K. ABE, AND Y. NAITO. 1989. A report on unusual mass occurrences of ribbon seal pups along the northeastern coast of Honshu and southern Hokkaido, Japan. *Proceedings of the National Institute of Polar Research* (Tokyo) *Symposium on Polar Biology* 2:139–45.

SCAMMON, C. M. 1874. *Marine mammals of the north-western coast of North America, described and illustrated: Together with an account of the American whale-fishery.* San Francisco, CA : J. H. Carmany. N.Y.: G. P. Putnam's Sons.

SCHEFFER, V. B. 1958. *Seals, sea lions, and walruses. A review of the Pinnipedia.* Stanford, CA: Stanford University Press.

SCHEFFER, V. B., C. H. FISCUS, AND E. I. TODD. 1984. History of scientific study and management of the Alaskan fur seal, *Callorhinus ursinus,* 1786–1964. *NOAA Technical Report NMFS SSRF-780.*

SERGEANT, D. E. 1974. A rediscovered whelping population of hooded seals *Cystophora cristata* Erxleben and its possible relationship to other populations. *Polarforschung* 44:1–7.

SERGEANT, D. E. 1976. History and present status of populations of harp and hooded seals. *Biological Conservation* 10:95–118.

SERGEANT, D. E. 1991. *Harp seals, man and ice.* Ottawa, Ontario, Canada: Canadian Special Publication of Fisheries and Aquatic Sciences 114.

SERGEANT, D. E., K. RONALD, J. BOULVA, AND F. BERKES. 1978. The recent status of *Monachus monachus,* the Mediterranean monk seal. *Biological Conservation* 14:259–87.

SHANE, S. H. 1983. Abundance, distribution, and movements of manatees (*Trichechus manatus*) in Brevard County, Florida. *Bulletin of Marine Science* 33:1–9.

SHAUGHNESSY, P. D. 1985. Interactions between fisheries and Cape fur seals in southern Africa. In J. R. Beddington, R.J.H. Beverton, and D. M. Lavigne, eds., *Marine mammals and fisheries.* London: Allen & Unwin, pp. 119–34.

SHAUGHNESSY, P. D., AND F. H. FAY. 1977. A review of the taxonomy and nomenclature of North Pacific harbour seals. *Journal of Zoology, London* 182:385–419.

SHAUGHNESSY, P. D., AND L. FLETCHER. 1987. Fur seals, *Arctocephalus* spp., at Macquarie Island. *NOAA Technical Report NMFS* 51:177–88.

SHAUGHNESSY, P. D., AND K. R. KERRY. 1989. Crabeater seals *Lobodon carcinophagus* during the breeding season: Observations on five groups near Enderby Land, Antarctica. *Marine Mammal Science* 5:68–77.

SHAUGHNESSY, P. D., AND R. M. WARNEKE. 1987. Australian fur seal, *Arctocephalus pusillus doriferus*. *NOAA Technical Report NMFS* 51:73–77.

SHIPLEY, C., M. HINES, AND J. S. BUCHWALD. 1981. Individual differences in threat calls of northern elephant seal bulls. *Animal Behaviour* 29:12–19.

SIEGFRIED, W. R., P. R. CONDY, AND R. M. LAWS, EDS. 1985. *Antarctic nutrient cycles and food webs.* Berlin: Springer-Verlag.

SIELFELD, W. H. 1988a. Densidad y tamaño poblacional de *Lutra felina* y *Lutra provocax* (Carnivora, Mustelidae) en el litoral de Chile austral. Unpublished report of Reunión Consultiva de Expertos para la Conservación de Mamíferos Marinos en el Pacífico Sudeste.

SIELFELD, W. H. 1988b. Presencia de *Lutra felina* y *L. provocax* (Carnivora, Mustelidae) en Sudamerica austral. Unpublished report of Reunión Consultiva de Expertos para la Conservación de Mamíferos Marinos en el Pacífico Sudeste.

SIELFELD K., W. 1983. *Mamiferos marinos de Chile.* Santiago: Ediciones de la Universidad de Chile.

SINIFF, D. B., D. P. DEMASTER, R. J. HOFMAN, AND L. L. EBERHARDT. 1977. An analysis of the dynamics of a Weddell seal population. *Ecological Monographs* 47:319–35.

SINIFF, D. B., AND S. STONE. 1985. The role of the leopard seal in the trophodynamics of the antarctic marine ecosystem. In W. R. Siegfried, P. R. Condy, and R. M. Laws, eds., *Antarctic nutrient cycles and food webs.* Berlin: Springer-Verlag, pp. 555–60.

SINIFF, D. B., I. STIRLING, J. L. BENGTSON, AND R. A. REICHLE. 1979. Social and reproductive behaviour of crabeater seals (*Lobodon carcinophagus*) during the austral spring. *Canadian Journal of Zoology* 57:2243–55.

SKINNER, J. D., AND R. J. VAN AARDE. 1983. Observations on the trend of the breeding population of southern elephant seals, *Mirounga leonina,* at Marion Island. *Journal of Applied Ecology* 20:707–12.

SKINNER, J. D., AND L. M. WESTLIN-VAN AARDE. 1989. Aspects of reproduction in female Ross seals (*Ommatophoca rossii*). *Journal of Reproduction and Fertility* 87:67–72.

SMITH, R.I.L. 1988. Destruction of antarctic terrestrial ecosystems by a rapidly increasing fur seal population. *Biological Conservation* 45:55–72.

SMITH, T. G. 1973. Population dynamics of the ringed seal in the Canadian eastern Arctic. *Bulletin of the Fisheries Research Board of Canada* 181.

SMITH, T. G. 1976. Predation of ringed seal pups (*Phoca hispida*) by the arctic fox (*Alopex lagopus*). *Canadian Journal of Zoology* 54:1610–16.

SMITH, T. G. 1987. The ringed seal, *Phoca hispida,* of the Canadian western Arctic. *Canadian Bulletin of Fisheries and Aquatic Sciences* 216.

SMITH, T. G., M. O. HAMMILL, AND G. TAUGBØL. 1991. A review of the developmental, behavioural and physiological adaptations of the ringed seal, *Phoca hispida,* to life in the arctic winter. *Arctic* 44:124–31.

STEJNEGER, L. 1887. How the great northern sea-cow (*Rytina*) became exterminated. *American Naturalist* 21: 1047–54.

STEJNEGER, L. 1936. *Georg Wilhelm Steller. The pioneer of Alaskan natural history.* Cambridge, MA: Harvard University Press.

STEWART, B. S. 1981. The Guadalupe fur seal (*Arctocephalus townsendi*) on San Nicolas Island, California. *Bulletin of the Southern California Academy of Sciences* 80:99–101.

STEWART, B. S., AND H. R. HUBER. 1992. *Mirounga angustirostris. Mammalian Species.* (In press).

STEWART, B. S., S. LEATHERWOOD, P. K. YOCHEM, AND M.-P. HEIDE-JØRGENSEN. 1989. Harbor seal tracking and telemetry by satellite. *Marine Mammal Science* 5:361–75.

STEWART, B. S., P. K. YOCHEM, R. L. DELONG, AND G. A. ANTONELIS, JR. 1987. Interactions between Guadalupe fur seals and California sea lions at San Nicolas and San Miguel islands, California. *NOAA Technical Report NMFS* 51:103–6.

STEWART, B. S., P. K. YOCHEM, R. L. DELONG, AND G. A. ANTONELIS. 1992. Status and trends in abundance of pinnipeds on the southern California Channel Islands. In F. G. Hochberg, ed., *Recent advances in California islands research. Proceedings of the third California Islands symposium.* Santa Barbara, CA: Santa Barbara Museum of Natural History. (In press).

STEWART, R.E.A. 1986. Energetics of age-specific reproductive effort in female harp seals, *Phoca groenlandica. Journal of Zoology, London* 208:503–17.

STEWART, R.E.A., AND D. M. LAVIGNE. 1984. Energy transfer and female condition in nursing harp seals *Phoca groenlandica. Holarctic Ecology* 7:182–94.

STEWART, R.E.A., B. E. STEWART, D. M. LAVIGNE, AND G. W. MILLER. 1989. Fetal growth of northwest Atlantic harp seals, *Phoca groenlandica. Canadian Journal of Zoology* 67:2147–57.

STIRLING, I. 1984. A group threat display given by walruses to a polar bear. *Journal of Mammalogy* 65:352–53.

STIRLING, I. 1986. Research and management of polar bears *Ursus maritimus. Polar Record* 23(143):167–76.

STIRLING, I. 1988. *Polar bears.* Ann Arbor: University of Michigan Press.

STIRLING, I., AND W. R. ARCHIBALD. 1977. Aspects of predation of seals by polar bears. *Journal of the Fisheries Research Board of Canada* 34:1126–29.

SUE, L. L. MOU, D. H. CHEN, R. K. BONDE, AND T. J. O'SHEA. 1990. Distribution and status of manatees (*Trichechus manatus*) in Panama. *Marine Mammal Science* 6:234–41.

TEMTE, J. L., M. A. BIGG, AND Ø. WIIG. 1991. Clines revisited: The timing of pupping in the harbour seal (*Phoca vitulina*). *Journal of Zoology, London.* 224:617–32.

TESTA, J. W., S.E.B. HILL, AND D. B. SINIFF. 1989. Diving behavior and maternal investment in Weddell seals (*Leptonychotes weddellii*). *Marine Mammal Science* 5:399–405.

TESTA, J. W., AND D. B. SINIFF. 1987. Population dynamics of Weddell seals (*Leptonychotes weddellii*) in McMurdo Sound, Antarctica. *Ecological Monographs* 57:149–65.

TESTA, J. W., D. B. SINIFF, J. P. CROXALL, AND H. R. BURTON. 1990. A comparison of reproductive parameters among three populations of Weddell seals (*Leptonychotes weddelli*). *Journal of Animal Ecology* 59:1165–75.

TESTA, J. W., D. B. SINIFF, M. J. ROSS, AND J. D. WINTER. 1985. Weddell seal–antarctic cod interactions in McMurdo Sound, Antarctica. In W. R. Siegfried, P. R. Condy, and R. M. Laws, eds., *Antarctic nutrient cycles and food webs*. Berlin: Springer-Verlag, pp. 561–65.

THOMAS, J., V. PASTUKHOV, R. ELSNER, AND E. PETROV. 1982. *Phoca sibirica*. *Mammalian Species* 188:1–6.

THOMAS, J. A., S. R. FISHER, W. E. EVANS, AND F. T. AWBREY. 1983. Ultrasonic vocalizations of leopard seals (*Hydrurga leptonyx*). *Antarctic Journal* 16:186.

THOMAS, J. A., AND I. STIRLING. 1983. Geographic variation in the underwater vocalizations of Weddell seals (*Leptonychotes weddellii*) from Palmer Peninsula and McMurdo Sound, Antarctica. *Canadian Journal of Zoology* 61:2203–12.

THOMPSON, P. 1988. Timing of mating in the common seal (*Phoca vitulina*). *Mammal Review* 18:105–12.

THOMPSON, P., AND P. ROTHERY. 1987. Age and sex differences in the timing of moult in the common seal, *Phoca vitulina*. *Journal of Zoology, London* 212:597–603.

THOMPSON, P. M. 1989. Seasonal changes in the distribution and composition of common seal (*Phoca vitulina*) haul-out groups. *Journal of Zoology, London* 217:281–94.

THORSTEINSON, F. V., AND C. J. LENSINK. 1962. Biological observations of the Steller sea lions taken during an experimental harvest. *Journal of Wildlife Management* 26:353–59.

TIMM, R. M., L. ALBUJA V., AND B. L. CLAUSON. 1986. Ecology, distribution, harvest, and conservation of the Amazonian manatee *Trichechus inunguis* in Ecuador. *Biotropica* 18:150–56.

TORRES N., D. 1987. Juan Fernández fur seal, *Arctocephalus philippii*. *NOAA Technical Report NMFS* 51:37–41.

TORRES N., D., P. E. CATTAN, AND J. YANEZ. 1985. First census of the Juan Fernández fur seal, *Arctocephalus philippii* (Peters) during the breeding season, 1978–1979. *National Geographic Society Research Reports* 18:733–43.

TOWNSEND, C. H. 1934. The fur seal of the Galápagos Islands. *Zoologica* 18:43–47.

TRILLMICH, F. 1981. Mutual mother-pup recognition in Galápagos fur seals and sea lions: Cues used and functional significance. *Behaviour* 78:21–42.

TRILLMICH, F. 1984. The natural history of the Galapagos fur seal, *Arctocephalus galapagoensis* Heller. In R. Perry, ed., *Key environments: Galapagos*. Oxford, England: Pergamon Press.

TRILLMICH, F. 1987. Galapagos fur seal, *Arctocephalus galapagoensis*. *NOAA Technical Report NMFS* 51:23–27.

TRILLMICH, F., AND P. MAJLUF. 1981. First observations on colony structure, behavior, and vocal repertoire of the South American fur seal (*Arctocephalus australis* Zimmermann, 1783) in Peru. *Zeitschrift für Säugetierkunde* 46:310–22.

TRILLMICH, F., AND K. ONO, EDS. 1991. *Pinnipeds and El Niño: Responses to environmental stress*. Berlin: Springer-Verlag.

VAN BLARICOM, G. R., AND J. A. ESTES, EDS. 1988. *The community ecology of sea otters*. Berlin: Springer-Verlag.

VAZ-FERREIRA, R. 1981. South American sea lion *Otaria flavescens* (Shaw, 1800). In S. H. Ridgway and R. J. Harrison, eds., *Handbook of marine mammals*, Vol. 1, *The walrus, sea lions, fur seals and sea otter*. London: Academic Press, pp. 39–65.

VAZ-FERREIRA, R., AND A. PONCE DE LEON. 1987. South American fur seal, *Arctocephalus australis,* in Uruguay. *NOAA Technical Report NMFS* 51:29–32.

VELTRE, D. W., AND M. J. VELTRE. 1987. The northern fur seal: A subsistence and commercial resource for Aleuts of the Aleutian and Pribilof islands, Alaska. *Etudes/Inuit/Studies* 11(2):51–72.

VIBE, C. 1950. The marine mammals and the marine fauna in the Thule district (northwest Greenland) with observations on ice conditions in 1939–41. *Meddelelser om Grønland* 150(6):1–115.

VLADIMIROV, V. A. 1987. Age-specific reproductive behavior in northern fur seals on the Commander Islands. *NOAA Technical Report NMFS* 51:113–20.

WALKER, G. E., AND J. K. LING. 1981a. Australian sea lion *Neophoca cinerea* (Péron, 1816). In S. H. Ridgway and R. J. Harrison, eds. *Handbook of marine mammals,* Vol. 1, *The walrus, sea lions, fur seals and sea otter.* London: Academic Press, pp. 99–118.

WALKER, G. E., AND J. K. LING. 1981b. New Zealand sea lion *Phocarctos hookeri* (Gray, 1844). In S. H. Ridgway and R. J. Harrison, eds., *Handbook of marine mammals,* Vol. 1, *The walrus, sea lions, fur seals and sea otter.* London: Academic Press, pp. 25–38.

WANG PEILIE AND SUN JIANYUN. 1986. [Distribution of the dugong off the coast of China.] *Acta Theriologica Sinica* 6:175–81. [In Chinese, with English abstract.]

WARD, H. L. 1887. Notes on the life-history of *Monachus tropicalis,* the West Indian seal. *American Naturalist* 21:257–64.

WARNEKE, R. M., AND P. D. SHAUGHNESSY. 1985. *Arctocephalus pusillus,* the South African and Australian fur seal: Taxonomy, evolution, biogeography, and life history. In J. K. Ling and M. M. Bryden, eds., *Studies of sea mammals in south latitudes.* Adelaide, Australia: South Australian Museum, pp. 53–77.

WATKINS, W. A., AND G. C. RAY. 1977. Underwater sounds from ribbon seal, *Phoca (Histriophoca) fasciata. U.S. Fishery Bulletin* 75:450–53.

WATKINS, W. A., AND G. C. RAY. 1985. In-air and underwater sounds of the Ross seal, *Ommatophoca rossi. Journal of the Acoustical Society of America* 77:1598–1600.

WESTLAKE, R. L., AND W. G. GILMARTIN. 1990. Hawaiian monk seal pupping locations in the northwestern Hawaiian Islands. *Pacific Science* 44:366–83.

WIIG, Ø. 1985. Morphometric variation in the hooded seal *(Cystophora cristata). Journal of Zoology, London* 206:497–508.

WIIG, Ø. 1986. The status of the grey seal *Halichoerus grypus* in Norway. *Biological Conservation* 38:339–49.

WIIG, Ø., M. EKKER, T. EKKER, AND N. RØV. 1990. Trend in the pup production of grey seals *Halichoerus grypus* at Froan, Norway, from 1974 to 1987. *Holarctic Ecology* 13:173–75.

WIIG, Ø., AND R. W. LIE. 1984. An analysis of the morphological relationships between the hooded seals (*Cystophora cristata*) of Newfoundland, the Denmark Strait and Jan Mayen. *Journal of Zoology, London* 203:227–40.

WILSON, D. E., M. A. BOGAN, R. L. BROWNELL, JR., A. M. BURDIN, AND M. K. MAMINOV. 1991. Geographic variation in sea otters, *Enhydra lutris. Journal of Mammalogy* 72:22–36.

WOODHOUSE, C. D., JR., R. K. COWEN, AND L. R. WILCOXON. 1977. A summary of knowledge of the sea otter, *Enhydra lutris, L.*, in California and an appraisal of the completeness of biological understanding of the species. *Report MMC-76/02.* Springfield, VA: U.S. Department of Commerce, National Technical Information Service, PB-270 374.

WORTHY, G.A.J., AND D. M. LAVIGNE. 1983. Energetics of fasting and subsequent growth in weaned harp seal pups, *Phoca groenlandica. Canadian Journal of Zoology* 61:447–56.

WORTHY, G.A.J., AND D. M. LAVIGNE. 1983. Changes in energy stores during postnatal development of the harp seal, *Phoca groenlandica. Journal of Mammalogy* 64:89–96.

YABLOKOV, A. V., AND OLSSON, M., eds. 1989. *Influence of human activities on the Baltic ecosystem. Proceedings of the Soviet-Swedish symposium Effect of toxic substances on dynamics of seal populations, Moscow, 14–18 April 1986.* Leningrad: Gidrometeoizdat.

YOCHEM, P. K., B. S. STEWART, R. L. DELONG, AND D. P. DEMASTER. 1987. Diel haul-out patterns and site fidelity of harbor seals (*Phoca vitulina richardsi*) on San Miguel Island, California, in autumn. *Marine Mammal Science* 3:323–32.

YORK, A. E. 1983. Average age at first reproduction of the northern fur seal (*Callorhinus ursinus*). *Canadian Journal of Fisheries and Aquatic Sciences* 40:121–27.

YORK, A. E., AND P. KOZLOFF. 1987. On the estimation of numbers of northern fur seal, *Callorhinus ursinus,* pups born on St. Paul Island 1980–86. *U.S. Fishery Bulletin* 85:367–75.

APPENDICES

Appendices

Appendix 1: Common and Scientific Names of Animals and Plants

THROUGHOUT THIS BOOK, we refer to various animals and plants other than pinnipeds, sirenians, otters, and polar bears. We have attempted to provide in this appendix the family, genus, or species names of each of them. However, some of our original sources contained insufficient information. Also, scientific classifications and names are not always universally accepted, and they change as new findings are published. This list, then, is neither complete nor authoritative. It is meant to assist readers who are interested in looking for further information.

FISHES

American plaice *Pleuronectes platessa*
Anchovy, family Engraulidae
Antarctic silver fish *Pleuragramma antarcticum*
Arctic char *Salvelinus alpinus*
Arctic cod *Boreogadus saida*
Atlantic cod *Gadus morhua*
Atlantic salmon *Salmo salar*
Australian salmon *Arripis trutta*
Barracouta *Leionura(=Thyrsites) atun*
Barramundi *Lates calcarifer*
Bearded goby *Barbulifer ceuthoecus*
Blue shark *Prionace glauca*
Blue tilapia *Tilapia aurea*
Bluechin parrot fish *Scarus ghobban*
Cape hakes *Merluccius capensis* and *M. paradoxus*
Capelin *Mallotus villosus*
Carp *Cyprinus carpio*
Chum salmon *Oncorhynchus keta*
Cod, family Gadidae
Cookie-cutter shark *Isistius* sp.
Eelpout *Lycodes* sp.
Eulachon *Thaleichthys pacificus*
Fiddler ray *Trygonorhina fasciata*
Flounders families Bothidae *and* Pleuronectidae
Gobies (Caspian Sea), including *Neogobius kesslerip* and *Knipowitschia longicaudata*
Golomyankas *Comephorus* spp.
Gray reef shark *Carcharinus amblyrhynchos*
Great white shark *Carcharodon carcharias*
Greenland halibut *Reinhardtius hippoglossoides*
Greenland shark *Somniosus microcephalus*
Greenling *Ophiodon elongatus*
Halibut *Hippoglossus* spp.
Hammerhead shark *Sphyrna* spp.
Herring *Clupea harengus*
Herring (Caspian), family Clupeidae
Hoki *Macruronus novaezelandiae*
Jack mackerel *Trachurus symmetricus*
Lamprey, family Petromyzontidae
Lanternfish, family Myctophidae
Lumpfish *Cyclopterus lumpus*
Mackerel, family Scombridae
Mackerel (So. Africa) *Trachurus* sp.
Mullet *Mugil* spp.
Northern anchovy *Engraulis mordax*
Omul *Coregonus autumnalis*
Pacific cod *Gadus macrocephalus*
Pacific mackerel *Scomber japonicus*
Pacific whiting (hake) *Merluccius productus*
Pike-perch (Caspian Sea) *Lucioperca lucioperca*

Pilchard *Sardinops ocellata*
Plainfin midshipmen *Porichthys notatus*
Polar cod *Arctogadus glacialis*
Pollack (Saithe) or Pollock *Pollachius virens*
Rainbow smelt *Osmerus mordax*
Redfish *Sebastes* spp.
Reef white-tip shark *Triaenodon obesus*
Remoras, family Echeneidae
Roach (Caspian Sea) *Rutilus rutilus*
Rockfish *Sebastes* spp.
Rosefish *Helicolenus* spp.
Saffron cod *Eleginus gracilis*
Salmon (North Atlantic) *Salmo salar*
Salmon (North Pacific) *Oncorhynchus* spp.
Salmon (southern South America) *Oncorhynchus* sp. (introduced)
Salmon (Caspian Sea) probably *Salmo* sp.
Sand lance (eel) *Ammodytes* spp.
Sand smelts (Caspian Sea), family Osmeridae
Sardines, family Clupeidae
Saury *Cololabis saira*
Sculpins, family Cottidae
Shiner perch *Cymatogaster aggregata*
Silver hake *Merluccius bilinearis*
Smelt *Osmerus* sp.
Snoek *Leionura (=Thyrsites) atun*
Spotted cusk eel *Chilara taylori*
Sprat (Caspian Sea) *Clupeonella delicatula*
Steelhead trout *Oncorhynchus mykiss*
Sturgeon (Caspian Sea), family Acipenseridae
Tiger shark *Galeocerdo cuvieri*
Toadfish *Opsanus* sp.
Walleye pollock *Theragra chalcogramma*
White pointer shark (Australia) *Carcharodon carcharias*
Wolffish *Anarhichas* sp.

BIRDS

Adelie penguin *Pygoscelis adeliae*
Andean condor *Vultur gryphus*
Bald eagle *Haliaeetus leucocephalus*
Cape gannet *Morus capensis*
Cape petrel *Daption capensis*
Chinstrap penguin *Pygoscelis antarctica*
Common grackle *Quiscalus quiscula*
Cormorants *Phalacrocorax* spp.
Gentoo penguin *Pygoscelis papua*
Giant petrels *Macronectes* spp.
Glaucous gull *Larus hyperboreus*
Jackass (Black-footed) penguin *Spheniscus demersus*
King penguin *Aptenodytes patagonica*
Little (Fairy) penguin *Eudyptula minor*
Macaroni penguin *Eudyptes chrysolophus*
Magellanic penguin *Spheniscus magellanicus*
Pintail *Anas georgica*
Pipit *Anthus antarcticus*
Puffin (Atlantic) *Fratercula arctica*
Raven *Corvus corax*
Rockhopper penguin *Eudyptes crestatus*
Skuas *Catharacta* spp.
Thayer's gull *Larus thayeri*
Wilson's storm petrel *Oceanites oceanicus*

REPTILES

Alligator *Alligator mississippiensis*
Caymans *Caiman* spp.
Crocodiles *Crocodylus* spp.
Green turtle *Chelonia mydas*

MAMMALS

African clawless otter *Aonyx capensis*
Amazon River dolphin *Inia geoffrensis*
American black bear *Ursus americanus*
Arctic fox *Alopex lagopus*
Asiatic black bear *Ursus (=Selenarctos) thibetanus*
Black-backed jackal *Canis mesomelas*

Bottlenose Dolphin *Tursiops truncatus*
Bottlenose whales *Hyperoodon* spp.
Bowhead whale *Balaena mysticetus*
Brown bear *Ursus arctos*
Brown hyena *Hyaena brunnea*
Burmeister's porpoise *Phocoena spinipinnis*
Caribou *Rangifer tarandus*
Common hippopotamus *Hippopotamus amphibius*
Coyote *Canis latrans*
European river otter *Lutra lutra*
Jaguar *Panthera onca*
Killer whale (Orca) *Orcinus orca*
Mink *Mustela vison*
Mountain lion *Felis concolor*
Narwhal *Monodon monoceros*
North American river otter *Lutra canadensis*
Peale's dolphin *Lagenorhynchus australis*
Platypus *Ornithorhynchus anatinus*
Pygmy hippopotamus *Choeropsis liberiensis*
Red fox *Vulpes vulpes*
Roe deer *Capreolus capreolus*
Sea mink *Mustela macrodon*
Skunk *Mephitis* spp.
Sperm whale *Physeter catodon*
Vampire bat *Vampyrum spectrum*
Weasels *Mustela* spp.
White whale (Beluga) *Delphinapterus leucas*
Wolf *Canis lupus*
Wolverine *Gulo gulo*

INVERTEBRATES

Abalone *Haliotis* spp.
Butter clam *Saxidomus giganteus*
Codworm ("Sealworm") *Pseudoterranova decipiens*
Crayfish *Panulirus ornatus*
Cuttlefish *Sepia* spp.
Heartworm *Dipetalonema (=Skrjabinaria) spirocauda*
Krill *Euphausia superba*
Lobster (Atlantic) *Homarus* spp.
Market squid *Loligo opalescens*
Octopus *Octopus* spp.
Pismo clam *Tivela stultorum*
Rock lobster *Munida* sp.
Seal louse *Echinophthirius horridus*
Spiny lobster, family Palinuridae

PLANTS

Hair grass *Deschampsia antarctica*
Hydrilla *Hydrilla verticillata*
Live oak *Quercus virginiana*
Marsh grass *Spartina alterniflora*
Tussock grass *Poa flabellata*
Water hyacinth *Eichhornia* spp.

Appendix 2: Dental Formulas of Pinnipeds

Species	Incisors	Canines	Post-canines	Total teeth
Odobenidae				
Walrus	1/0	1/1	3/3	18
Otariidae				
Northern fur seal	2/2	1/1	5–6/5	32–34
Juan Fernández fur seal	3/2	1/1	6/5	36
Guadalupe fur seal	3/2	1/1	6/5	36
Galápagos fur seal	3/2	1/1	6/5	36
South American fur seal	3/2	1/1	6/5	36
South African and Australian fur seal	3/2	1/1	6/5	36
New Zealand fur seal	3/2	1/1	6/5	36
Antarctic fur seal	3/2	1/1	6/5	36
Subantarctic fur seal	3/2	1/1	6/5	36
Northern sea lion	3/2	1/1	5/5	34
California (and Galápagos) sea lion	3/2	1/1	5–6/5–6	34–38
South American sea lion	3/2	1/1	6/5	36
Australian sea lion	3/2	1/1	5/5	34
New Zealand sea lion	3/2	1/1	6/5	36
Phocidae				
Harbor seal	3/2	1/1	5/5	34
Spotted seal	3/2	1/1	5/5	34
Ringed seal	3/2	1/1	5/5	34
Baikal seal	3/2	1/1	5/5	34
Caspian seal	3/2	1/1	5/5	34
Harp seal	3/2	1/1	5/5	34
Ribbon seal	3/2	1/1	5/5	34
Bearded seal	3/2	1/1	5/5	34
Hooded seal	2/1	1/1	5/5	30
Gray seal	3/2	1/1	5–6/5	34–36
Crabeater seal	2/2	1/1	5/5	32
Ross seal	2/2	1/1	5–6/4–6	30–36
Leopard seal	2/2	1/1	5/5	32
Weddell seal	2/2	1/1	5/5	32
Northern elephant seal	2/1	1/1	5/5	30
Southern elephant seal	2/1	1/1	5/5	30
Mediterranean monk seal	2/2	1/1	5/5	32
Caribbean monk seal	2/2	1/1	5/5	32
Hawaiian monk seal	2/2	1/1	5/5	32

Appendix 3: Breeding Ranges By Major Ocean Areas (adapted from Scheffer 1958)

Key: Principal Breeding Range •; Peripheral Breeding Range ○

Species	Atlantic Arctic/ Subarctic	Pacific Arctic/ Subarctic	Temperate North Atlantic	Temperate North Pacific	Tropical Atlantic	Tropical Pacific	Tropical Indian Ocean	Temperate South Atlantic	Temperate South Pacific	Temperate Indian Ocean	Sub-Antarctic	Antarctic	Other
Walrus	•	•											
Northern fur seal		•		•									
Guadalupe fur seal				•									
Juan Fernández fur seal									•				
Galápagos fur seal						•							
South American fur seal								•	•				
South African and								•		•			
Australian fur seal									•	•			
New Zealand fur seal									•				
Antarctic fur seal											•	•	
Subantarctic fur seal											•		
Northern sea lion				•									
California (and Galápagos) sea lion				•		○							
South American sea lion								•	•				
Australian sea lion									•	•			
New Zealand sea lion									•				
Harbor seal	○	○	•	•									

Baikal seal													A
Caspian seal													B
Harp seal	●												
Ribbon seal		●											
Bearded seal	●	●											
Hooded seal	●												
Gray seal	○		●										
Crabeater seal											○	●	
Ross seal												●	
Leopard seal												●	
Weddell seal											○	●	
Northern Elephant seal				●									
Southern Elephant seal								○	○		●	○	
Mediterranean Monk seal					●								
Caribbean Monk seal					●+								
Hawaiian Monk seal						●							
West Indian manatee					●*								
Amazon manatee					●**								
West African manatee					●*								
Dugong						●	●						
Steller's sea cow				●+									
Sea otter				●									
Marine otter									●				
Polar bear	●	●											

Footnotes: Restricted to Lake Baikal (A) and Caspian Sea (B); Extinct (+); Principally fresh water (*); Entirely fresh water (**)

Index

DISCARD
LIBRARY
VT TECHNICAL COLLEGE
RANDOLPH CTR VT 05061